# Cultural Turns/ Geographical Turns: Perspectives on Cultural Geography

# Cultural Turns/ Geographical Turns: Perspectives on Cultural Geography

### Edited by
### Ian Cook, David Crouch, Simon Naylor and James R. Ryan

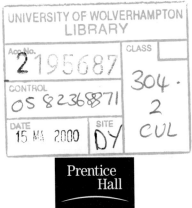
**Prentice Hall**

*An imprint of* **Pearson Education**

Harlow, England · London · New York · Reading, Massachusetts · San Francisco
Toronto · Don Mills, Ontario · Sydney · Tokyo · Singapore · Hong Kong · Seoul
Taipei · Cape Town · Madrid · Mexico City · Amsterdam · Munich · Paris · Milan

**Pearson Education Limited**
Edinburgh Gate
Harlow
Essex CM20 2JE
England

and Associated Companies throughout the World.

*Visit us on the World Wide Web at:*
www.pearsoneduc.com

First published 2000

© Pearson Education Limited 2000

ISBN 0 582 36887 1

*British Library Cataloguing-in-Publication Data*
A catalogue record for this book can be obtained from the British Library

*Library of Congress Cataloging-in-Publication Data*
A catalogue record for this book can be obtained from the Library of Congress

10 9 8 7 6 5 4 3 2 1
05 04 03 02 01 00

Typeset by 35 in 10/12pt Sabon
Produced by Pearson Education Asia Pte Ltd.
Printed in Singapore

# Contents

# List of Figures

# List of Contributors

Jacquie Burgess, Department of Geography, University College London, England
David B. Clarke, School of Geography, University of Leeds, England
John Clarke, School of Social Policy, Open University, England
Ian Cook, School of Geography, University of Birmingham, England
Mike Crang, Department of Geography, University of Durham, England
Philip Crang, Department of Geography, University College London, England
David Crouch, University of Derby, England
Marcus A. Doel, Department of Geography, University of Loughborough, England
Sheila Hones, Department of Area Studies, University of Tokyo, Japan
Keith D. Lilley, School of Geography, The Queen's University of Belfast, Northern Ireland
George E. Marcus, Department of Anthropology, Rice University, USA
David Matless, School of Geography, University of Nottingham, England
Linda McDowell, Department of Geography, London School of Economics, England
Daniel Miller, Department of Anthropology, University College London, England
Simon Naylor, School of Geographical Sciences, University of Bristol, England
Chris Philo, Department of Geography and Topographic Science, University of Glasgow, Scotland
George Revill, Geography Unit, School of Social Sciences, Oxford Brookes University, England
James R. Ryan, School of Geography, The Queen's University of Belfast, Northern Ireland
Andrew Sayer, Department of Sociology, University of Lancaster, England
David Sibley, School of Geography and Earth Resources, University of Hull, England
James D. Sidaway, School of Geography, University of Birmingham, England
Mark Thorpe, Simons Priest and Associates, London, England
Nigel Thrift, School of Geographical Sciences, University of Bristol, England
Judith Tsouvalis, School of Geography, University of Oxford, England
Melanie Wall, Department of Geography, University of Edinburgh, Scotland
Sarah Whatmore, School of Geographical Sciences, University of Bristol, England

# Acknowledgements

This book is the culmination of a long, drawn-out project. It began at a committee meeting in the Geography Department at University College London, went through a conference at Rhodes House at Oxford University, various editorial meetings at Pearson's London Office, and through the Geography Departments where we have worked over the past four years (at the University of Wales Lampeter, Anglia Polytechnic University, Keele University, Oxford University, Queen's University Belfast, and the University of Birmingham). Very little would have happened without the Social and Cultural Research Group of the Institute of British Geographers/Royal Geographical Society, who funded and led us to organize this follow-up to its ground breaking New Words/New Worlds initiative; those who helped us run the Cultural Turns/Geographical Turns conference, which was stage one of our project (the speakers, audience members, support staff at Oxford University's School of Geography and at Rhodes House, as well as the latter's trustees); and those who helped us work through stage two of the project, the production of this book (the anonymous referees of our initial proposals, those who have given permission for the use of various photos and figures, all of the contributors, Pearson's Matthew Smith, Magda Robson and the people who create the magnificent sandwich feasts at their London office). We are grateful to Music Sales Ltd. for permission to reproduce an excerpt from 'Caractacus' by Sir Edward Elgar, published by Novello & Co. Ltd. and John Hugh McKenzie Goodacre for 'Land of the Ridge and Furrow' The Pageant Song by Hugh Goodacre (music by Walter Groocock) © 1932. Particular thanks go to Matthew for keeping us enthusiastic and to reasonable deadlines, to Magda for being so cooperative and helpful with the copy-editing, to the authors for getting their work to us on time, to Pat Woodward of Oxford's School of Geography for preventing things from going pear-shaped when organiser illness struck, to Stuart Franklin and Hamish Fulton of Magnum for their generous allocations of time and photographic knowledge (not to mention the photos), to the Lanyon family for permission to use Peter Lanyon's 'Turn Again' work on the cover, and to all of the others who have kept us going throughout the project, each other included.

# Foreword

## Ian Cook, David Crouch, Simon Naylor and James R. Ryan

Over the last twenty or so years geography has followed and absorbed work from across the social sciences and the humanities, to become a discipline thoroughly inculcated in debates over and around 'culture'. From landscape paintings to consumption, from Foucault to Empire, from maps to sexuality, geography has engaged with a wide variety of issues, concerns, theories and methods of relevance to the study of 'culture' in all its forms. In some ways marking the developments in the nascent field that came to be known as ' "new" cultural geography' came a series of conferences: at University College London in 1987, at Edinburgh in 1991 (from which the volume *New Words, New Worlds* emerged), and at Oxford in September 1997. Whilst the arguments contained within the realms of this particular volume track across numerous fora they became focused through the 1997 conference organized under the auspices of the Social and Cultural Geography Research Group of the RGS/IBG.

This volume brings together 31 authors from many different intellectual backgrounds and disciplines. Within it there are a number of approaches and standpoints, varying in length and style. The balance of empirical and theoretical work also ranges widely amongst chapters and parts. Moreover, the chapters do not 'toe' a single 'line', yet common threads run through each contribution and the part in which they are placed – the investigation of the relationship between culture and space.

This book stretches the idea of 'the cultural' in geography to embrace large areas across the discipline. Moreover, it breaks the boundaries of geography itself and overlaps with work in anthropology, cultural studies, sociology and the humanities. Yet it holds all of these aspects in a relationship with geography and what geography does and what geographers contribute. The book is divided into five parts and the part titles demonstrate something of this range: 'Popular culture and cultural texts', 'Culture and political economy', 'Nature and society', 'Spaces and subjectivities'. Within these parts lie further categories concerning, for instance, nations and nationalism, cities and urban spaces, regions, ethnicity, sexuality and gender, but it has been the editors' intention to avoid categorizations as much as possible and provide parts that are inclusive. However, a groundwork of issues is contained in the first part, 'Cultural turns/ geographical turns', whilst the introductions to each part outline engagements of four arenas of geographical debate and present key fulcra around which these debates are likely to evolve in the immediate future.

Just from this short foreword it is already obvious that there has been less a *cultural turn* than a series of cultural insights, turns, multiple circuits. Geography has become, perhaps, embodied by cultural discourse. However, in this book there is a critical awareness of the dilemmas of cultural discourse in geography that will hopefully set the scene for numerous further discussions and investigative work.

## Reference

Philo, C. (ed.) (1991) *New Words, New Worlds. Reconceptualising Social and Cultural Geography*, Lampeter, Social and Cultural Geography Study Group.

# Introduction

## Dead or alive?

## *Nigel Thrift*

Surely there are few commentators who would want to deny that the cultural turn in the social sciences and humanities – including in human geography – has paid enormous intellectual dividends. More than this, it has simply made things a lot more *interesting*.

Yet, this is a particularly propitious time to be examining the state of this cultural turn. Why? Because there are signs of a thaw in the standard accounts of its emergence and maintenance: a number of critics have started to criticize both the foundational stories and the propositions upon which the turn was founded, and not always in bad-tempered ways.[1] The charge sheet can be distilled down into six related propositions which run as follows.

First, the cultural turn is charged with having become theoreticist. Not only does cultural work too often, in Sivanandan's (1993) phrase, turn the world into the word, but this is a specific case of a more general problem – that of letting theory outrun the data presented (when a more modest approach might be to cleave to a 'certain brand of empiricism, making the data so presented apparently outrun the theoretical effort to comprehend it' (Strathern, 1999, p. 199)). In apparent contradiction, a second often-voiced criticism is that the cultural turn has become a hard-hearted empiricism with a soft veneer; journalism without immediacy but with added sentiment. The world becomes a lucky dip, a source of innumerable case studies waiting to be plucked, suitably agonized over – in ever so reflexive ways, of course – and published. This is the cultural studies academic as world traveller.

Then, third, the cultural turn is charged with having become culturalist. This charge is often given voice by suggesting that cultural work neglects 'political economy', is insufficiently 'sociological', and the like. Usually this is a coded way of implying that cultural work too often ignores the gritty aspects of life like poverty that are still the lot of far too many. Fourth, the cultural turn is charged with abrogating all responsibility for asking questions concerning value, quality and truth. At best, cultural work has become an 'anything goes' eclecticism with no sense of political project; at worst, an illustration of Oscar Wilde's aphorism that 'modern morality consists in accepting the standards of one's age'.

Fifth, the cultural turn is then charged with a lack of political bite. Whilst ostensibly pursuing political goals, there is an increasing gap between political mission and cultural practice – there is plenty of hype but no action (McChesney, 1995). Perhaps, so critics argue, this is because the cultural turn has been captured by the expanding disciplinary apparatus of cultural studies and, in this moment of institutionalization (Cultural Studies, 1998), what it has gained in legitimacy, it has lost in relevance.

Sixth, the cultural turn is caught up in its own pretensions. Written into its very fabric is a sense that it is both avant-garde – ahead of the game – and, by extension, on the margins of society. This is even though much of the cultural turn still consists of intellectual habits that might not have appeared out of place at the end of the last century, or before: scholastic residue has become radical gloss.[2]

Insofar as these charges deflate the pretensions of the cultural turn, they can only do good. They can help us to see that the cultural turn will probably prove to be primarily concerned with 'allowing everyday life and cultural experience to be fashioned into instruments of government via their inscription in new forms of teaching and training' (Bennett, 1998, p. 51; see also Barnett, 1998a, 1998b). They can help us to see that the cultural turn is still dominated by tired constructionist themes. They can help us to see that the prevailing academic mode of operation still depends on the figure of irony and so distance. They can help us to see how easy it is to slip into a too easy politics in which inquiry is 'cast as critique understood as the uncovering of relations of domination and, consequently, [is] rarely far from denunciation. Denunciation [is] commonly elided to politics and, in the end, victory is more often than not spiritual and discursive' (Rabinow, 1999, p. 9).

*But*, all this said, I think that we have to be careful not to lose that sense of *engagement with the emergent* which is so clearly a reason why the cultural turn has been so successful[3]. I think that this will be a much more modest sense of engagement borne out of joint curiosities as much as revolutionary programmes. I think this will also be a more performative sense of engagement, intended to identify

> those forces or potencies where origins and outcomes cannot be specified independently of the open and necessarily incomplete series of their actualization. Such is their multiplicity that it can never be reduced to a set of discrete elements or to the different parts of a closed or organic whole. (Rajchman, 1998, p. 116)

And I think this will be a mode of operation which moves to new performative figures which do not view knowledge as just dominative power and/or protean desire. Knowledge as sight but also cite and site (Spinosa, Flores and Dreyfus, 1997).

This, in turn, requires valuing some related disclosive skills more than we do now. I think that these skills are of three kinds.[4] The first of these skills is *methodological*. Cultural geographers have, over time, allied themselves with a number of qualitative methods, and most notably in-depth interviews and ethnographic 'procedures'. But what is surprising is how narrow this range of skills still is, how wedded they still are to the notion of bringing back the 'data', and then re-presenting it (nicely packaged up as a few supposedly illustrative quotations), and the narrow realms of sensate life they register.

This is all very odd for, since the 1960s, a whole plethora of methods has become available which have within them the potential to take us beyond this (repressed) place (Thrift, 1999). I am thinking here of all the methods fostered in the performing arts which attempt to co-produce the world – street theatre, community theatre, legislative theatre, and so on. I am thinking also of all those techniques which work with bodies – various forms of dance and music therapy, contact improvisation, and so on. I am thinking of the large number of initiatives in social psychology, aimed at boosting relational responsibility by focusing on group interaction (e.g. McNamee and Gergen, 1998). I am thinking of various forms of performative writing which have sprung up from disciplines as diverse as anthropology, feminism and performance studies. I am thinking, too, of the increasing use of interactive web sites which reform and perform on the internet (e.g. Hill and Paris, 1998).

Further, each and every one of these developments involves, centrally and intimately, the production of spaces, for example, both through altering the conditions of possibility of extant spaces, and by producing new spaces. These are not just gestures to new ground.

The emphasis on world-making which informs this methodological breakout, also informs another set of skills that are important. These are the skills of 'fabulation', a term used by Bergson to refer to a visionary faculty and taken up by Deleuze (1995, p. 17) to refer to 'creative story telling that is, as it were, the obverse of the dominant myths and fictions, an act of resistance that creates line of flight on which a minority discourse and a people can be constituted'. Such fabulations can be used to generate new models of government which resonate more fully with experience than some of the totalizing designs of old. I admit to being utterly taken, for example, by Charles Fourier's often bizarre account of Harmony, an imaginary community founded on the circulation of all the senses. There is no space here to go into all the details of Fourier's synaesthetic union (but see Beecher and Bienvenu, 1971; Beecher, 1986; Barthes, 1976; Mattelart, 1996; Classen, 1998) in which the profound and the burlesque cohabit.[5] But what is startling is how few parallels there are to these kinds of imaginative political projects today. It is as if utopian thought is off the agenda, replaced by the work of technicians (as in Giddens' (1998) *The Third Way*). Yet it is not as if the longing for these stories is not there. I think of the renewed interest in Geddes, whose often bizarre inventions and diagrams are a part of the integral character of his work, of some feminist science fiction which clearly is, in part, a political project and, most recently, of Michael Taussig's extraordinary (1997) *The Magic of the State*, an attempt to conjure up a performance of government which, at the same time, provides clues as to how it might be out-performed. In other words, what we can see here is a politics of desiring, taking shape as an immediation (Grossberg, 1998).[6]

These thoughts lead to one more set of skills which seem to me to be in short supply, what might be called ethical expertise. These skills have been given a very high profile of late but as a kind of technology of decision. The problem is how we can make ethics into something more general. I therefore go along with Tom Osborne (1998, p. 193) when he argues that:

> My intention in using the vocabulary of ethics has not been so as to join the decisionist bandwagon by saying that ethics is all, in effect, that we have left. It is rather to isolate the ethical dimension that is the background to all our coherent human activities, including those associated with the pursuit of faith. That means a critique of enlightenment would be completely useless in formulating any kind of ethical theory. It is more deliberately critical than the grandiose projects of a MacIntyre or a Bauman in that it is concerned not with a general diagnosis of the possibility of ethics in the present, but with, so to speak, acting directly on an ethical propensity. This is not a discourse *about* ethics, but an ethical form of discourse in its own right, one which takes the task to be to stretch the limits of what it is possible to think, not in the interest of this or that project of enlightenment or reform, but in the interests of an exercise in judgement itself.

It is this exercise of judgement which seems to me to be such a crucial element of three different but utterly related literatures which I believe to be some of the most interesting currently in this regard. One is the derivation of a 'trans-human' constitution of the kind to be found in actor-network theory, and some of the work on animals, and on new technologies. This literature is attempting to produce an ethics which will be able to take in the expanding ecology of mind. The second is the work of feminist theorists like Nussbaum and others who are attempting to expand the horizon of care. And the third is the growth of work on the modern university as a place where the capacity for judgement is (or at least should be) fostered through an openness which prevents academics from clustering in small inward-looking groups which superannuate and narrow (Perl, 1998). In each case, what these literatures incorporate is a sense of an ethics of possibility which seems to me to be the only stance we can now take.

To conclude, it seems to me that we can shake off some of the illusions of the cultural turn and still generate a sense of intense engagement by recognizing that practice does not 'soil the pure expectation of theoretical possibility with its very groundedness' (Kirby, 1997, p. 155). Further, this sense of engagement insofar as the cultivation of cultural and practical *generosity* is one of the key principles, can also provide a renewed sense of personal engagement (just as long as we realize that the person is precisely one of the terms that is now most contested). So perhaps there is no need to pack up and go home quite yet, especially since, in so many parts of the world, people still find the energy to dance in chains. To the extent that the risk is there, there is no reason why the cultural turn has to turn bad.

# Notes

1. See, for example, the collections edited by Ferguson and Golding (1997) and McRobbie (1997), books such as Bennett (1998) and responses such as Morley (1998), or Kline (1998). Bennett (1998) is particularly good on how being 'in the true' in cultural studies entails employing narratives based on highly romantic stereotypes of both politics and persons.

2. For example, in cultural geography nearly all work in areas like landscapes and postcolonialism is still based on inadvertently Cartesian models of representation and interpretation which have been junked elsewhere in the social sciences and humanities. Similarly in research on consumption, ethnicity, and other areas much effort is still devoted to models of 'reading' (even if the reader is allowed to become more active, in itself, as Gell (1998) notes, an assumption with massive cultural presuppositions) which are increasingly seen as inappropriate elsewhere in the social sciences and humanities.

   What does all this mean for cultural geography? The problem is that cultural geography needs to be both more theoretical *and* more empirical. More theoretical certainly. Most cultural geographers are not trained in theory, and use it as though it were a technique (hence the bizarre calls for applying particular theories to politics). On the whole, cultural geographers' use of theory therefore remains resolutely eucomistic and representational; and therefore in danger of simply retracing steps already made by others. But cultural geography is not empirical enough either. Its range of methods is remarkably small and, underneath all the rhetoric, really quite conservative. Why might this be? There are no doubt many reasons, but I think one of the most important is probably training. Most undergraduate and postgraduate degrees simply do not give sufficient depth or challenge to enable prospective geographers to see the possibilities.

3. For me, such a sense of engagement comes from three main sources. The first is that Nietzschean strain to be found in so many modern works from Foucault to Deleuze, a gay science which attempts to conserve and create the positive, 'a form of expansive knowledge that will heat up the universe and render it conducive for the mixing of all kinds of foreign elements and the explosion of new sparks' (Ansell-Pearson, 1998, p. 55). The second is some modern feminist theory which tries to engage the limits of the possible, especially the work of Cixous and Irigaray. The third is distributed theories of practice (see Thrift, 1996). Each and every one of these strands of work is distinguished by their commitment to non-representational issues.

4. I become a little frustrated with academics who believe that they can simply read off politics and, indeed, the urge to be political as though people simply took a decision to be political on the basis of rational deliberation. In contradiction, each of these three skills are, it should be noted, skills of *intense involvement* (Spinosa, Flores and Dreyfus, 1997).

5. Fourier based his cosmos on the model of music:

   > on earth, humans constitute musical notes, with the particular note of each individual dominated by his or her dominant character traits. In Fourier's plan for the perfect society, human 'notes' are grouped together in work and social units to fashion 'major and minor scales', and 'choirs'. This plan for harmonizing personality types through social organization gave Fourier's utopia its name, Harmony. (Classen, 1998, p. 25)

6. And a desperate need to teach skills such as creative writing.

# References

Ansell-Pearson, K. (1998) *Viroid Life*. London, Routledge.

Barnett, C. (1998a) 'Cultural twists and turns.' *Environment and Planning D: Society and Space*, 16, 631–4.

Barnett, C. (1998b) 'The cultural worm turns: fashion or progress in human geography?' *Antipode*, 30, 379–94.

Barthes, R. (1976) *Sade, Fourier, Loyola*. Baltimore, Johns Hopkins University Press.

Beecher, J. (1986) *Charles Fourier. The Visionary and His World*. Berkeley, University of California Press.

Beecher, J. and Bienvenu, R. (1971) *The Utopian Vision of Charles Fourier. Selected Texts on Work, Care and Passionate Attraction*. Boston, Beacon Press.

Bennett, T. (1998) *Culture. A Reformer's Science*. London, Sage.

Classen, C. (1998) *The Colour of Angels. Cosmology, Gender and the Aesthetic Imagination*. London, Routledge.

Cultural Studies (1998) Special Issue on 'The Institutionalisation of Cultural Studies.' *Cultural Studies*, 12 (4).

Deleuze, G. (1995) *Negotiations*. New York, Columbia University Press.

Ferguson, M. and Golding, P. (eds) (1997) *Cultural Studies in Question*. London, Sage.

Gell, A. (1998) *Art and Agency. An Anthropological Perspective*. Oxford, Blackwell.

Giddens, A. (1998) *The Third Way*. Polity Press, Cambridge.

Grossberg, L. (1998) 'The Victory of Culture Part I: Against the logic of mediation.' *Angelaki*, 3 (3), 3–30.

Hill, L. and Paris, H. (1998) 'I never go anywhere I can't drive myself.' *Performance Research*, 3 (2), 102–8.

Kirby, V. (1997) *Telling Flesh. The Substance of the Corporeal*. London, Routledge.

Kline, S. (1998) 'W(h)ither cultural studies?' *European Journal of Cultural Studies*, 1, 436–41.

McChesney, R.W. (1995) *Telecommunications, Capitalism and the Media*. New York, Oxford University Press.

McNamee, K. and Gergen, K. (1998) *Relational Responsibility*. London, Sage.

McRobbie, A. (ed.) (1997) *Back to Reality*. Manchester, Manchester University Press.

Mattelart, A. (1996) *The Invention of Communication*. Minneapolis, University of Minnesota Press.

Morley, D. (1998) 'So-called cultural studies: dead ends and reinvented wheels.' *Cultural Studies*, 12, 476–97.

Osborne, J. (1998) *Aspects of Enlightenment. Social Theory and the Ethics of the Truth*. London, UCL Press.

Perl, J. (1998) 'Editorial.' *Common Knowledge*, 7, 1–3.

Rabinow, P. (1999) 'Ratios: an anthropology of reason is to the social studies of knowledge as a counter-sublime is to irony' (Forthcoming).

Rajchman, J. (1998) *Constructions*. Cambridge, Mass., MIT Press.

Sivanandan, A. (1993) *Race and Class*.

Spinosa, C., Flores, F. and Dreyfus, H.C. (1997) *Disclosing New Worlds. Entrepreneurship, Democratic Action and the Cultivation of Solidarity*. Cambridge, Mass., MIT Press.

Strathern, M. (1999) *Property, Substance and Effect*. London, Athlone Press.

Taussig, M. (1997) *The Magic of the State*. New York, Routledge.

Thrift, N.J. (1996) *Spatial Formations*. London, Sage.

Thrift, N.J. (1999) 'Afterwords.' *Environment and Planning D. Society and Space* (Forthcoming).

# PART I

# Cultural turns, geographical turns

Paris. Bois de Vincennes. 1987. © Josef Koudelka/Magnum

# Introduction

*James R. Ryan*

Few commentators would disagree with the claim that the cultural turn in contemporary geography has represented and produced new ways of thinking about culture and geography. There is, however, much less consensus on the processes underpinning this trend as well as its long-term consequences (Barnett, 1998b; Castree, 1999). Warnings of the theoreticist, journalistic, culturalist, irrelevant, pretentious tendencies of the currents which constitute the cultural turn might deflate its grander posturing. Yet at the same time its critical currency at the centre of such debate further establishes it as a fixture in a range of inter-secting intellectual landscapes. As Nigel Thrift points out in his preceding Introduction in this volume, this is an opportune moment in geography, and in the social sciences and humanities more generally, for the refashioning of critical cartographies of the cultural turn. Having said this, we should not posit the situation as one of either–or choices; the cultural turn as a single track, downhill path which is either taken or rejected in favour of some alternative, more level-headed route. Indeed, it is profitable intellectually and politically to travel on a number of routes and roundabouts; to support and critique simul-taneously this absorption with the cultural (Butler, 1998).

This opening part consists of three essays which engage in different ways with the general themes of this volume and set the scene for many of the debates which follow. Each essay makes an important contribution to the generation of more nuanced accounts of the distances covered and directions taken thus far on the cultural turn. They should not, however, either individually or collectively, be taken as attempts at comprehensive overviews of some uniform, singular and complete development. To map fully the interwoven twists and turns of 'culture' and 'geography' would be a mammoth and a complex task, involving negotiation of the complex genealogies of both these terms, as well as their inflection within dynamic disciplinary and geographical settings. In any case, useful introductions to many of these fields, from cultural geography (Crang, 1998) to cultural theory and popular culture (Storey, 1993), as well as more comprehensive assessments of the cultural turn in geography are readily available elsewhere (Barnett, 1998a; Matless, 1995, 1996; McDowell, 1994). There is neither the space nor the intention to produce full surveys here. Rather, the essays in Part I – like others throughout this book – are offered as deliberately selective and careful accounts of aspects and effects of the cultural turn from particular perspectives.

Whilst the essays in Part I refer inevitably to disciplinary boundaries and foci, notably of anthropology, geography and sociology, they also emphasize their porosity and the fruitful outcomes which stem from their blurring. For whilst ideas of culture in geography have drawn on work in cultural studies, anthropology, sociology and related fields, geographical metaphors and techniques for 'mapping' the 'topographies' of culture have become incorporated into the language and practice of disciplines such as cultural studies (see, for example, Baldwin *et al.*, 1998).

Part I begins with an essay by anthropologist George E. Marcus in which he considers the more recent cultural interweavings of geography and anthropology. Conceptualizations of culture within geography have long been influenced by parallel developments in anthropology. Early cultural geography drew on ideas developed in cultural anthopology for its definitions of culture which focused on the material productions of particular social groups. More recently, the lively interest across human geography in retheorizing culture – central to the whole 'cultural turn' – owes much to the important work by anthropologists such as Marcus on the interpretation of cultures and politics of cultural representation (see, for example, Geertz, 1973; Clifford and Marcus, 1986).

In his essay Marcus considers the potentials of geographical and anthropological dialogue within the setting of what he terms 'the sobering wake of the great awakening', where the 'theoretical imaginaries' which characterized the beginnings of the cultural turn have given way to more grounded and applied work in a second wave of interdisciplinary research strategies. Taking 'multi-sited ethnography' as his focus, Marcus argues that both 'obvious' strategies (tracking movements and exchanges of transnational communities, or the circulation of objects) and 'non-obvious' strategies (exploring the unknown networks and disjunctions between particular places and emergent social relations) pose important reformulations of the idea and practice of 'fieldwork' and suggest new dialogues between and within geographical and anthropological enquiry. Whilst Marcus is keen to endorse such interdisciplinary connections and developments he ends his essay in cautious mode with an interview with Wlad Godzich in which the latter argues for the significance of the temporal, as opposed to the spatial, in ethnographic enquiry.

The effects of the cultural turn upon embedded research practice is also the concern of Chris Philo who, in the second essay, reflects critically upon the cultural turn within social and cultural geography. As an enthusiastic pioneer of the cultural turn within British geography Philo is well placed to offer an avowedly personal account of aspects of this trend and the problems which its hegemony poses for the theory and practice of human geography. In particular, Philo shows how a narrow preoccupation with the 'cultural' can result in 'dematerialized' and 'desocialized' geographies. In his concern with the neglect of the 'social' and his critique of abstract, un-'earthed' enquiry Philo has much in common with Marcus. Whilst Marcus puts forward multi-site fieldwork as one means of renewing 'social' questions whilst keeping the interpretative gains of the cultural turn, Philo points to recent studies in geography which, blending theoretical sophistication with empirical grounding, do not simply redress the balance away from flighty 'cultural' concerns to proper 'social' questions, but actively recast categories of 'social' and 'cultural', 'material' and 'immaterial', revisioning the relations between them.

Questions of the fate of 'the social' under the shadow of the cultural turn also provide the focus for the third and final essay in Part I, by the sociologist John Clarke who takes a more 'applied' domain than either geography or anthropology, namely social policy. Clarke explores the place of the cultural turn within dynamic currents in social policy and a transforming welfare state. As Clarke reminds us, welfare policies are shaped decisively by conceptions of social groupings and cultural identity, from 'the family' to 'the nation'. Yet what is actually meant and understood by concepts of 'social' and 'cultural' within fields such as social policy has changed considerably in the last half century.

The insistence in recent years upon the 'socially constructed' nature of identities previously thought of as fixed or natural – of 'race', disability, sexuality – within professional welfare practice has been closely connected to conceptual and methodological debates within social policy. Here a localized manifestation of the cultural turn, incorporating a range of theoretical perspectives and methodological influences as well as individuals and social movements, has nourished new conceptions of social policy study and practice. Clarke shows that whilst the rise of interest in the cultural and social construction of identity can offer important new ways of thinking about social policy, it can render welfare issues vulnerable to attempts to 'de-socialize' them. Such attempts stem from a range of factors, including political tendencies towards individualism; redefinitions of social problems as 'moral' ones; the rediscovery of biology (notably in genetics); and the definition of social issues (such as 'race') as matters of geography. The results, as Clarke shows, are both complex and contradictory, and yet it is within such a landscape of the cultural turn that the most provocative and difficult questions arise.

These three essays provide an important opening on to the issues developed throughout this book by posing a range of difficult questions of the cultural turn from different disciplinary and theoretical perspectives. In their accounts of particular moments and settings within anthropology, geography and social policy, these essays show some of the complexity obscured within that catch-all term 'the cultural turn' and show how it has particular and localized versions and effects. Nevertheless, the essays also show the virtue of tracking some of the shared, intersecting and interdisciplinary domains in the social sciences which have been informed by the cultural turn and which continue to shape its myriad effects.

All three essays in Part I are concerned with making sober assessments of the cultural turn in the light of renewed interest in theoretically and empirically grounded enquiry. In doing so they connect directly to wider calls within both geography (Barnett, 1998a) and cultural studies (Grossberg, 1993, 1998) for a renewed need to subject culture and cultural analysis to further theoretical scrutiny, and associated moves in human geography towards a more informed engagement with philosophical debate on issues such as ethics (Smith, 1997; Thrift, this volume). Taken together these contributions warn of any easy acceptance of theoretical and methodological assumptions characteristic of much of the cultural turn. Declarations that all manner of things are 'socially and culturally constructed' roll off many tongues and printers, yet the theoretical and analytical complexity behind such assertions are more rarely discussed (see Hacking, 1999). Furthermore, all of these essays stress the serious difficulty of combining critical, analytical awareness of the constructions of culture with attention to the 'solidification' of such constructions across a range of geographies. It is the potentials and problems of the kinds of engagement with the cultural turn discussed in these introductory essays which inform many of the essays in the remainder of this book.

# References

Baldwin, E., Longhurst, B., Smith, G., and Ogborn, M. (1998). *Introduction to Cultural Studies*, Hemel Hempstead: Prentice Hall.

Barnett, C. (1998a). 'The cultural worm turns: fashion or progress in human geography?', *Antipode*, 30, 379–94.

Barnett, C. (1998b). 'Cultural twists and turns', *Environment and Planning D: Society and Space*, 16, 631–4.

Butler, J. (1998). 'Merely cultural', *New Left Review*, 227, 33–44.

Castree, N. (1999). 'Situating cultural twists and turns', *Environment and Planning D: Society and Space*, 17, 257–60.

Clifford, J. and Marcus, G.E., eds. (1986). *Writing Culture: The Politics and Poetics of Ethnography*, Berkeley, CA: University of California Press.

Crang, M. (1998). *Cultural Geography*, London: Routledge.

Geertz, C. (1973). *The Interpretation of Cultures*, New York: Basic Books.

Grossberg, L. (1993). 'Cultural studies and/in new worlds', *Critical Studies in Mass Communications*, 10, 1–12.

Grossberg, L. (1998). 'The victory of culture', *Angelaki*, 3, 3–29.

Hacking, I. (1999). *The Social Construction of What?*, London: Harvard University Press.

McDowell, L. (1994). 'The transformation of cultural geography', in Gregory, D., Martin, R. and Smith, G., eds. *Human Geography: Society, Space and Social Science*, London: Macmillan, 146–73.

Matless, D. (1995). 'Culture run riot? Work in social and cultural geography, 1994', *Progress in Human Geography*, 19, 395–403.

Matless, D. (1996). 'New material? Work in cultural and social geography, 1995', *Progress in Human Geography*, 20, 379–91.

Smith, D. (1997). 'Geography and ethics: a moral turn?', *Progress in Human Geography*, 21, 583–90.

Storey, J. (1993). *An Introductory Guide to Cultural Theory and Popular Culture*, London: Harvester Wheatsheaf.

# The twistings and turnings of geography and anthropology in winds of millennial transition

*George E. Marcus*

Several of the papers in this volume seem to operate from the assumption that the so-called cultural turn is mature, or even spent, as a purely theoretical enterprise. As an enterprise primarily of the 1970s and 1980s (also associated with overlapping labels like 'the postmodern turn' or 'the poststructuralist turn'), there was great ferment and stimulation in English-speaking academia over the study of culture, and the theory and practice of interpretation associated with it. I believe that the sense of the waning of the interdisciplinary fervour over the working out of new ideas at a sometimes indulgently theoretical level is very widespread at present. There now seems to be as widespread a preoccupation with developing the stock of ideas and styles of analysis as distinctive or experimental research programmes within those disciplinary traditions that were most affected by the cultural turn. While certain disciplines like geography (and less so anthropology) might have kept the balance between theoretical exegesis and empirical research traditions all along, there is a sense in which innovation for the moment has been recentred within the bounds of disciplinary authority rather than in their interstices – the sites of earlier interdisciplinary spaces built upon the critique and even scorn of disciplinarity. Yet, in the sobering wake of the great awakening, so to speak, that the cultural turn was for many scholars in the social sciences and humanities, I think there is immense potential for a second wave of interdisciplinary fusions (partnerships, even) more fine-tuned and defined, along certain borders that crystallize as a result of affinities in how certain disciplines or disciplinary fractions responded to and assimilated the cultural turn. Reality, so to speak, signalled in the pervasive use of the yet poorly understood framing trope of globalization, is pressing upon the storehouse of theoretical imaginaries explored in the 1970s and 1980s. The task now is to experiment with research practices that both are rooted in distinctive disciplinary styles and arise from partnerships in how the cultural turn has defined affinities in modes of inquiry.[1]

Here I want to explore the possibilities for this sort of partnership between anthropology and geography in the applied aftermath of the cultural turn as theoretical fervour, by offering to this volume a discussion of an emerging development in the practice of anthropology – what I have termed multi-sited ethnography – that brings it even closer to the kinds of research in geography most influenced by the cultural turn. Of course the working relations of the two disciplines have long been entwined while their professional fortunes have differed quite markedly, at least in the United States. Areal and regional geography has always been the base for the coherence of ethnographic research in anthropology. But as will be seen, the basis of this relationship in the present moment of so-called globalization, and after the heavily theoretical postmodern and cultural turns, is much less clear, but no less strong.

To sketch their current relationship briefly, while both are marginal disciplines in the United States, the disciplinary authority and presence of anthropology are better established than are geography's in the university (this of course has its advantages and disadvantages in terms of the freedom to innovate and synthesize). What the cultural turn has meant for anthropology is a strong reinforcement of its already strong valorization of hermeneutic depth in ethnography, and consequently its concerns with the interpretation of symbols, meanings and representations, at the cost of its treatment of social relations, social structures and systematic differences in political economy. As I will discuss, the necessity of multi-sited ethnography can be understood as a corrective to the inattention to the social while staying with the commitment to hermeneutic depth.

What the cultural turn has meant for geography is a strong intervention of interpretive theories, methods and ideas in a field heavily influenced by tasks of mapping, describing societies spatially, and by economistic thinking. Perhaps more broadly open to considering and exploring the range of theories and ideas that the cultural turn offered than anthropology has been, geography also has not lost its own sense of the social to the same degree, nor has it pursued hermeneutic issues and problems with the culture concept as obsessively as anthropology has.

As will be seen below, the basis for a new sort of partnership between the current applied cultural turns in geography and anthropology arises from the tendencies and limits that the preceding period of rather freewheeling interdisciplinarity left each of these disciplines. Multi-sited ethnography, out of the cultural turn, certainly needs dialogue with the venturesome sort of culturally inflected consideration of contemporary processes of political economy that geography has evolved (and that the wing of anthropology concerned with political economy failed to produce). More importantly, the consideration of multi-sized ethnography, or at least a multi-sized field of research, has already been anticipated and strongly pursued in geography (see, for example, Cook and Crang, 1996; Jackson and Thrift, 1995; Katz, 1992). And geography might benefit from the hermeneutic depth to cultural analysis that ethnography insists upon even as it becomes more geography-minded as it moves problematically from the intensely observed and engaged loci that have constituted the traditional *mise-en-scène* of fieldwork.

## The emergence of multi-sited ethnography

Ethnography is in the midst of a second wave of reassessment, following upon the critique of traditional strategies of ethnographic representation, the so-called 'writing culture' critique of the 1980s (see Clifford and Marcus, 1986). This second wave focuses upon the sacred ground of fieldwork practice and is motivated by the ongoing changing pragmatics and conditions of fieldwork. Situating ethnography in relatively stable geographical space is giving way to the necessity of tracking and even defining cultural formations across a number of related sites of fieldwork. This is what the macro-processes keyed to the ubiquitous trope of globalization are forcing upon the traditional *modus operandi* of ethnographic fieldwork in anthropology. In a series of recent writings (Marcus, 1999a, 1999b), I have tried to articulate the considerable changes in the regulative assumptions and idealized norms that would be required in setting up a multi-sited ethnographic object of study for fieldwork. This is on the surface a plea for the consideration of methodology as the most important focus of theoretical attention now. But this is to some degree illusory. It is really a plea to look at the legacies of the so-called postmodern moment or the cultural turn not in theory, but as embedded in research practices as these finally try to come to terms with a self-consciously conceived world of great transformation and transition. At stake is how many of the traditional presuppositions of ethnography will be left standing and in what forms.

There are both obvious and non-obvious applications of multi-sited strategies. The obvious cases these days follow well marked pathways of process defined by macro-social theories or historical metanarratives, or they follow previously local subjects and objects of ethnography as they literally move in space and time under altered conditions of historical political economy. Tracking movements of migrants transnationally in diaspora and exile, or the history of the circulation of objects and techniques, or studying the relationships of dispersed communities and networks that define well designated macro-processes in the global flow of capital and expertise – these are the obvious contexts of contemporary ethnographic work that are challenging traditional norms of fieldwork in productive ways and are also opening new conduits of exchange with past interdisciplinary partners in other social sciences. Arjun Appadurai's influential 'scape' essays of the early 1990s (see Appadurai, 1997), for example, are a metaphorical map for this now strong wave of research. When anthropologists think of multi-sited work, most think of projects like these whose contexts are defined by given lines of macro-social theory and historical narrative. Macro-processes may be changing, but there is nothing particularly problematic or unclear – untrackable – about the relationships or connections of peoples or objects ethnographically probed within the framework of these processes. This is where the border between anthropology and human geography is most clearly productive. The intensity of ethnography combined with the scoping political economy of geography with a heightened sensitivity to cultural interpretation shows much promise.

The other alternative is non-obvious applications of multi-sited strategies. It poses somewhat different challenges and suggests more radical alternatives than the above to the norms that have traditionally regulated fieldwork. These are the cases where processes are not given by contextualizing macro-theories and historical narratives, where the metaphors of tracking or following therefore do not work as well in constituting multi-sited objects of ethnographic study, where relationships or connections between sites are indeed not clear, the discovery and discussion of which are precisely in fact the main problem, contribution and argument of ethnographic analysis.

This frankly more mysterious kind of ethnography is often posed as that which operates in the realm of emergent social and cultural phenomena. Such multi-sited ethnography arises from putting questions to an emergent object of study whose contours, sites and relationships are not known beforehand, but are themselves a contribution of making an account that has different, complexly connected real-world sites of investigation. These strategies raise the question of the nature of relationships between sites of activity and social locations that are disjunctive in space or time and perhaps in terms of social category as well (say, elites juxtaposed to subalterns, middle class to the poor, experts to non-experts, institutions to communities) whose relations are not obvious as given in any framing macro-social theoretical discourse or historical narrative, and may or may not be posed by ethnographic subjects themselves.[2] This problem for ethnography is signalled on the one hand by the increasing occurrence of virtuality in social relations and the problem of how to study it, and on the other hand, by the recent interest in circulation and flows in anthropology trying to come to terms with an idea of globalization at the level of ethnography, consistent with its past concerns with sites in which fieldwork with hermeneutic depth can be done.

Superficially, non-obvious strategies involving disjunctions and juxtapositions with speculations about relations retain the same use of geographical metaphors, and there is the strong potential of association with the work of human geography as is clearly the case with obvious strategies. Non-obvious strategies merely require more speculative geographies, insular to what they are trying to associate (and partly track). But there is also something challenging in this kind of multi-sited work to the geographical categories themselves on which it still depends. This is the kind of cautionary challenge that Wlad Godzich, a major figure responsible for bringing much of the European intellectual capital of the cultural turn during its theoretical heyday to English-speaking scholars, has articulated, and whose thinking is sampled in the final section of this paper.

# The loss and reclamation of the social in recent ethnography

What is most importantly and most generally at stake for anthropology – and its connection to human geography – in the emergence of multi-sited ethnography is indeed a reorientation to the very idea of the social in situated cultural analysis inspired by the cultural turn of the past two decades. There is a general feeling among anthropologists today, more or less articulated in professional 'corridor talk', that the interest in all of those things that would be classed under the 'social' – social relations, processes, structures, systems, institutions, matters of political economy – have been relatively neglected in favour of attention to, for example, subject positions, identity construction, dialogic exchange and micro-examinations of embedded practices, restricted to the intimate traditional scene of fieldwork. Indeed, it might be argued that this finely wrought preoccupation with the micro-cultural is about the social (*à la* Anthony Giddens' location of structuration in situated agency, and Pierre Bourdieu's location of what is systemic in the situation in the habitus) – just a different way of constructing it. But there is no doubt that within the production of ethnography the description of the terrain of the macro-social has suffered in its materialities, attention to scale, regimes of exchange, and resulting exposures of and concerns for inequalities. While it may be that what has happened has not been truly a loss of the social, at least there is a widespread sense among anthropologists of a relative inattention to it as a focusing concern in terms of classic questions of social theory. Instead there has been a tendency in ethnography to let the constructs, theories and work of other kinds of academics (including human geographers, political economists and postcolonial theorists, among others) to stand in, so to speak, for the macro, patterned sense of the social that contextualizes ethnographic work while itself probes voice, discourse, subjectivity and identity as its primary concerns.

At present, there seems to be a desire among anthropologists to find new ways to extend specifically ethnographic method and analysis into the contextualizing zone of the frames and constructs that now stand in for the macro-social, while also preserving the real advances in micro-cultural analysis that have been made over the past two decades. Certainly there is a concern to avoid the 'swing of the pendulum' situation – that the only way to refocus attention on the social would be to move back to older political economy perspectives, to allow the inevitable swing of fashion between the two poles of the symbolic and the material, the cultural and the political economic, to take its course. Indeed, since the 1980s there have been efforts by those trained in existing styles of political economy research in anthropology (e.g. the students of Eric Wolf), from their side, to take culture more seriously and especially, to incorporate cultural historical narratives into their accounts of global systems. But to those interested in the anthropology of the present, and who are primarily invested in the complex questions of cultural analysis raised by theories of representation, the micro-production of subjectivity, and the ever present reflexive effects of particular regimes of constituting knowledge, these efforts have not been satisfying. Thus, it is for those who have been most vested in cultural analysis – those who have reinforced the study of culture in anthropology through the influences of the more general cultural turn – to produce their own reclamation of the social. The broad perspectives of writers such as Pierre Bourdieu, David Harvey, Fredric Jameson, and the many theorists of postcoloniality have provided provocative macro-views of late modern or postmodern social structural processes that for much contemporary ethnography has been allowed to stand for this dynamic contextualizing function of representing the social, but in some sense, this manner of contextualizing ethnography contradicts the basic impulse of the ethnographer to know the insides of whatever he or she studies – including its contexts – intimately and from subjects' perspectives. In a sense, letting constructs of the social define the ethnographic centre of contemporary anthropological research constrains the interpretation of ethnographic materials by essentially non-ethnographic perspectives. One major aim then of trying to theorize the already occurring multi-sited transformation of fieldwork is to encourage this move of a certain nomadic or rhizomic tendency in ethnography into a motivated change in the way that much contemporary ethnography constructs the space of the social as the contexts for its intimate eye and ear. The world of finance, markets, politics, and their institutions are ethnographic objects of study implicated in every fieldwork project these days and cannot be left to other constructions, if anthropology is to be responsible for its own contexts of meaning and the forging of its arguments from inside the ethnographic process of research itself. This reclamation of the social context of ethnography is for me the most important stake for the current reconstruction of anthropology through multi-sited ethnographic projects.

The emerging norms of multi-sited fieldwork are the means of growing a renewed attention to the social out of the cultural turn's revolution in thinking about what goes on in the site of ethnography while still preserving the intensity and complexity of that revolution. Different paradigms of doing multi-sited work – and there are many detectable – would develop the social in different ways, but they all would move from the site of initial fieldwork into a contextualizing consideration of what Douglas Holmes in his study of the circulation of European right wing discourse (in press) has called 'the social within reach' as an integral dimension of the design of any multi-sited fieldwork project.

To the question 'Is this kind of always limited probing of the social in multi-sited ethnography enough?', I would answer 'Certainly not'. But this constitution of the social through the strategized movement of ethnography among different sites is always at least implicitly, and should end up explicitly, in dialogue or tension with non-ethnographically composed constructions, narratives and representations of the social in the work of historians, social theorists, and for my special interest here, of human geographers. The critical point, though, is that multi-sited research hesitates to work immediately with these constructions as the ground of ethnography.

The alternative, and a distinctly anthropological style, is to build the social out of local knowledge of it in the sites of fieldwork. There is a crucial and agile methodological suspicion here that is deeply embedded in the tradition of cross-cultural translation at the core of anthropology. Thus, rather than the social being given or already defining of the terms of ethnography, there will always be a gap between other constructs and those generated by ethnography – and this is a very productive gap. Meditations on this gap are what generates the social in anthropological research deeply affected by the cultural turn. It is never more constructed or less problematic than this play in the gap between the interpreted situated knowledge of subjects and more distanced expert constructs, and there is no more valid way for it to return. Anthropologists themselves do not construct macro-social theory. In the past, they often have had reason to contest such theory. So why should they take such frames, narratives and macro-visions for granted in constructing their own projects of fieldwork? Not to do so, but to use the social within reach in any project is a way to re-engage representations of the social as processes which the ethnography documents in multi-sited space and then uses to challenge the existing theories, constructs, narratives and problematics that it would otherwise have precipitously embraced as the context (and limits of interpretation, as well) of its work.

Finally, there may be the fear that deriving the social from what the informant thinks (especially when the ethnographer starts with the situated imaginary within an empowered space of elite or expert practice) will just be native sociology, so to speak, albeit of natives who are in a sense counterparts of the anthropologist. And this, too, is not enough. Just as grasping the native point of view was never the sole point of ethnography, so ethnographic knowledge cannot be only native knowledge. The growing of a perspective on the social from intensive ethnography in multiple sites cannot be synonymous with elaborated native models of the social. As always, understanding native knowledge holds the key to how anthropologists achieve an independence of perspective. In multi-sited work, the anthropologist does take the native construction of the social seriously, as something to absorb, critique and extend, but in moving beyond an initial site where situated knowledge is very literally probed in relationship to its referent elsewhere (see note 2), the anthropologist produces his or her own construction of the social within reach. This construction based on fieldwork in other sites is played back eventually within an initial one, often as a critical challenge or intervention to so-called native sociology found there. A critical engagement with particular elite or expert representations of the social, understood itself through acts of ethnographic fieldwork, by the anthropologist who eventually returns to elite sites with the experience of grounded ethnography among referents that are only imagined or abstractly constructed in elite visions of the social, is the most fulfilling end of multi-sited ethnographic projects.

In a sense, this is what anthropologists have always done in their traditional terrain – undermining the west's view of primitives, exotic others through bringing back detailed, empirical knowledge of such referent peoples. Only now the terrain and geography of this traditional task have changed considerably, reconfiguring and fragmenting the communities among whom anthropologists move and define their work.

So, on the one hand multi-sited ethnography faces macro-theoretical and historical narrative efforts to represent what it develops in complicity with social actors – the social within reach of ethnographic projects – and on the other hand, builds its knowledge through strategies of work with these actors that has its moments of complex interdependence with as well as independence from them. All the social will ever be again for anthropologists most affected by the cultural turn is in these productive gaps, dialogues and engagements, on two fronts. This is by no means an unfavourable place for anthropologists to be after the cultural turn.

How, then, does this vision and implication of an even more deeply cultural anthropology after the cultural turn, but that is striving also to reconnect to issues of the society, intertwine with the ferment and possibilities in human geography as surveyed in the papers of this volume? Well, clearly the emergence of a multi-sited ethnography makes anthropological enquiry more geographic in nature, and not just in locating itself in relation to the conventional cartography of the globe, but in the very rhizomic constitution of its subjects of study amid sites, the relationships among which are the object of ethnographic interpretation and speculation.

## Wlad Godzich's cautionary note

Much of the necessary and ongoing remaking of ethnographic method that I have described above along with its implication for the return of a kind of contingent but committed concern with the social depends on the use of spatial concepts, resting on metaphors like mapping, tracking, cartography, and imaginaries as keys to what sorts of perspectives and knowledge is now desired from subjects in fieldwork settings. This of course would appear to strengthen anthropology's connections to the sorts of developments in geography described in the diverse papers of this volume under the same influences of the 1980s and 1990s cultural turn. Undoubtedly, and happily so. But it is also useful to end on a cautionary note, since while spatial and geographic conceptual metaphors are irresistible, and in terms of the nineteenth century storehouse of social and cultural theories on which we still depend, they seem natural to what we confront empirically in our work, the very emphasis on the spatial may no longer work. Temporalities, for example, may now be more important in orienting social action at particular locations. A stimulating expression of this caution, for me, emerged in an interview that my colleagues and I conducted with Wlad Godzich[3] on his visit to Rice in the autumn of 1998. His is a strong statement for the emphasis on the temporal rather than the geographic as required by emerging shifts in technologies of form-giving. There is much imprecision and informality in the way our interview unfolded and much to argue with, but I leave the following cuttings of Godzich's responses as a constructive provocation. They do not banish the related analytic styles of the ethnographic and the geographic of the cultural turn that I have evoked in this paper, but offer them challenges and probings that are well worth considering.

*Wlad*:   I think the kinds of questions that you are raising are to be found in many disciplines at present, and this is in large measure a testament to the fact that we are going through an important change. A paradigm shift is taking place in a number of disciplines. One can see, for example, how a number of Marxists are returning to Gramsci's notion of the interregnum, as a way of describing this particular moment – which I think in part is very good, and in part is a cop-out in as much as an interregnum is precisely a moment when people cannot tell what is going on. What I think is fundamentally problematic for anthropology is indeed this major shift from the geographical to the temporal. I mean, at the present time it is not where you are located in space that matters, it's where you are located in time . . . and how do you think about time where you are. What interests me is how to identify how people all over the globe now are thinking about the moment in which they live. That is, what is their conception of their own historicism? What is happening with time within the conditions of globalization, with people referring to notions such as real time and virtual time that are so important in the ways we now operate. However, when we are thinking in terms of space, we used to think of geometrical categories of distance . . . we would be calculating how far this is from that, and so on. The nature of time that we have to think about now is a much more topological notion of time . . . for example, take your handkerchief and just crumple it up, and the crumpling of the handkerchief does not alter the geometrical relationships of the things within the handkerchief, but it produces rather different configurations. There are different folds that bring things together. This is what is happening certainly for various people in different parts of the globe. The handkerchief has been crumpled up in a different way, or in some instances, torn, so that suddenly experiences which they thought could be mediated by very long distances become immediate . . . they have to find ways of talking about that, and describing it, and there is little evidence that there is an available vocabulary for what's happening. It seems to me there are two things that we ought to be doing: one is simply tracking these things and seeing how people are reacting, how they are putting it in their own language, what kind of little stories they are telling themselves. But the other one is asking 'what are the invariant geometric organizations within the topology and then what are the prophecies that this topology is being subjected to at present?' If we can keep track of these two things then we have a possibility of grasping what's going on . . . What is at the heart of all of this is that under conditions of globalization, there is a revaluation of what the local is. The primary concern with one's historicity, placing oneself within historical experience, changes the nature of the sense of location. Location is no longer a spatial notion, but rather an indication of how the invariant of the geometry has been altered in order to constitute the sense of a particular kind of place . . . One should reject all of these discussions that say these are the exclusions, these are the inclusions within processes of globalization. That's nonsense. Everybody is within this process. Globalization means that there is nothing outside of the frame. It means there is no margin. And so all we have to be preoccupied with is what are the forms of the figuration that are taking place within that. And I think that it really doesn't matter whether you do this as an anthropologist and I do this as a comparativist . . . that's why I didn't use 'locality', but rather location – location is always in relation to something else . . . Let me tell you, for example, how I think of Geneva, because that might be a good way of describing this: Geneva is a very small city of about 180,000 people, and yet it is really a global city. It has a very peculiar function in the world because of the international organizations but also the large number of NGOs that are there, plus the kind of banking that is going on in the city. It is a place where there is forged an international language of reference. In other words, what are claims in specific fields have to come to Geneva to get validated at conferences . . . there are seven expert conferences per day in Geneva. That's the most important activity in that city. And so all claims to knowledge that are dispersed now come to converge . . . and in Geneva this suddenly becomes the language of reference that governments and other international institutions try to introduce and use. I think we need an understanding of various other places in roughly these terms – what they do, what flows through them, how a globally inflected organization takes shape, etc. Geneva has an elegance to it, but really that is not very important. So there is a tourist industry, but what we really need to accommodate are those 735 experts concurrently on different days. That means that there has to be a printing industry, a translation industry, all kinds of things that are organized in that way. And that becomes reproduced in the rest of the society . . . and that produces then other kinds of opportunities . . . you know, secondary or tertiary, and so on. . . .

I think that the kinds of contemporalities that we are dealing with at present are not going to depend on notions of mediation, because notions of mediation essentially try to regulate certain kinds of flows and to reorganize them. That would prove simply inadequate for what we are doing. And I'm afraid that an ethnographic approach is going to finesse this simply by being descriptive at this very moment where in fact it's the moment that requires the most analysis. I think the best model that we have at the moment is that of arbitrage. Arbitrage is the handling of the contemporalities, that's all it is. It is in fact making decisions and it's not negotiating . . .

This is a post-hermeneutic situation, it's not a question of understanding what the causes are . . . but rather, what do you do? The paradigm shift means not that you are actually going to be describing things, but that you are going to start looking at things differently. Methodology is no use right now because a methodology is essentially a blueprint of a road that you are going to follow. Instead, you can only pose certain kinds of goals, expectations, or to give an idea of the extent of the area that you are trying to saturate. All you can do is say, I'm going to take these, these and these factors and I'm going to look at their various combinations and this is where I am going to place my intervention. It is within these moves that you have to be concerned with issues of arbitrage, having to do with the recognition that there are different temporalities. And temporalities in this instance means, in a Deleuzian way, intensities, because what is at the heart of talking about events is, as Deleuze said, disjunctive syntheses. Current events do indeed conjoin and disjoin things, but the significant thing produced in events are new dispersals. You cannot predict what is going to be, or the spread of things. But what you can do by looking at the kinds of intensities and by trying to measure the kinds of intensities is to get a sense of the potential productivity of an event. And this is what those who engage in arbitrage do. If they predict correctly, they survive and make money; if not, they go down. And in some sense, we who are engaged in thinking in and about the present, whether we want it or not, are cast in a very similar role as the arbitrageur . . .

In conventional cultural analysis with its valorization of the everyday, events were not being considered, the focus on the everyday is a way even of avoiding the notion of the event. But events are these disjunctive syntheses. And they have to be considered. We have to bear in mind that an event is not something that jumps at you, but that you have to construct . . . so it's a relationship. And it is for that reason that anthropology was able to focus on spatial and geographic relations, but the moment it focuses on temporal ones then it has to take into account that notion of event and also avoid falling into the trap of simply hierarchizing events . . . this is an event of world importance . . . whereas this is something about the dog that bit the postman, or something like that . . . We no longer live in a world that can afford perspectivalism of a Nietzschean sort, where perspective was born of this very strong notion of locality – simply, this is my deixis, this is where I'm anchored, this where I do things from. We find ourselves for the time in a situation where, in terms of linguistic categories, we have no deixis. So that language is not taking place . . . this is what virtual means . . . that things no longer take place. What does it mean to no longer have access to that kind of deixis, where when you read an e-mail message, you don't really know what the deixis of that e-mail message is, because it can be routed from so many locations? When I send an e-mail message to people, they think it comes from Geneva, but I'm sending it from South Africa. So what is the deixis of that e-mail message? They have no idea. Increasingly, we will have these abstract addresses where things reach us, but they don't locate us. So we have to think of a system of address. And what we need to do then is to allow what used to be called the native perspective to be a mode of address, a way of entering into a conversation. That's really what a mode of address is, it's a way of entering into a conversation. The best example of that kind of address is in the Odyssey, when the dead address the living. It is a moment when Odysseus does not know where he is, and what they are telling him is don't worry where you are on the surface of this water here, that is not what matters. You need to know where you are in time . . . what is your genealogy. And that is what I mean by this mode of addressing. Addressing, again, is a temporal mode of locating oneself, rather than this kind of geographical mode . . .

# Notes

1.  The sense of this more precise and segmented interdisciplinarity is very evident now. For example, even though in this paper I am evoking the borders between anthropology and geography, I have recently been approached by sociologists who have pursued ethnography in their discipline and who have suggested fusion with the tradition of ethnography in anthropology in the name of the changes impelled by globalization. I would say this sort of proposal to form micro, precise interdisciplinary partnerships is typical of this moment. In general what the cultural turn has legitimated pragmatically are modes of research out of particular disciplinary trainings that are experimental, result in 'messy' texts (see Marcus, 1999a), and put theoretical resources together in unusual ways. The emergence of multi-sited ethnography, discussed in this paper, is very much in this trend. In a review of the recent book by Mary Poovey,

*A History of the Modern Fact*, John Brewer gives an elegant characterization of this trend that I think is a second wave of interdisciplinarity, grounding the cultural turn in enquiries:

> Poovey's wide-ranging and erudite study seems to me to exemplify a growing body of American scholarship that is united not by its subject matter but by what Poovey herself calls 'its mode of argumentation.' Such works written by scholars over the tenure barrier but still far from the twilight zone, share a family resemblance, a common 'poetics.' They are lengthy and learned, formulate an object of analysis that has no obvious place in one discipline, and experimentally apply critical theory rather than produce it. (Brewer, 1999, p. B22)

2. Tangentially, I offer a description of one paradigm for non-obvious multi-sited research. The presentation is abstract but several projects in science studies resemble it (e.g. see Galison, 1997), as do studies of the circulation of activist and political discourses (see Holmes, in press; and Fortun, in press). More specifically, it reflects my own experience of studying dynastic families and fortunes in the United States (see Marcus, 1992). Schematically posed, the paradigm has these features: there is very little actual contact or exchange between two sites but the functioning of one of the sites depends on a very specific imaginary of what is going on elsewhere. The complex nature of the relation between the disjunctive sites, how they are co-ordinated, if they are, is the main objective puzzle of the ethnography. Fieldwork in an initial site is interested primarily in a shared imaginary of a set of subjects derived from attention to situated discourse, as so much ethnography is these days. It is their social cartography or social referents elsewhere within this imaginary that become particularly important. The initial fieldwork in the interest of its multi-sited strategy is looking for a current preoccupation with an elsewhere that bounds a situated imaginary. It is the referent of an imaginary discovered in initial fieldwork that suggests where the fieldwork will literally move and contextualize itself by enquiry in a related site. The fieldwork literally moves then to what the imaginary within the first site refers to or constructs as its dynamic other, and makes a second ethnographic object of study of that referent. Literalness, as a kind of naive realism, in making this move, is, I would argue, both a virtue and provocation of the project. The fieldwork in the second site is often different in nature than in the first site. It is perhaps less intensive than in the first, but always with the first site in mind. It is fieldwork more in the manner of Malinowski than, say, Geertz. The second site is probed for its relation to the first site, and this becomes the foremost question. Is there a reciprocal relation at the level of imaginary or not? Is there a material relation, one of periodic exchange, or is the relation totally virtual? In the second site does the social imaginary there point elsewhere? In what manner? Most interestingly, how does an ethnographic project whose fieldwork stops here at this second site leave itself open, satisfying the practical need to bound the potentially unbounded while leaving the unbounded imagined not as some abstract system but as a further network of sites that could be investigated? The project might end with this analysis and interpretation of the relation or connection of the two sites across disjunction where the imaginary at one site is juxtaposed to the ethnography of its literal referent, but a third phase of such a multi-sited project would define itself as intervention, with some strategy of bringing back the ethnography at the second site to the first site, as a practice of critique, involving re-engagement with one's original subjects from whose imaginaries and regimes of representation the impetus and strategy for moving the project literally elsewhere was derived.

This paradigm for multi-sited ethnography – distanced from conventional geography – makes its own geography in collaboration with one's subjects. This is a geography forged from the found (and cartographic) imaginaries of subjects and the interests of the mobile curiosity of the ethnographer in relation to them.

3. Wlad Godzich is a very interesting person to be making such claims. He was a key and instrumental figure in making available the crucial theoretical texts to the inter-disciplinary movements of the 1970s and 1980s, through the *History and Theory of Literature* series, published by the University of Minnesota Press, and produced through the unique programme in comparative literature that he created at the same university. In the early 1990s, he moved away from a very active role in the 'scene' of the cultural turn in North America (holding simultaneous faculty appointments at Harvard, Minnesota and Montreal), to a professorship in Geneva, understanding perhaps precociously the end of the cultural turn's theoretical moment. During his time in Geneva, he has become a true global scholar, moving between consultancies to 'masters of the universe' – writing reports and position papers for corporations, banks, financiers and policy experts – and producing projects with local scholars in places like South Africa, Brazil, China and Eastern Europe, that gauge the state of emergent literatures, a concept that he first developed at Minnesota. He has thus been developing points of access to contemporary change on the surface of events, in ways that are rhizomic rather than totalistic, that create a symptomology that constitutes a grounded sense of the social and that challenges more orderly perspectives, and improves upon them in so doing.

# References

Appadurai, A. (1997). *Modernity at Large*. Minneapolis: University of Minnesota Press.

Brewer, J. (1999). 'The Cult of the Particular.' A review of *A History of the Modern Fact* by Mary Poovey. *Lingua Franca* 9(3): B20–B22.

Clifford, J. and Marcus, G.E. (eds) (1986). *Writing Culture: the Poetics and Politics of Ethnography*. Berkeley: University of California Press.

Cook, I. and Crang, P. (1996). 'The world on a plate: culinary culture, displacement, and geographical knowledges.' *Journal of Material Culture*, 1(2), 131–53.

Fortun, K. (in press). *Advocating Bhopal*. Chicago: University of Chicago Press.

Galison, P. (1997). *Image and Logic*. Chicago: University of Chicago Press.

Holmes, D. (in press). *Integral Europe*. Princeton: Princeton University Press.

Jackson, P. and Thrift, N. (1995). 'Geographies of Consumption', in Miller, D. (ed.) *Acknowledging Consumption: a review of new studies*, London, Routledge.

Katz, C. (1992). 'All the world is staged: Intellectuals and the projects of ethnography', *Environment and Planning D: Society and space*, 10, 495–510.

Marcus, G.E. (1992). *Lives in Trust*. Boulder: Westview Press.

Marcus, G.E. (1999a). *Ethnography Through Thick and Thin*. Princeton: Princeton University Press.

Marcus, G.E. (ed.) (1999b). *Critical Anthropology Now*. Santa Fe: School of American Research Press.

# More words, more worlds

## Reflections on the 'cultural turn' and human geography

*Chris Philo*

## *Introduction*

> Increasingly, it seems, there is a tendency among those closely associated with
> the 'new' cultural geography to eschew the use of the phrase 'cultural turn',
> just at the moment it begins to take on a certain solidity within the discipline.
> (Barnett, 1998a, p. 379)

Clive Barnett is only half right in this statement, since a number of human geographers, including those contributing to this volume, *are* prepared to talk about
a 'cultural turn' now occurring within the discipline.[1] Even so, he is probably
correct to identify a certain reticence in this respect, given that talk of a cultural
turn implies that one can discern a modicum of unity, of coherence, to this
emerging shape within the overall field of human geography. Ron Johnston
chooses to use this term for the title of a chapter in the recent edition of his text
*Geography and Geographers* (Johnston, 1997, Chap. 8), but he swiftly admits
that, rather than there being any obvious 'core' to the developments involved,
they are best described simply as 'a series of sometimes interwoven strands
coming out of previous approaches' (Johnston, 1997, p. 271).[2] From the chapters which follow in this book, it will be possible to gain a rich sense of these
many 'interwoven strands', and it is very much in the details of these chapters
that the value of human geography's cultural turn (and also of cultural studies'
'geographical turn') should become apparent. Nonetheless, it is important that
some attempts are made to appreciate the overall direction being taken by this
cultural turn, and where appropriate to offer (constructive) criticisms of what
has or has not been achieved. And, foolishly, this is a task that I wish to undertake here in helping to set a context for what follows, although in so doing

I must immediately admit to borrowing heavily from the views of others such as Gregson (1993).

I am not intending to provide anything like a panoptic survey of this uneven and shifting intellectual landscape, although relatively comprehensive surveys can be found in pieces already mentioned (Barnett, 1998a; Johnston, 1997, Chap. 8) as well as in Linda McDowell's thoughtful overview (McDowell, 1994)[3] and in regular progress reports published in the journal *Progress in Human Geography*. It is important, I feel, to resist the impression of sitting in some kind of satellite circling human geography's cultural turn, claiming the 'scopic power' to see clearly all that is taking place but which others closer by cannot themselves comprehend. In the spirit of what Cindi Katz (1996) has argued about 'minor theory', I wish to offer reflections which are themselves somewhat hesitant, asking questions as much as making definitive judgements, and in so doing I will also be less in praise of self-consciously important position statements and more in sympathy with those very many human geographers with a cultural interest who are now mixing up conceptual elaboration with substantive detail. In addition, my discussion will refer to certain projects familiar to me personally, one of which is particularly significant because it influenced the cultural turn within British human geography during the opening years of the 1990s, and I will also touch upon one or two of my own writings in the process. Given this personal involvement in much that I will cover, the text will inevitably include some first person commentary, observation and judgement, and I hope that this personalized tone will not detract from the broader arguments being advanced. I have also decided to retain the more 'conversational' style originally adopted for the conference paper from which this essay is now derived (see the editors' introduction).

Back in the early 1990s I did have occasion to identify a cultural turn emerging within human geography (Philo, 1991a, p. 3), and the prompt for this remark was a project which became known as the 'New Words, New Worlds' initiative. This initiative was co-ordinated by what was then the Social and Cultural Geography Study Group of the Institute of British Geographers,[4] and it entailed enlisting inputs from numerous British human geographers of varying backgrounds, career stages, conceptual orientations and substantive concerns. Following through several phases, the initiative culminated in a sizeable conference held in Edinburgh during September 1991, and then in the compiling together of the various inputs (position papers, a discussion document, conference papers and postscripts) for a small desk-top publication entitled *New Words, New Worlds: Reconceptualising Social and Cultural Geography* (Philo, 1991b). One or two authors at the time commented that this initiative bore witness to a new departure in how geographers, or at least British geographers, were dealing with matters of culture (e.g. J.S. Duncan, 1993a, pp. 372–6), and in retrospect similar claims are commonly made about its contribution to a broader cultural turn within the discipline (e.g. Barnett, 1998a, p. 381; Johnston, 1997, pp. 314–15).

A first point to notice, however, is that the initiative was undoubtedly but one straw in a rather larger wind. It certainly did not by itself kick start a wholly new orientation, since in many respects it simply picked up on a slate of changes (of approach, subject matter, emphasis, style, politics) which were then coursing through (and beyond) the discipline. Indeed, it emerged from the recognition of the Study Group Committee that its members were beginning to think differently about the very objects of 'the social' and 'the cultural' which featured in the name of the Group (and it was only in the late 1980s that the Group changed its name from just 'Social Geography' to 'Social and Cultural Geography': Cosgrove, 1988). The original purpose of the initiative was hence to cast a critical eye over this reconceptualizing of both 'the social' and 'the cultural', and in the process to tease out the implications for what might be taken as the subdisciplinary fields of both 'social geography' and 'cultural geography'. Elements of this purpose can still be traced in the *New Words* compilation, but with hindsight the main achievement was arguably to heighten our senses of how all things *cultural* might be raised to a much more prominent position in studies throughout the corpus of human geography (and not just in one or two neatly parceled-off subdisciplines). It was to take much more seriously than hitherto all manner of things that might be construed as constituting the cultural 'stuff' of human life, not just phenomena routinely designated as cultural (e.g. 'highbrow' arts and 'lowbrow' media), but also the complete panorama of meaning systems both collective (e.g. religions and nationalisms) and more individual (built up in personal psychic economies).[5]

This was to anticipate the amazingly rich arc of the cultural turn within human geography as it has now arisen: a turn which has obviously had great ramifications for both social and cultural geography, but one too which has undoubtedly sent shockwaves throughout the length and breadth of human geography, leading to debates (more or less explicit, more or less heated) about the merits of a cultural turn within (say) economic geography, political geography, population geography, environmental geography and elsewhere. This is not the place to detail these debates (although see Parts III and IV of this collection), but what I will insist is that it was never the intention of the original initiative to indulge in some kind of 'empire-bilding' on behalf of a reformulated cultural geography. I know that there have been some mutterings of complaint along these lines,[6] but we are talking here about 'unintended consequences' of our project, albeit consequences that I personally would agree have not always been entirely positive. Let me therefore acknowledge my worry that in certain respects the cultural turn has been *too* successful, has become *too* hegemonic, and has led to the realms of (say) economic and political geography making too many accommodations with a cultural orientation.[7] This has perhaps meant that human geographers have lost interest in certain domains of (say) economy, politics, demography and nature which arguably should still command our attention because they remain so inescapable in how today's human geographies are pieced together in the world outside of the academy. Moreover, I would go so far as to suggest that the success of the cultural turn is actually now posing a few problems for the study of *social* geography, the subdisciplinary realm which, in Britain at least, initially spawned the present cultural turn. And this is actually a key claim of my paper here: namely, the threat to social geography posed by the cultural turn.

There are a number of other introductory points which I really ought to make, since passing mention should be made of differences between the British experience and that of both North American geography (where the development of a new cultural geography has struggled with the baggage of an older Sauerian cultural geography: see Price and Lewis, 1993, and the responses from Cosgrove, 1993; J.S. Duncan, 1993b; Jackson, 1993) and perhaps too of contrasts with how culture has been treated in other national or regional schools of geography (in France the Vidalian heritage is obviously crucial: see Claval, 1998; a sense of the Australian perspective can be gained from Anderson and Gale, 1992).[8] Attention might also be given to the possible causes of the cultural turn, in which case reference should be made to Barnett's (1998a, 1998b; see also Castree, 1999) account of how the discipline has been affected by the growing popularity of cultural studies as propelled by an academic publishing industry looking for products attractive to growing student markets. And further notice might be given to the reciprocal moves whereby geography's cultural turn has diffused into wider, inter- and multidisciplinary realms, cultural studies included, while arguably being invigorated in return by new currents flowing in from such realms (and see Part II of this collection). Rather than expanding upon such points, crucial though they are, let me instead simply underline my own enthusiasm regarding the changes accompanying the cultural turn in human geography. Indeed, my basic response is one of excitement at the explosion of new studies emerging under the loose umbrella of the so-called turn. I think that all of this newness is highly stimulating, and is creating a new ambience for human geography that has blown away many cobwebs of convention, conservatism and even downright prejudice. Any critical remarks that I make in what follows should be seen in the context of my underlying sense that what is happening in this respect is a progressive thing for the discipline, which is of course *not* to say that it is then immune from criticism.

Before getting into the heart of my paper, there are one or two additional worries that I should first express. For instance, there are just so many good new things to read which look interesting in the literature, packing out the pages of even the supposedly more 'staid' geographical journals, that it is incredibly difficult to 'keep up': to retain either a sense of the overall intellectual landscape or a knowledge of the more detailed excursions within this landscape. What is also difficult is the pace of the changes occurring: the rapidity with which new possibilities for thought and practice come tumbling into the literature. To give one tiny personal example, just as I was coming to terms with arguments about the value of 'giving voice' to research subjects, particularly those who are powerless, disadvantaged and marginalized, serious questions have begun to be raised about the assumption of unitary self-present subjects who can somehow 'voice' them-*selves* (note in particular work by Gillian Rose and others drawing upon psychoanalysis: e.g. Rose, 1997; see also Part V of this collection). Furthermore, other geographers have begun to question the very politics of forcing 'subaltern' peoples to have a voice when maybe strategies of silence are more relevant to them (note in particular work by Clive Barnett and others drawing upon strains of postcolonial and critical literary theory: e.g. Barnett, 1997). It might be objected that this pace of change is simply a matter of 'being trendy', of continually trumping one another by always being one step ahead in the theoretical marketplace, and I suppose that this is an element of the cultural turn (Mohan, 1994; see also Barnett's, 1998a, discussion of academic 'fashions'). Even so, I genuinely think that the older combative model of academic 'cut and thrust' has been largely erased in the work of today's human geographers, so that papers by the likes of Rose and Barnett are presented in the spirit of additional (if important) issues needing to be stirred into the mix of endeavour, rather than in the vein of saying that all previous treatments of given issues are completely misguided and in need of total renewal. (I doubt that everyone will agree with my assessment here, though, and there is definitely scope for thinking further about the current state of what might be called the 'politics' of disciplinary debate: see Short, 1998, Part 2.)

Having given a thumbnail outline of the cultural turn in human geography, and having also cleared some preliminary ground, in the remainder of the paper I now wish to offer a handful of more focused reflections which *do* contain criticisms of what an overplayed cultural turn can instil into our theorizing and researching of human geography. My reflections are organized into two sections which I entitle, respectively, 'dematerialized geographies' and 'desocialized geographies', in the course of which I will develop an argument of sorts about certain dangers attaching to how the cultural turn has drained some of the substance out of what many of us used to conceive of as *social* geography. This is not a clarion call for a return to a social geography *without* culture, however, but rather a plea for some of our studies – but certainly not all of them – to retain a feel for the place of the cultural within those various social shapes, forces and structures that social geographers themselves have only introduced to a disciplinary audience relatively recently (from the 1960s onwards).

## *Dematerialized geographies*

In my opening piece from the *New Words* compilation (Philo, 1991a), I reviewed a long-term trajectory whereby human geographers have gradually overcome their *fear of the immaterial*, of things without obvious material expression in the world, and have thereby opened up the possibilities for the kind of cultural turn which we are considering in this collection. I indicated that traditional human geography, covering the first half of this century, generally took the end-point of its enquiries to lie in patient accounts of obvious, tangible, countable and mappable phenomena present to the senses (primarily sight) of the geographical researcher. It thereby concentrated on observable human modifications of the natural environment (e.g. clearing the forest) or equally observable human productions in the landscape (e.g. farms, factories, settlements, transport routes), and it might consider these individually (perhaps tracing their environmental determinations) or it might consider them more collectively in the shape of distinctive regional assemblages. I also hinted that geography as spatial science, emerging in the 1950s and 1960s, replicated this obsession with the material world, albeit translating the configurations of material phenomena into more abstract geometric representations supposedly disclosing the deeper truths of humanity's spatial organization. Very rarely did such versions of human geo-graphy consider the possibility that there might be other sorts of phenomena, defiantly immaterial ones, that could possess their own distinctive (and significant) geographies or be highly relevant in explaining the more material patterns normally studied. We might wonder about this aversion to the immaterial. Was it simply a by-product of a particular view of scientific enquiry, a positivist one with its strict criteria about what might count as legitimate candidates for existence, for the attribution of ontological status, which would then be worthy of scientific attention?[9] Or is there some deeper story to tell about an anxiety on the part of earlier geographers when encountering the mysteries of the unseen, the mysterious, whether it lies 'out there' in (say) the political-economic machinations of a capitalism that they dare not question or 'in here' in (say) the impulses, desires and hauntings of the human psyche?[10]

Now, this narrative is certainly partial, in that it is possible to find examples of earlier geographers *not* shying away from the immaterial or, if not exactly embracing it, clearly recognizing it as something which must on occasion leak into the discussion. Think of Peter Kropotkin (the wonderful early anarchist geographer) nodding towards the role of such not-immediately-materially-present forces as state bureaucracies, national jealousies and racial prejudices in moulding uneven human geographies (Kropotkin, 1885; Mac Laughlin, 1986); think of J.K. Wright's tireless efforts to bring the *terrae incognitae* of human imaginings, from those of the Crusaders to those of North American Indians, into the picture as 'geosophical' adjuncts to traditional human geography (Wright, 1925, 1947); or think of Jean Brunhes's provocative claims about what he termed 'the psychological factor', including collective wills with their 'spontaneous or deliberate appetite', intervening between 'natural phenomena and human activity' (Brunhes, 1920). All of these early experiments in geo-graphical thought, along with J. Wreford Watson's (esp. Wreford Watson, 1951, p. 469: see Philo, 1991a, pp. 8–11) mid-century urgings that we should take seriously 'a great number of heterogeneous things . . . , immaterial as well as material', evidently anticipated the later shift towards a human geography alerted to the realms of the immaterial.

Again, to offer sketchy but I think not entirely indefensible generalizations, the twin routes out of spatial science after *circa* 1970 signposted by, respectively, radical/Marxist geography (with its focus on social structure) and behavioural/humanistic geography (with its focus on human agency) *both* offered startling new voyages away from an obsession with the material into the complications of the immaterial.[11] Some might object that the first-mentioned route, into a radical and increasingly Marxist geography (e.g. Harvey, 1973; Peet, 1978), was avowedly *materialist* and as such did not entail the brush with the immaterial just claimed. Given that we are here talking about a move towards a Marxist 'historical materialism' (subsequently reframed as an 'historical-geographical materialism' by many authors: esp. Harvey, 1984, 1990), an orientation pro-claiming its analytical foundations in the elemental operations of humans on nature, then such an objection seems justified. However, it might be replied that the whole architecture of Marxist thought – conducting theoretical work to unpack the hidden structures of social relations as bound up with control over the means of production – is all about bringing into the consciousness of the academic (and maybe too of the downtrodden) the reality of structures which are themselves immaterial (in the sense of not being immediately available to human sensory apprehension) but which nonetheless have dramatic material effects in the well-being or otherwise of everyday people.[12] For behavioural and more so for humanistic geographers (e.g. Ley and Samuels, 1978), the entertaining of the immaterial was of course much more obvious, since these scholars readily began detailed investigations into the shadowy recesses of human perception, cognition, interpretation, emotion, meanings and values, creating a rich vein of enquiry for which terms such as 'mental maps' and 'senses of place' stood as helpful if inadequate shorthands. The details as such do not matter for my purposes here however, for the crucial thing is simply that the fear of the immaterial was breached. In all sorts of ways, from a diversity of perspectives and reflecting a diversity of conceptual, methodological, political and other motivations, the immaterial began to be more fully released into the studies of human geographers.

My next claim is that these valuable manoeuvres paved the way for the cultural turn of more recent vintage. I am again oversimplifying, but it strikes me that the wider horizons of late 1980s human geography which gave birth to the cultural turn included a variety of developments which extended, and in many cases recast, the Marxist and humanistic revolutions of the previous decade (for overviews, if partial ones, of the many developments involved here, see Barnes and Gregory, 1997; Gregory, 1994; Massey, 1994; Peet, 1998; WGSG, 1997). The reworkings of Marxist geography – the spatializing of Marxist concepts; the restructuring, spatial divisions of labour and locality studies debates; the advent of regulation and regime theories; the critical realist encounter; the agitations around post-Marxism[13] – have all created a receptiveness to the ways in which immaterial cultural processes become implicated in political-economic spaces. It is true that some commentators do not see the cultural turn as having anything to do with Marxist geography, and I realize that some who might see themselves as heirs of Marxist geography can be hostile to this cultural turn (Sayer, 1994),[14] but I would suggest that much of what has occurred within the cultural turn does have crucial reference points back in ongoing dialogues between Marxism and geography (a point which I will revisit from another angle in Box 1 on pages 38–40). More obviously, though, the reworkings of humanistic geography – the reappraisal of the landscape tradition; the growing attention to intersubjective meaning systems; the fragmenting of the singular figure of 'Man' in recognizing the sheer diversity of peoples occupying this planet; the new mapping of the human subject(s); the alertness to the psychodynamics of gender and sexuality[15] – have all fostered a sensitivity to the many dimensions of immaterial culture which enter into the making of virtually all human spaces imaginable. This furore within what a few writers are now calling 'post-humanistic geography' (Barnes and Gregory, 1997, pp. 359–60), as fuelled by inputs from sources as diverse as feminism, postcolonial theory, psychoanalysis, social psychology and cultural anthropology, has been central to the cultural turn (which has of course then acted as a spur to further borrowings from, and adaptations of, such intellectual sources). Indeed, a different way of putting things is simply to say that what I have so gesturally outlined in this paragraph *is* the stuff of the cultural turn.

Whatever the details, though, the logic of my argument is that this gradual overcoming of the fear of the immaterial within human geography can be traced from a few older antecedents, through the excitements of Marxist and humanistic revolutions, to a fully-fledged embracing of the immaterial in the fires of the recent cultural turn. I also want to reiterate my applause for this outcome: it marks an important maturation of certain trends set in train by generations of geographers fumbling to forge something more inclusive and challenging than what has been offered by traditional and spatial-scientific versions of the discipline (accepting that there are still useful aspects of these older geographies worth retaining or revisiting). Yet, what I wish to signpost now are some concerns that I (and others) have about this *dematerializing* of human geography: this preoccupation with immaterial cultural processes, with the constitution of intersubjective meaning systems, with the play of identity politics through the less-than-tangible, often-fleeting spaces of texts, signs, symbols, psyches, desires, fears and imaginings. I am concerned that, in the rush to elevate such spaces in our human geographical studies, we have ended up being less attentive to the more 'thingy', bump-into-able, stubbornly there-in-the-world kinds of 'matter' (the material) with which earlier geographers tended to be more familiar.

A number of people who might be identified as new cultural geographers have already said quite a lot in this respect, Don Mitchell (e.g. Mitchell, 1995) being maybe the most obvious example. Moreover, my impression is that a *re*-introduction of the material into human geography, and even into the new cultural geography, is something that is already starting to occur (hence deflecting my critique before I have even finished making it). One possibility can be found in the moves being urged by writers drawing upon Henri Lefebvre's reworkings of Marxism, and in this respect I like Virginia Blum and Heidi Nast's recent recovery of Lefebvre's 'emphasis on embodied agency' ('He asks who is holding the pen, who is thrusting the fist, who occupies the towers, etc.': Blum and Nast, 1996, p. 560) and also Derek Gregory's (1994, 1995, 1997; see also Stewart, 1995) writings on Lefebvre's long-term history of bodies and spaces. This mention of bodies is also instructive, since a number of studies are now emerging which examine what occurs to the 'fleshly' geographies of human bodies (as material eruptions in space) when they are disciplined into certain postures and conducts, inscribed with myriad values and expectations, and mobilized as possible sites of resistance and repositories of counter-hegemonic meanings. Various references could be provided in this respect (see, for instance, the contributions to Nast and Pile, 1998, and the survey by Longhurst, 1997; see also Part V of this collection), but I specifically want to mention John Bale's fostering of a sports geography interested in the making and breaking of 'body cultures' through the highly spatialized demands of different sports regimes (e.g.

Bale, 1996, 1999; Bale and Sang, 1996; Bale and Philo, 1998). It is apparent that for all of the geographers cited here it is not a simple question of stressing the material *or* the immaterial, however, for the strength of what they are doing is to re-vision the relations between the material and the immaterial: in fact, they would probably insist upon the need to see the one in the other, and vice versa, to the point where the binary opposition has been dissolved. (The recent 'Latourian' turn in human geography, inspired by the social studies of science and technology associated with Bruno Latour and others, is surely trying to achieve something similar in stressing the inescapable hybridity of nature and culture, practices and discourses, things and (no)things: e.g. Bingham, 1996; Demeritt, 1996; Hinchliffe, 1996, 1999; Murdoch, 1997a, 1997b; Thrift, 1996; Whatmore, 1997; see also Part IV of this collection.)

This is to run a little ahead of myself, though, and to risk making claims that I cannot follow up any further in this paper. Instead, let me be more specific in wondering about this dematerialization of human geography, and let me do this by considering two papers running one after the other in a recent issue of *Society and Space*. My choice of these is pretty much random, merely reflecting the fact that I read them together a few days before writing the original conference paper from which the present chapter is derived. They are both in their own ways excellent pieces, almost certainly containing more depth and subtleties than I can detect, yet I am intrigued by certain key differences of content and emphasis which lie beneath their surface similarity as instances of a sophisticated, theoretically informed new cultural geography (and both of which consider questions of utopia, 'heterotopia' and 'ageographia').

The first paper is by Richard Smith, entitled 'The end of geography and radical politics in Baudrillard's philosophy' (R.G. Smith, 1997), and it provides a sustained enquiry into Jean Baudrillard's ideas about late capitalism's annihilation of 'the real' in the virtual spaces of the 'sim-territory', drawing out various lessons for any attempt to conceive either of 'geography' or of 'radical politics'. Smith actually criticizes some remarks of my own (in Philo, 1992)[16] about what Baudrillard might mean by the stray term, 'the geography of things', which surfaces in one of his writings (Baudrillard, 1987), and Smith argues that my own interpretation – which positions Baudrillard alongside Michel Foucault as resisting the abuses of totalizing 'grand theories' or 'metanarratives' in the name of awkward worldly 'substantive geographies' – is really untenable. I certainly defer to Smith on this count, but the ramifications of what he ends up arguing about the apparent irrelevance of thingy, fleshly, mucky stuff in the world (however faithful a rendition of Baudrillard) does disturb me greatly. I am probably not understanding him fully, but I am puzzled by Smith's happy acquiescence in the apparent loss of any substantive content from both geography and radical politics – from how these might be conceived – as seemingly sanctioned by Baudrillard's rhetorical stress on 'a multiplicity of ends or liquidations' (R.G. Smith, 1997, abstract, p. 305). I am aware that there are many who quarrel with Baudrillard on such grounds, as in reactions to his infamous claims about the Gulf War only happening on television, but my unease in this regard remains with how Smith himself develops his reasoning about

Baudrillard. He seems content to take quotes from Baudrillard as if they are accurate accounts of what is really going on in today's 'actual hyperreal society'. If Baudrillard has said that the real world has disappeared because everyone is sitting at home watching television, then that must be true. Similarly, Smith repeats claims from both Virilio and Debord about the 'cancellation of town and country' (in R.G. Smith, 1997, p. 308), and from Sorkin (1992) about the American city becoming an 'ageographical' no-place of nothing but simulations (one big Disneyland), and does so without noting that such claims might come as rather surprising to countless people struggling to get by in the middle of very material cities and countrysides whose everyday mundane reality (the buildings, the distances, the deprivations) is never in doubt to them. I realize that these are obvious criticisms, but I am convinced that they remain ones worth pressing when confronted with a paper from someone who – his brilliance at discussing Baudrillard notwithstanding – appears to inhabit a strange dematerialized world of texts, thought-games, electronic transmissions and simulations (*and imagines that everyone else does too*). It is hence revealing that the opening sentence of Smith's paper reads as follows: 'There has been an explosion in the jargon of space after the rapid development of the various virtual reality technologies from which *we* increasingly gain our spatial experience' (R.G. Smith, 1997, p. 305: my emphasis). But who, and where, are *we*?

The second paper is by Loretta Lees, entitled 'Ageographia, heterotopia and Vancouver's new public library' (Lees, 1997), and it provides a lengthy examination of this new public space for Vancouver in the light of both Foucault's notion of 'heterotopia' (as a potentially subversive site of difference: Foucault, 1986) and Sorkin's notion of 'ageographia' (as a basically conservative site of simulated sameness). In the course of a piece which I certainly enjoyed – a piece of 'minor theory' which cheerfully mixes up theoretical debates, more personal statements of belief and a detailed exposition of everything about the Vancouver public library (from the 'wrecked grandeur' of its Colosseum-like exterior to the intimate design of its children's section) – Lees ranges freely across a host of different material and immaterial realms. The brute thereness of the building and its contents is described and in effect mapped out for the reader, while attention is paid to precisely who can access the library, with reference to both the bodily exclusion of 'undesirable' homeless people by a private security force and the clear divide between the library and the neighbouring Granville Street district (which contains many 'street kids and homeless'). At the same time more immaterial aspects of the library's services are documented, and in particular Lees reports on the 'intellectual spaces' (the 'cyberwaves') of the internet which can be freely accessed by whoever turns up and in partcular by anyone who reaps the benefit of the library's free computer literacy classes. Intriguingly, given my own line of reasoning in this chapter, Lees writes as follows about this intellectual space within the library: 'Though abstract and immaterial, this site is directly related to the material site of the library by the very fact that the resources themselves, or the resources for accessing other resources (that is, computers), are situated in the library building' (Lees, 1997, pp. 341–2). Moreover, she observes that the library is 'standing astride the material–immaterial dualism' (Lees, 1997, p. 342) in part because the spaces of thought which are accessible through its printed and electronic media do themselves have an inescapable material dimension in being anchored in books and computers (themselves positioned in certain parts of the building rather than in others) which are not automatically or instantaneously available to all. This means that Lees's approach is very much more 'earthed' than is Smith's: she is under no illusion about the gaps, privilegings and exclusions endemic to who can access the material sites from which it is possible to 'surf the cyberwaves', and as such hers is a much more materialized geography full of substance (it does retain a sense of being the 'earth-writing' that Smith so dislikes). This does not make her account any less attuned to difficult theoretical issues, and in practice she debates matters neatly parallel to those preoccupying Smith, although it does perhaps make her less prepared to give up on asking questions about substantive geographies and radical politics (including ones about the role of public spaces in empowering people who are disadvantaged). Whatever precise judgements might be made about the two papers, both of which would certainly be positioned as representatives of the cultural turn in human geography, it should be evident that the first (R.G. Smith's) dematerializes his geography while the second (Lees's) shows that the new cultural geography can easily *rematerialize* geography (albeit in the context of a vision sensitive to the complex fusings of material and immaterial).

## *Desocialized geographies*

Looking back once more at my opening piece in the *New Words* compilation, I notice that, immediately prior to my discussion of human geographers overcoming their fear of the immaterial, I had also speculated about some drawbacks that might be occasioned in the process. I wrote as follows: 'I am sure that some would feel . . . that a process originating in rethinking "the social" had ended up (ironically enough) spawning a "cultural turn" in which "the social" itself becomes completely deconstructed' (Philo, 1991a, p. 3). This being said, I was indeed hardly the only one, and elsewhere in *New Words* itself it is possible to find this point rehearsed in different ways: with Stephen Daniels (1991, p. 165) warning us that '[t]here is a world out there, of physical as well as cultural difference, that we are in danger of forgetting'; with Nigel Thrift (1991) wondering about the trap of descending into a preoccupation with 'over-wordy worlds'; and with both Michael Keith (1991) and Peter Jackson (1991) clearly stating the need for dimensions of (for instance) social struggle and urban racism to remain securely in the frame of our enquiries. In Keith's paper, in particular, it is possible to see a remarkably prescient discussion of Spivak's 'strategic essentialism' as one way of retaining a materialist moment (but no simplistic 'romance of the real': Keith, 1991, p. 191) in the face of 'the sort of metaphoric spatialisation of social theory that Anglo-American cultural understandings of the urban exemplify' (Keith, 1991, p. 186: see also Smith and Katz, 1993). Such a view then led Keith to accept that 'it is essential to return analytical categories to the firmly social contexts of racism' (Keith, 1991, p. 190). Soon after reading *New Words*, Nicky Gregson (1993, pp. 528–9) arrived at a view which in effect extended such worries into a more trenchant critique of how, '[r]ather than rethinking the social, then, it is the social itself which is being deconstructed in the current reflexive phase'. As she added, there is a risk here of slipping into an 'in-house dialogue' entailing 'the retreat from the empirical world', with the geographers concerned 'redefining their project[s] so as to exclude empirical social worlds and the others who are seen to constitute them' (Gregson, 1993, p. 529).

I wish now to echo such sentiments, and in a more basic sense I wish to underline the need – in our formulations post the cultural turn – *not* to evacuate both the social and social geography of their substantive 'guts'. What I mean by this is that we need to keep an eye open to the processes – we might call them more material processes, even if they are not directly observable in the fashion of, say, trees, roads and libraries – which are the stuff of everyday social practices, relations and struggles, and which underpin social group formation, the constitution of social systems and social structures, and the social dynamics of inclusion and exclusion. More concretely (or even more materially), it is to continue paying urgent attention to the mundane workings of families and communities (however we understand such phenomena); it is to register the battles to get by on a daily basis, to earn a crust, to keep the house warm, to cope with the neighbours, to walk down the street without being afraid; it is to take a stab at sharing the happiness and the sadness of being people with or without friends, groups to hang out in, things to do, to share, to enjoy, to complain about; it is to pay attention to the child crying in the road, the old man shuffling to the pub, the young mother and her pram negotiating kerbstones; and so on and so on. I am sure that readers will know what I am getting at, perhaps a 'romance of the real', a wish to access some kind of 'gritty' real social world from which many academics end up feeling wholly alienated, and I immediately realize that objections of all kinds might be raised to my comments here.[17] Yet I cannot help feeling that somehow the cultural turn in human geography *has* risked emptying out much of this stuff from our lenses, from both the approaches that we adopt and the subject matters that we tackle. This is certainly not entirely the case, and I am delighted with the many exciting projects currently afoot to study the everyday socio-spatial lives of groupings such as children, elderly people, people with alternative sexual orientations, people with physical disabilities, people with mental health problems, and the like;[18] but even here I sometimes become concerned that the researcher's focus all too quickly becomes one hung up on identity politics and cultural representations, rather than patiently excavating the grain of component social lives, social worlds and social spaces. Such comments are maybe more provocative than I mean them to be, but I do still have a concern, a lingering sense of unease, about what strikes me as a *desocializing* of geography (in tandem with the demateri-alizing of geography mentioned previously) which arises even when enquiries are being directed at topics that seem irreducibly 'social', and even when qualitative research methods are being used which one would expect to burrow into the social grains and knots of the situations under study (e.g. Cook and Crang, 1995).

An additional thought revolves around whether or not human geographers have passed on all too swiftly from any sustained encounter with the discipline of sociology, and it might be argued that insufficient time and energy has been expended by human geographers in coming to terms with the classic texts of sociology – Karl Marx in his more grounded analyses of class and the labour process (see Box 1), but what about Durkheim, Weber, Simmel, Parsons, Merton? – which means that our appreciation of the 'social geographies' to be found therein has actually been quite limited. I would even suggest that there are surprisingly few works which explicitly set out to explore the possible overlaps between human geography and sociology, and I can only think of a handful of pieces by geographers which have set themselves this goal: Daryll Forde's 'Human geography, history and sociology' (1939); J. Wreford Watson's 'The sociological aspects of geogaphy' (1951); and Ray Pahl's 'Sociological models in geography' (1967). It is true that various social geographers have thought carefully about what can be recovered from the corpus of Chicago urban sociology (e.g. Ley, 1974; Jackson and Smith, 1984, Chap. 4[19]), and it is also true that the 1980s saw a lengthy dalliance with the ideas of Anthony Giddens, who certainly regarded his deliberations around the 'theory of structuration' as contributing to the remaking of sociology (e.g. J.S. Duncan, 1985; Gregory, 1989a). Mention might also be made of Benno Werlen's *Society, Action and Space: An Alternative Human Geography* (Werlen, 1993), with its thoroughgoing attempt to secure the lineaments of a human geography which adopts the stance of 'methodological individualism' as refracted through the sociological writings of Parsons, Popper, Schutz and others. But, I would still stand by the assertion that human geography's encounter with sociology (and hence with Mills's 'sociological imagination': see Harvey, 1973, pp. 23–7) has actually been quite patchy, leading to serious gaps in the foundations of social geography, whereas the current encounter with cultural studies (with the whole

---

**Box 1: Marx's 'social geography'**

In a small cottage at Gauxholme, near Todmorden (West Riding of Yorkshire), there lived a father and two daughters; the father was old and feeble, the girls earned their living as workers at Halliwell's cotton mill. They lived in a miserable room on the ground floor, a few feet from a filthy little brook, and past their window a staircase, used by the people who lived upstairs cut off the light from their dreary habitation. At the best of times, they earned just enough to 'keep body and soul together', but for the last fifteen weeks they had lost the only source of their livelihood. The factory had been closed down; the family could no longer earn the means to buy food. Step by step, poverty dragged them into the abyss. Every hour brought them nearer to the grave. (Marx, 1862, in Marx and Engels, 1984, p. 241)

One of the daughters finally dragged herself to the head of the local poorhouse, asking for some parish relief for her starving sister and sick father, but, when the local Poor Law officials finally deigned to visit the house five days after the daughter's plea, they found that one of the sisters had died of starvation:

> Stretched out on a wretched plank bed, among the signs of the most terrible poverty, lay the corpse of the starved girl, while her father, worn and helpless, sobbed on his bed and the surviving girl had just enough strength to tell the story of her woe. (Marx, 1862, in Marx and Engels, 1984, p. 242)

In this extract from the Viennese liberal newspaper, *Die Presse*, Marx carefully sketches out a small fragment of social geography: to do with a particular household whose wretched conditions of existence were greatly exacerbated by where the family lived, from the unavailability of alternative local employment once Halliwell's cotton mill had closed, to the distance from the local workhouse, to the filthy stream outside and the staircase cutting out light. This is certainly not a well known part of Marx's corpus, but it is arguably typical of his alertness to the details of social life, problems and spaces of ordinary people struggling to get by in the face of the emerging, largely unremitting demands of industrial capitalism across nineteenth century Europe.

Consider now Marx's account in *Capital*, Vol. I of labour power, wherein he explains how many working-class households are forced to eat cheap, adulterated bread from their local bakers because they do not have the time to travel further afield *and* because often they can only buy on credit, particularly if their wages are only paid fortnightly or monthly. The latter circumstance prevailed '[i]n many English and still more Scotch [sic] districts', meaning that the labourer here 'is in fact tied to the shop which gives him [sic] credit' (Marx, 1867, in Elster, 1986, footnote 2, p. 144). More famously, *Capital*, Vol. I includes passages discussing 'the struggle for the normal working day', and refers in passing to a contrast between the rural and the urban situations with respect to Sunday working: more specifically, Marx contrasts the situation whereby 'even now occasionally in rural districts a labourer is condemned to imprisonment for desecrating the Sabbath' (Marx, 1867, in Elster, 1986, footnote 2, p. 150) with, for instance, '[a] memorial . . . in which the London day-labourers in fish and poultry shops asked for the abolition of Sunday labour' (Marx, 1867, in Elster, 1986, footnote 2, pp. 150–1). Marx also discusses here the insatiable demand of capital for labour power, for the bodies (the muscles, sinews, dexterity) of labourers, explicitly observing how 'the capitalist mode of production . . . has seized the vital power of the people by the very root' (Marx, 1867, in Elster, 1986, p. 155: see also Harvey, 1998). He explains too the fate of labourers from the countryside recruited into the 'industrial population', as well as recording that even in agricultural

regions the demands of an 'improving' capitalist agriculture were prompting the bodily and everyday social decline of inhabitants. One example
that he gives was Sutherland in the Scottish Highlands, where '[i]n the
healthiest situations, on hill sides fronting the sea, the faces of their
famished children are as pale as they could be in the foul atmosphere of
a London alley' (a report on the Highlands quoted by Marx, 1867, in
Elster, 1986, footnote 2, p. 155).

Often found in newspaper articles or in footnotes to his major texts,
commonly quarried from the voluminous British Parliamentary 'Blue Books',
there is nonetheless ample evidence in his writing of what can be termed
a thoroughly grounded *social geography*: a dwelling on the minutiae of
particular groups of people in particular places, labouring in particular
ways under particular systems, experiencing all manner of deprivations in
their everyday social lives to do with nutrition, health and mental well-
being. Such claims are obviously framed by the broader logic of political-
economic reasoning for which Marx is rightly better known, and also by
the precepts of an 'historical materialist' philosophy energized by a sustained
enquiry into the '*earthly* basis for history' (Marx, 1846, in Elster, 1986,
p. 175), but the social geographical evidence unearthed in his writing – as
enriched by these respective framings – surely warrants further inspection
as an input to the (re)thinking of social geography.

There has of course been a deep-seated Marxist 'revolution' in geographical thought since the early 1970s (e.g. Harvey, 1973, 1982, 1984, 1990,
1996; see also note 13), but I would contend that the studies involved have
concentrated principally on the political-economic and more philosophical
dimensions of Marx's writing, with little systematic consideration being
given to what his more sociological reflections might imply for social geography. This is perhaps surprising given that in what is perhaps human
geography's most significant early exchange with Marxism (Harvey, 1972;
reprinted in Harvey, 1973, Chap. 4), the stated intention is 'nothing more
nor less than the self-conscious and aware construction of a new paradigm
for *social* geographic thought' (Harvey, 1972, p. 10, my emphasis). Harvey's
chief manoeuvre here is to borrow from Frederick Engels's description of
1840s Manchester (Engels, 1958; see also Marcus, 1973), with its fine-
grained probing of the social spaces of this growing industrial city, as a
route into (the seeds of) a Marxist enquiry into how capital's 'law' of
'surplus value extraction' can aid in explaining the uneven social geography
of urbanization under capitalism. Subsequent geographers, particularly
historical geographers (e.g. Billinge, 1984; Dunford and Perrons, 1983;
Gregory, 1984), have drawn upon Marx's ideas about (and substantive
examples of) class formation, consciousness and struggle when researching the social spaces of town and country, but there is still much to be
done in continuing Harvey's outline project of critically reconstructing
Marx's social geography as a point of departure for social geographers
(and maybe also cultural geographers[20]) working in the present.

field from the most straightforward of cultural analysis to the most esoteric of literary-cultural criticism) has been much more full-blooded, comprehensive and productive in the sense of delivering a new cultural geography. This is perhaps why the cultural turn has led relatively easily to an evaporation of what many of us once understood by social geography, a dismissal of it (if not all that consciously and certainly not maliciously) as something which should be central to the endeavours of contemporary human geography.

There is much that I could say in elaborating these claims. For instance, while finding much of value in a recent major text such as Steve Pile and Nigel Thrift's edited volume *Mapping the Subject: Geographies of Cultural Transformation* (Pile and Thrift, 1995), it is instructive that the subtitle speaks of 'cultural transformation' rather than of 'social transformation'. Moreover, I guess that the chapter written by myself with Hester Parr (Parr and Philo, 1995) is one of the few in the collection which is very much a work of social geography, tackling what might be cast as quite mundane sociological issues to do with how people in the social spaces of the city construct their fragile self-identities out of the resources made available to them in and through the rough-and-tumble of everyday social situations, interactions, inclusions and exclusions (which is certainly not to suggest that our chapter is in any way superior to the other chapters). David Sibley's essay (Sibley, 1995a) on 'families and domestic routines' in the construction of childhood also contains such elements of the mundanely sociological, discussing the micro-social geographies of home spaces (of bedrooms, living rooms and favourite chairs), although it is interesting that Sibley provides an explanatory framework which owes much to psychoanalysis (to Kristeva and to 'object relations theory') and not all that much to the discipline of sociology. This is again in no way a criticism, and I (like many others) have found much that is convincing and exciting in the turn to psychoanalysis which has informed Sibley's investigations both here and in his *Geographies of Exclusion* book (Sibley, 1995b). Yet, I do wonder if there was not also something else present in his earlier work, notably in the path-breaking *Outsiders in Urban Societies* (1981), where he drew upon a mix of ideas from sociology – aspects of Bernstein's sociology of education, traces of Weberian sociology as adapted by Parkin, a critical take on modernization theory, elements of social anthropology such as Mary Douglas on the 'sacred/profane' coupling – to give shape to his thinking about the worlds of travellers, 'Gypsies' and other social 'outsiders' (both as structured from without and as felt from within). Very recently I picked up a text on Weberian theories of 'social closure' (Murphy, 1988), and also read a text in which the geographer Jim Mac Laughlin (1995; see also Mac Laughlin, 1998, 1999) deploys notions of 'closure' in studying the related geographies of both Irish travellers and anti-traveller racism,[21] and it led me to recall these more sociological ingredients of Sibley's earlier work. Indeed, it prompted me to wonder if there would be mileage in retrieving such ingredients, brushing them off and seeing what they could do in the light of his more recent psychoanalytic turn. It led me to wonder the extent to which Sibley as a leading exponent of the cultural turn in British human geography regards his earlier social geographical formulations as crucial building blocks in his present work, or whether he supposes that there were fatal flaws in some of these earlier formulations which needed to be overcome by shifting his theoretical lenses away from more sociological territory to that of a psychoanalytically inflected cultural studies (see also Sibley's chapter in this collection, Part V).

Interestingly, there *are* some scholars expressly identifying themselves as social geographers who do not share the above-voiced concerns about the cultural turn as a threat to the integrity of a social geography alert to what Gregson (1993) termed those 'empirical social worlds' full of 'others'. To give an example, Ceri Peach's (1999) recent review of the subdiscipline is really quite assertive in responding to Gregson's fears about the death of social geography. He indicates that Gregson's claims might be true 'of that branch of the subject whose titles include playful parentheses to indicate a knowing deconstruction of their subject matter' (Peach, 1999, p. 282), and he thereby offers a thinly veiled criticism of those geographers who have become involved in the cultural turn, and whose academic co-ordinates have taken them towards the difficult issues of cultural identity politics, the crisis of representation, the constitution of the self, and so on. At the same time, he insists that 'in that area that is based more strongly on the empirical side, there has been a lively outpouring of material in the last few years' (Peach, 1999, p. 282), and he concludes by contrasting 'the Hamlet tendency of self-doubt' (in which he lumps both Gregson and those in the cultural turn who she, in turn, is criticizing) with 'the Caliban school of brutish energy' (Peach, 1999, p. 286) comprising those working 'on the empirical side'. More precisely, by the latter he means those social geographers who have been actively engaged in highly focused research on topics such as ethnic segregation, and more specifically still who have been involved in a four-volume analysis of 'the ethnic question data' (Peach, 1999, p. 282) available from the 1991 Census. I have no doubts regarding the quality and utility of such research, and indeed of Peach's own contributions in this respect (e.g. Peach, 1996), but I would object to his judgement on the superiority of 'grounded' research by 'social geographers', which supposedly recognizes 'real' divisions within the ethnic social spaces of western cities, over the tendency of 'cultural geographers' to remain locked in definitional wranglings. Such a judgement rather misses the significance of the 'social constructionist' move within social geographical studies of race and ethnicity (e.g. Jackson, 1987), in that the concern here is to tease out how complex definitional disputes – together with the often tacit underpinnings of racist and ethnic stereotyping, as impregnating the operations of countless institutions implicated in, for instance, housing and labour markets (e.g. S.J. Smith, 1989) – enter into every pore of routine social life in 'grounded' social spaces for so many late twentieth century people. In this context, I would want to defend the gains of the cultural turn *within* studies of what is usually taken as social geography, and to clarify that my own call for a *re*socializing of human geography is not meant to sanction any return to an unthinking empiricism obsessed with mappable patterns and devoid of interest in the constitution, contestation and lived meanings of such material geographies. While respecting aspects of what Peach is arguing in his 1999 review, I do think that the way forward requires a more inclusive reinsertion of the cultural into the social – of insights from the cultural turn being brought into the orbit of a social geography – than appears to be envisaged by Peach when he talks of recombining cultural and social geography.

## Concluding thoughts

Some final thoughts in this connection arise in relation to Andrew Leyshon's useful *Environment and Planning A* editorial (Leyshon, 1995), which asked 'whatever happened to the geography of poverty?', and which by implication suggested that the British Social and Cultural Geography Study Group, with its *New Words* compilation, was running too far with a certain type of approach (the cultural turn) which was leading it to neglect what many would regard as fundamental, unavoidable and uncomfortable issues such as the geography of poverty. The broader point raised by Leyshon is a valid one, and it chimes with my own arguments here, although in the specifics it was a touch awry because 1995 also saw the publication of *Off the Map: The Social Geography of Poverty in the UK* (Philo, 1995), a collection edited by myself on behalf of the Study Group and published by the Child Poverty Action Group (who commissioned the project from the Group).[22] I am not about to suggest that this was a particularly remarkable collection, although it did a reasonable job in reclaiming a geographical perspective on studying poverty, bringing together a handful of well constructed essays addressing poverty, deprivation and inequality at different spatial scales and in different types of environment across Britain. In my view, this basic exercise of social geography does remain an absolutely crucial component of the overall human geographical enterprise. I know that back in the early 1970s David Harvey complained about the continual documenting of socio-spatial inequalities as an exercise in 'moral masturbation' (Harvey, 1972, p. 10), his point being that we had enough descriptive 'welfare' studies and needed instead to begin offering more theoretically driven (i.e. Marxist) explanations for revealed patterns of inequality, but there surely remains an indispensable task for geographers to keep on charting such inequalities (after all, every society is perpetually reinventing its inequalities and their spatial disposition, and so continual monitoring is essential to provide a society with what might be termed its 'geographical conscience'). Whatever the precise claims to be made, however, all I really want to note is the enthusiasm of British social and cultural geographers to get involved in this project with the Child Poverty Action Group, and their conviction in the value of such a relatively simple use of conventional social geographical skills in teasing out the 'map' of rich and poor Britain. What this should caution against is any unthinking criticisms of today's culturally turned human geographers for adopting an irresponsible, indulgent, headlong dive into the unfathomable depths of cultural studies and 'clever' French theory. It should caution against knee-jerk accusations directed at such geographers for being feckless in forgetting about everything but their own egos, career paths and institutional power games (and there are a number of such savage assaults out there, albeit as yet chiefly conveyed in conference whispers rather than in published papers).

What it also signals is that, notwithstanding my own fears about an impulse towards both dematerializing and desocializing human geography which I think *does* accompany the cultural turn, in practice the overall programme which might be referred to as 'cultural *and* social and cultural geography' continues to be about much more than its critics might suggest. It continues to be a diverse, living, lively entity, and indeed a tolerant one in which (at least as I understand it) those veering towards the more extreme ends of the cultural turn are pleased that work continues to be done in the vein of a text such as *Off the Map*. From a personal position, therefore, I do want 'to let a thousand flowers bloom': to allow and even actively to promote a contemporary human geography which encompasses a whole range of possibilities for dealing with the material and the immaterial, the social and the cultural, often at one and the same time. Furthermore, so long as this span of human geographical enquiry includes within it studies which continue – while drawing inspiration from the whole sweep of the cultural turn – to embrace the material and the social, thereby resisting any dogmatic dematerializing or desocializing of the discipline, then I might be tempted to conclude that the 'one good turn' under review here will probably beget a few more in the near future.

## Acknowledgements

Thanks are due to various people for reading and commenting on the original version of this paper, notably Eric Laurier, Hester Parr and Joanne Sharp, but in particular I wish to thank James R. Ryan for his very helpful editorial advice. Thanks are also due to participants at the Oxford conference for their supportive remarks and constructive criticisms, notably to Dolores Garcia-Ramon. A rather different version of this paper is now being published in Spanish, in the journal *Documents d'Anàlisi Geogràfica*.

## Notes

1. I am tempted to place 'cultural turn' in quote marks throughout the paper, given how amorphous and contested this term has quickly become. In the interests of appearance on the page, though, I have resisted this temptation.
2. The strands discussed here by Johnston (1997) are: 'postmodernism'; 'feminism'; 'positionality, difference and identity politics'; 'language, texts and discourse'; and 'images, consumption and cultural geography'. This list embraces a variety of conceptual claims, methodological moments and substantive studies, often with little in common. Johnston particularly stresses postmodernism, although its connection with the cultural turn is arguably less straightforward than Johnston implies. Postmodernism may have originated in the sphere of what might loosely be termed 'cultural production', in relation to architecture, art, music and literature, but in its guise as an assault on modernist forms of intellectualizing – as an attack on 'grand theories' and 'metanarratives' of all sorts – it does something much more far-reaching than simply asking disciplines to take seriously the stuff of culture (see

Cloke, Philo and Sadler, 1991, Chap. 6; Gregory, 1989b; Strohmayer and Hannah, 1992). As such, I will say nothing more in this chapter about postmodernism.

3. McDowell (1994) usefully distinguishes between: (i) geographers drawing upon a 'cultural materialism' anchored in the Marxist cultural theory of Raymond Williams as then extended in the work of Stuart Hall and the Birmingham 'Centre', and growing into what might be called a geographical inflexion of cultural politics (e.g. Jackson, 1989); and (ii) a 'new landscape school' seeking to excavate the power relations etched into human-made environments and their many representations (e.g. Cosgrove, 1984; Cosgrove and Daniels, 1988; Daniels, 1993).

4. Now the Social and Cultural Geography Research Group of the Royal Geographical Society with the Institute of British Geographers.

5. The chief ingredient here has probably been the shift to a 'semiotic' view of cultures as 'webs of meaning', generated and conveyed both discursively and through other signs and symbols, which are shared intersubjectively (up to a point) by members of social groups, large and small, who are usually able to be spatially co-present with one another on a regular basis. Crucial in all of this, I would suggest, were the 'symbolic interactionist' (after Mead) and 'cultural anthropology' (after Geertz) streams which various geographers brought into their initial rethinkings of social and cultural geography (e.g. Duncan, 1978; Jackson and Smith, 1984, esp. Chap. 2).

6. For a taster, see Peet (1998, Chap. 6), but the tone of Johnston (1997, Chap. 8) is also fairly sceptical about a cultural turn which has over-reached itself, and perhaps even ended up reinventing what many other geographers had long been doing anyway.

7. Which is certainly not to question the valuable gains which have clearly been made by studies which strive to 'enculture' the other human geography subdisciplines.

8. At the 'regional' scale, reference might be made to the example of a vibrant and distinctive Welsh 'school' of human geography associated with the decidedly cultural concerns of scholars such as E.G. Bowen, H.J. Fleure and C. Daryll Forde (see Bowen, 1976; Fleure, 1919; Forde, 1934; Peate, 1930: see also Gruffudd, 1994).

9. Lengthy critiques of geography as spatial science, with its positivist leanings, were developed during the 1970s: a common complaint was the failure to take seriously all manner of phenomena not immediately available for 'sense verification', and for regarding such phenomena – whether social classes or human emotions, the imperatives of capitalism or the alienation of souls – as mere chimera unworthy of scientific investigation (e.g. Gregory, 1978, Chaps 1 and 2; Olsson, 1980, esp. Chap. 3; Sack, 1980).

10. A recent statement of some importance, I think, is Sibley's (1998) questioning of the repressed fears and desires which contributed to the obsession for making visible the simple spatial order of the world that, arguably, afflicted many spatial scientists.

11. For commentaries, see Cloke, Philo and Sadler, 1991, Chaps 2 and 3; Gregory, 1978, Chaps 3 and 4, 1981; Peet, 1998, Chaps 2, 3 and 4.

12. The discussion of 'structural explanation' in geography, particularly in Marxist geography, has itself been complex and contested: see Gregory (1978, Chap. 3) for an insightful account of the move to conceptualizing 'invisible' structures of economy, polity and society.

13. There are too many elements here to discuss or to reference fully, but see, as just a small sample, Castree, 1996; Cooke, 1989; Corbridge, 1993; Massey, 1984, 1994; Smith, 1984, 1994; Swyngedouw, 1992; and Watts, 1991, as well as all of the Harvey references dotted throughout this chapter.

14. Harvey (1989), moreover, regards postmodernism as a 'cultural change' which is itself the product of transformations within advanced capitalism, thus reworking Fredric Jameson's (1984) well known thesis about postmodernism comprising the 'cultural logic of late capitalism'.

15. There are again too many elements here to discuss or to reference fully, but see, as just a small sample, Bell and Valentine, 1995; Cosgrove and Daniels, 1988; N. Duncan, 1996; Fincher and Jacobs, 1998; Gregory, 1994; Jackson and Smith, 1984; Pile, 1996; Pile and Thrift, 1995; and Rose, 1993.

16. This is genuinely *not* the reason why I have selected this paper for attention here.

17. Pamela Shurmer-Smith insightfully critiques my points here, as originally made at the Oxford conference, while also accepting some import to what I claimed:

> I'm not quite as enthusiastic as Chris about the need to give voice to a list of disadvantaged categories within British society, because I fear the potentially exploitative outcomes of 'charity' work, but I do believe that the cultural without the social (political and economic) is meaningless, and that one of the most important aspects of cultural work is to examine the ways in which social groups and categories are constituted. (Shurmer-Smith, 1997, p. 8)

Her latter observation here also speaks back against the renewed 'empirical turn' seemingly being championed by Peach (1999): see main text.

18. As another small sample, see Bell and Valentine, 1995; Butler and Parr, 1999; Matthews, 1992; Rowles, 1978; Skelton and Valentine, 1998; C.J. Smith and Giggs, 1988; and Warnes, 1982.

19. It should be noted that Jackson and Smith (1984, Chap. 5) *do* also consider in some detail the ideas of both Georg Simmel and Max Weber, and in fact their *Exploring Social Geography* is arguably the best example available of geographers 'exploring' what the classic texts of sociology contain of value to social geography.

20. The relevance of Marx to more obviously cultural geographical studies has long been recognized through an engagement with 'cultural materialism', but it has also featured in the work of the 'new landscape school' (see McDowell's distinction, as laid out in note 3). Of particular interest in this respect, though, is Daniels (1989).

21. It occurs to me that an interesting exercise would be to compare the very similar intellectual trajectories, conceptual borrowings, substantive interests and political motivations (e.g. their anti-racism) of Mac Laughlin and Sibley. It also strikes me as strange how little they cite each other's work.

22. Rather nicely, in a review of *Off the Map*, Leyshon (1998) acknowledges the unfortunate coincidence of his editorial with this book's publication. Nonetheless, he does add the crucial rider that: 'While a welcome addition to the literature, it would be a mistake to conclude that the appearance of this book is testament to a thriving programme of work on the geography of poverty' (Leyshon, 1998, p. 500).

# *References*

Anderson, K. and Gale, F. (eds). (1992). *Inventing Places: Studies in Cultural Geography*. Longman Cheshire: Melbourne.

Bale, J. (1996). 'Rhetorical modes, imaginative geographies and body culture in early twentieth century Rwanda'. *Area*, 28, pp. 289–97.

Bale, J. (1999). 'Sport as power: running as resistance?' In Sharp, J.P., Routledge, P., Philo, C. and Paddison, R. (eds). *Entanglements of Power: Geographies of Domination/Resistance*. Routledge: London, pp. 148–63.

Bale, J. and Philo, C. (eds). (1998). *Body Cultures: Essays on Sport, Space and Identity by Henning Eichberg*. Routledge: London.

Bale, J. and Sang, J. (1996). *Kenyan Running: Movement Culture, Geography and Global Change*. Frank Cass: London.

Barnes, T. and Gregory, D. (eds). (1997). *Reading Human Geography: The Poetics and Politics of Inquiry*. Arnold: London.

Barnett, C. (1997). ' "Sing along with the common people": politics, postcolonialism and other figures'. *Environment and Planning D: Society and Space*, 15, pp. 137–54.

Barnett, C. (1998a). 'The cultural worm turns: fashion or progress in human geography?'. *Antipode*, 30, pp. 379–94.

Barnett, C. (1998b). 'Cultural twists and turns'. *Environment and Planning D: Society and Space*, 16, pp. 631–4.

Baudrillard, J. (1987). 'Forget Baudrillard: an interview with S. Lotringer'. In Baudrillard, J. *Forget Foucault*. Columbia University Press: New York, pp. 65–137.

Bell, D. and Valentine, G. (eds). (1995). *Mapping Desire: Geographies of Sexualities*. Routledge: London.

Billinge, M. (1984). 'Hegemony, class and power in late-Georgian and early-Victorian England: towards a cultural geography'. In Baker, A.R.H. and Gregory, D. (eds). *Explorations in Historical Geography*. Cambridge University Press: Cambridge, pp. 28–67.

Bingham, N. (1996). 'Object-ions: from technological determinism towards geographies of relations'. *Environment and Planning D: Society and Space*, 14, pp. 635–57.

Blum, H. and Nast, H. (1996). 'Where's the difference? the heterosexualisation of alterity in Henri Lefebvre and Jacques Lacan'. *Environment and Planning D: Society and Space*, 14, pp. 559–80.

Bowen, E.G. (1976). *Geography, Culture and Habitat: Selected Essays (1925–1975) of E.G. Bowen*. Gomer Press: Landysul.

Brunhes, J. (1920: trans.). *Human Geography: An Attempt at a Positive Classification (with) Principles and Examples*. Harrap: London.

Butler, R. and Parr, H. (eds). (1999). *Mind and Body Spaces: New Geographies of Illness, Disability and Impairment*. Routledge: London.

Castree, N. (1996). 'Birds, mice and geography: Marxisms and dialectics'. *Transactions of the Institute of British Geographers*, 21, pp. 342–62.

Castree, N. (1999). 'Situating cultural twists and turns'. *Environment and Planning D: Society and Space*, 17, pp. 257–60.

Claval, P. (1998). *Histoire de la Géographie française de 1870 à nos jours*. Nathan: Paris.

Cloke, P., Philo, C. and Sadler, D. (1991). *Approaching Human Geography: An Introduction to Contemporary Theoretical Debates*. Paul Chapman: London.

Cook, I. and Crang, M. (1995). *Doing Ethnographies*. CATMOG: Department of Geography, University of Durham.

Cooke, P. (ed.). (1989). *Localities: The Changing Face of Urban Britain*. Unwin Hyman: London.

Corbridge, S. (1993). 'Marxisms, modernities and moralities: development praxis and the claims of distant strangers'. *Environment and Planning D: Society and Space*, 11, pp. 449–72.

Cosgrove, D. (1984). *Social Formation and Symbolic Landscape*. Croom Helm: London.

Cosgrove, D. (1988). 'The cultural in human geography'. *Newsletter of the Social and Cultural Geography Study Group*, Spring, pp. 2–3.

Cosgrove, D. (1993). 'On "The reinvention of cultural geography", by Price and Lewis'. *Annals of the Association of American Geographers*, 83, pp. 515–17.

Cosgrove, D. and Daniels, S. (eds). (1988). *The Iconography of Landscape: Essays on the Symbolic Representation, Design and Use of Past Environments.* Cambridge University Press: Cambridge.

Daniels, S. (1989). 'Marxism, culture and the duplicity of landscape'. In Peet, R. and Thrift, N. (eds). *New Models in Geography, Vol.2.* Unwin Hyman: London, pp. 196–220.

Daniels, S. (1991). 'Demoralising social and cultural geography'. In Philo, C. (comp.). *New Words, New Worlds: Reconceptualising Social and Cultural Geography.* St David's University College, Lampeter: Social and Cultural Geography Study Group, pp. 164–5.

Daniels, S. (1993). *Fields of Vision: Landscape Imagery and National Identity in England and the United States.* Polity: Cambridge.

Demeritt, D. (1996). 'Social theory and the reconstruction of science and geography'. *Transactions of the Institute of British Geographers,* 21, pp. 484–503.

Duncan, J.S. (1978). 'The social construction of unreality: an interactionist approach to the tourist's cognition of environment'. In Ley, D. and Samuels, M.S. (eds). *Humanistic Geography: Prospects and Problems.* Croom Helm: London, pp. 269–82.

Duncan, J.S. (1985). 'Individual action and political power: a structurationist perspective'. In Johnston, R.J. (ed.). *The Future of Geography.* Methuen: London, pp. 174–89.

Duncan, J.S. (1993a). 'Landspaces of the self/landscapes of the other(s): cultural geography, 1991–1992'. *Progress in Human Geography,* 17, pp. 367–77.

Duncan, J.S. (1993b). 'Commentary'. *Annals of the Association of American Geographers,* 83, pp. 517–19.

Duncan, N. (ed.). (1996). *Bodyspace: Destabilising Geographies of Gender and Sexuality.* Routledge: London.

Dunford, M. and Perrons, D. (1983). *The Arena of Capital.* Macmillan: London.

Elster, J. (ed.). (1986). *Karl Marx: A Reader.* Cambridge University Press: Cambridge.

Engels, F. (1958). *The Condition of the Working Class in England in 1844.* Oxford University Press: Oxford.

Fincher, R. and Jacobs, J.M. (eds). (1998). *Cities of Difference.* Guilford: New York.

Fleure, H.J. (1919). 'Human regions'. *Scottish Geographical Magazine,* 35, pp. 94–105.

Forde, D. (1934). *Habitat, Economy and Society: A Geographical Introduction to Ethnology.* Harcourt, Brace & Co.: New York.

Forde, D. (1939). 'Human geography, history and sociology'. *Scottish Geographical Magazine,* 55, pp. 217–35.

Foucault, M. (1986). 'Of other spaces'. *Diacritics,* Spring, pp. 22–7.

Gregory, D. (1978). *Ideology, Science and Human Geography.* Macmillan: London.

Gregory, D. (1981). 'Human agency and human geography'. *Transactions of the Institute of British Geographers,* 7, pp. 1–18.

Gregory, D. (1984). 'Contours of crisis? Sketches for a geography of class struggle in the early Industrial Revolution in England'. In Baker, A.R.H. and Gregory, D. (eds). *Explorations in Historical Geography.* Cambridge University Press: Cambridge, pp. 68–117.

Gregory, D. (1989a). 'Presences and absences: time–space relations and structuration theory'. In Held, D. and Thompson, J.B. (eds). *Social Theory of Modern Societies: Anthony Giddens and his Critics.* Cambridge University Press: Cambridge, pp. 185–214.

Gregory, D. (1989b). 'Areal differentiation and post-modern human geography'. In Gregory, D. and Walford, R. (eds). *Horizons in Human Geography.* Macmillan: London, pp. 67–96.

Gregory, D. (1994). *Geographical Imaginations*. Blackwell: Oxford.

Gregory, D. (1995). 'Lefebvre, Lacan and the production of space'. In Benko, G.B. and Strohmayer, U. (eds). *Geography, History and the Social Sciences*. Kluwer: Dordrecht, pp. 15–44.

Gregory, D. (1997). 'Lacan and geography: the production of space revisited'. In Benko, G.B. and Strohmayer, U. (eds). *Space and Social Theory: Interpreting Modernity and Postmodernity*. Blackwell: Oxford, pp. 203–31.

Gregson, N. (1993). ' "The initiative": delimiting or deconstructing social geography'. *Progress in Human Geography*, 17, pp. 525–30.

Gruffudd, P. (1994). 'Back to the land: historiography, rurality and the nation in inter-war Wales'. *Transactions of the Institute of British Geographers*, 19, pp. 61–77.

Harvey, D. (1972). 'Revolutionary and counter-revolutionary theory in geography and the problem of ghetto formation'. *Antipode*, 4(2), pp. 1–13.

Harvey, D. (1973). *Social Justice and the City*. Arnold: London.

Harvey, D. (1982). *The Limits to Capital*. Blackwell: Oxford.

Harvey, D. (1984). 'On the history and present condition of geography: an historical materialist manifesto'. *Professional Geographer*, 36, pp. 1–11.

Harvey, D. (1989). *The Condition of Postmodernity: An Enquiry into the Origins of Cultural Change*. Blackwell: Oxford.

Harvey, D. (1990). 'Between space and time: reflections on the geographical imagination'. *Annals of the Association of American Geographers*, 80, pp. 418–34.

Harvey, D. (1996). *Justice, Nature and the Geography of Difference*. Blackwell: Oxford.

Harvey, D. (1998). 'The body as an accumulation strategy'. *Environment and Planning D: Society and Space*, 16, pp. 401–21.

Hinchliffe, S. (1996). 'Technology, power and space – the means and the ends of geographies of technology'. *Environment and Planning D: Society and Space*, 14, pp. 659–82.

Hinchliffe, S. (1999). 'Entangled humans: specifying powers and their spatialities'. In Sharp, J.P., Routledge, P., Philo, C. and Paddison, R. (eds). *Entanglements of Power: Geographies of Domination/Resistance*. Routledge: London, pp. 219–37.

Jackson, P. (ed.). (1987). *Race and Racism: Essays in Social Geography*. Allen & Unwin: London.

Jackson, P. (1989). *Maps of Meaning: An Introduction to Cultural Geography*. Unwin Hyman: London.

Jackson, P. (1991). 'Repositioning social and cultural geography'. In Philo, C. (comp.). *New Words, New Worlds: Reconceptualising Social and Cultural Geography*. St David's University College, Lampeter: Social and Cultural Geography Study Group, pp. 193–5.

Jackson, P. (1993). 'Berkeley and beyond: broadening the horizons of cultural geography'. *Annals of the Association of American Geographers*, 83, pp. 519–20.

Jackson, P. and Smith, S.J. (1984). *Exploring Social Geography*. Allen and Unwin: London.

Jameson, F. (1984). 'Postmodernism, or the cultural logic of late capitalism'. *New Left Review*, 146, pp. 53–92.

Johnston, R.J. (1997). *Geography and Geographers: Anglo-American Human Geography Since 1945 (Fifth Edition)*. Arnold: London.

Katz, C. (1996). 'Towards minor theory'. *Environment and Planning D: Society and Space*, 14, pp. 487–99.

Keith, M. (1991). 'Knowing your place: the imagined geographies of racial subordination'. In Philo, C. (comp.). *New Words, New Worlds: Reconceptualising Social and*

*Cultural Geography.* St David's University College, Lampeter: Social and Cultural Geography Study Group, pp. 178–92.

Kropotkin, P. (1885). 'What geography ought to be'. *The Nineteenth Century*, 18, pp. 940–56.

Lees, L. (1997). 'Ageographia, heterotopia and Vancouver's new public library'. *Environment and Planning D: Society and Space*, 15, pp. 321–47.

Ley, D. (1974). *The Black Inner City as Frontier Outpost: Images and Behaviour of a Philadelphia Neighbourhood.* Association of American Geographers: Washington, DC.

Ley, D. and Samnels, M.S. (eds.) (1978) *Humanistic Geography: Prospects and Problems.* Croom Helm, London.

Leyshon, A. (1995). 'Missing words: whatever happened to the geography of poverty?'. *Environment and Planning A*, 27, pp. 1021–8.

Leyshon, A. (1998). 'Review of Philo, C. (ed.), *Off the Map*'. *Journal of Rural Studies*, 14, p. 500.

Longhurst, R. (1997). '(Dis)embodied geographies'. *Progress in Human Geography*, 21, pp. 487–501.

McDowell, L. (1994). 'The transformation of cultural geography'. In Gregory, D., Martin, R. and Smith, G. (eds). *Human Geography: Society, Space and Social Science.* Macmillan: London, pp. 146–73.

Mac Laughlin, J. (1986). 'State-centred social science and the anarchist critique: ideology in political geography'. *Antipode*, 18, pp. 11–38.

Mac Laughlin, J. (1995). *Travellers and Ireland: Whose Country, Whose History?* Cork University Press: Cork.

Mac Laughlin, J. (1998). 'The political geography of anti-Traveller racism in Ireland: the politics of exclusion and the geography of closure'. *Political Geography*, 17, pp. 417–35.

Mac Laughlin, J. (1999). 'Nation-building, social closure and anti-Traveller racism in Ireland'. *Sociology*, 33, pp. 129–51.

Marcus, S. (1973). 'Reading the illegible'. In Dyos, H.J. and Wolff, M. (eds). *The Victorian City: Images and Realities, Vol.II – Shapes on the Ground and a Change of Accent.* Routledge & Kegan Paul: London, pp. 257–76.

Marx, K. and Engels, F. (1984). *Karl Marx and Frederick Engels: Collected Works, Vol.19, Marx and Engels, 1861–1864.* Lawrence & Wishart: London.

Massey, D. (1984). *Spatial Divisions of Labour: Social Structures and the Geography of Production.* Macmillan: London.

Massey, D. (1994). *Space, Place and Gender.* Polity Press: Cambridge.

Matthews, M.H. (1992). *Making Sense of Place: Children's Understandings of Large-Scale Environments.* Harvester Wheatsheaf: Hemel Hempstead.

Mitchell, D. (1995). 'There's no such thing as culture: towards a reconceptualisation of the idea of culture in geography'. *Transactions of the Institute of British Geographers*, 20(ns), pp. 102–16.

Mohan, G. (1994). 'Destruction of the con: geography and the commodification of knowledge'. *Area*, 26, pp. 387–90.

Murdoch, J. (1997a). 'Towards a geography of heterogeneous associations'. *Progress in Human Geography*, 21, pp. 321–37.

Murdoch, J. (1997b). 'Inhuman/nonhuman/human: Actor-network theory and the prospects for a nondualistic and symmetrical perspective on nature and society'. *Environment and Planning D: Society and Space*, 15, pp. 731–56.

Murphy, R. (1988). *Social Closure: The Theory of Monopolisation and Exclusion*. Clarendon Press: Oxford.

Nast, H.J. and Pile, S. (eds). (1998). *Places Through the Body*. Routledge: London.

Olsson, G. (1980). *Birds in Egg/Eggs in Bird*. Pion: London.

Pahl, R.E. (1967). 'Sociological models in geography'. In Chorley, R.J. and Haggett, P. (eds). *Models in Geography*. Methuen: London, pp. 217–42.

Parr, H. and Philo, C. (1995). 'Mapping "mad" identities'. In Pile, S. and Thrift, N. (eds). *Mapping the Subject: Geographies of Cultural Transformation*. Routledge: London, pp. 198–226.

Peach, C. (1996). 'Does Britain have ghettos?'. *Transactions of the Institute of British Geographers*, 21, pp. 216–35.

Peach, C. (1999). 'Social geography'. *Progress in Human Geography*, 23, pp. 282–8.

Peate, I. (ed.). (1930). *Studies in Regional Consciousness and Environment (Essays Presented to H.J. Fleure)*. Oxford University Press: Oxford.

Peet, R. (ed.). (1978). *Radical Geography: Alternative Viewpoints on Contemporary Social Issues*. Methuen: London.

Peet, R. (1998). *Modern Geographical Thought*. Blackwell: Oxford.

Philo, C. (1991a). 'Introduction, acknowledgements and brief thoughts on older words and older worlds'. In Philo, C. (comp.). *New Words, New Worlds: Reconceptualising Social and Cultural Geography*. St David's University College, Lampeter: Social and Cultural Geography Study Group, pp. 1–13.

Philo, C. (comp.). (1991b). *New Words, New Worlds: Reconceptualising Social and Cultural Geography*. St David's University College, Lampeter: Social and Cultural Geography Study Group.

Philo, C. (1992). 'Foucault's geography'. *Environment and Planning D: Society and Space*, 10, pp. 137–61.

Philo, C. (ed.). (1995). *Off the Map: The Social Geography of Poverty in the UK*. London: Child Poverty Action Group.

Pile, S. (1996). *The Body and the City: Psychoanalysis, Space and Subjectivity*. Routledge: London.

Pile, S. and Thrift, N. (eds). (1995). *Mapping the Subject: Geographies of Cultural Transformation*. Routledge: London.

Price, M. and Lewis, M. (1993). 'The reinvention of cultural geography'. *Annals of the Association of American Geographers*, 83, pp. 1–17.

Rose, G. (1993). *Feminism and Geography: The Limits of Geographical Knowledge*. Polity: Cambridge.

Rose, G. (1997). 'Situating knowledges: positionality, reflexivities and other tactics'. *Progress in Human Geography*, 21, pp. 305–20.

Rowles, G.D. (1978). *Prisoners of Space? Exploring the Geographical Experience of Older People*. Westview Press: Boulder.

Sack, R.B. (1980). *Conceptions of Space in Social Thought: A Geographic Perspective*. Macmillan: London.

Sayer, A. (1994). 'Cultural studies and "the economy, stupid"'. *Environment and Planning D: Society and Space*, 12, pp. 635–7.

Short, J.R. (1998). *New Worlds/New Geographies*. Syracuse University Press: Syracuse.

Shurmer-Smith, P. (1997). 'Report on the "Cultural Turns/Geographical Turns" Conference'. *Newsletter of the Social and Cultural Geography Research Group*, November, pp. 7–8.

Sibley, D. (1981). *Outsiders in Urban Societies*. Blackwell: Oxford.

Sibley, D. (1995a). 'Families and domestic routines: constructing the boundaries of childhood'. In Pile, S. and Thrift, N. (eds). *Mapping the Subject: Geographies of Cultural Transformation*. Routledge: London, pp. 123–42.

Sibley, D. (1995b). *Geographies of Exclusion: Society and Difference in the West*. Routledge: London.

Sibley, D. (1998). 'Sensations and spatial science: gratification and anxiety in the production of ordered landscapes'. *Environment and Planning A*, 30, pp. 235–46.

Skelton, T. and Valentine, G. (eds). (1998). *Cool Places: Geographies of Youth Cultures*. Routledge: London.

Smith, C.J. and Giggs, J.A. (eds). (1988). *Location and Stigma: Contemporary Perspectives on Mental Health and Mental Health Care*. Unwin Hyman: London.

Smith, N. (1984). *Uneven Development: Nature, Capital and the Production of Space*. Blackwell: Oxford.

Smith, N. (1994). 'Marxist geography'. In Johnston, R.J., Gregory, D. and Smith, D.M. (eds). *The Dictionary of Human Geography (Third Edition)*. Blackwell: Oxford, pp. 365–73.

Smith, N. and Katz, C. (1993). 'Grounding metaphor: towards a spatialised politics'. In Keith, M. and Pile, S. (eds). *Place and the Politics of Identity*. Routledge: London, pp. 67–83.

Smith, R.G. (1997). 'The end of geography and radical politics in Baudrillard's philosophy'. *Environment and Planning D: Society and Space*, 15, pp. 305–20.

Smith, S.J. (1989). *The Politics of 'Race' and Residence: Citizenship, Segregation and White Supremacy in Britain*. Polity Press: Cambridge.

Soja, E.W. (1989). *Postmodern Geographies: The Reassertion of Space in Critical Social Theory*. Verso: London.

Sorkin, M. (ed.). (1992). *Variations on a Theme Park: The New American City and the End of Public Space*. Hill & Wang: New York.

Stewart, L. (1995). 'Bodies, visions and spatial politics: a review essay on Henri Lefebvre's *The Production of Space*'. *Environment and Planning D: Society and Space*, 13, pp. 609–18.

Strohmayer, U. and Hannah, M. (1992). 'Domesticating posmodernism'. *Antipode*, 24, pp. 29–55.

Swyngedouw, E. (1992). 'Economic geography in the 1980s: the perplexing geography of uneven development'. In Rogers, A., Viles, H. and Goudie, A. (eds). *The Student's Companion to Geography*. Blackwell: Oxford, pp. 86–96.

Thrift, N. (1991). 'Over-wordy worlds? thoughts and worries'. In Philo, C. (comp.). *New Words, New Worlds: Reconceptualising Social and Cultural Geography*. St David's University College, Lampeter: Social and Cultural Geography Study Group, pp. 144–8.

Thrift, N. (1994). 'Taking aim at the heart of the region'. In Gregory, D., Martin, R. and Smith, G. (eds). *Human Geography: Society, Space and Social Science*. Macmillan: London, pp. 200–31.

Thrift, N. (1996). *Spatial Formations*. Sage: London.

Warnes, A.M. (ed.). (1982). *Geographical Perspectives on the Elderly*. Wiley: Chichester.

Watts, M.J. (1991). 'Mapping meaning, denoting difference, imagining identity: dialectical images and postmodern geographies'. *Geografiska Annaler*, 73B, pp. 7–16.

Werlen, B. (1993). *Society, Action and Space: An Alternative Human Geography*. Routledge: London.

WGSG (Women and Geography Study Group). (1997). *Feminist Geographies: Explorations in Diversity and Difference*. Longman: Harlow.

Whatmore, S. (1997). 'Dissecting the autonomous self: hybrid cartographies for a relational ethics'. *Environment and Planning D: Society and Space*, 15, pp. 37–54.

Wreford Watson, J. (1951). 'The sociological aspects of geography'. In Taylor, G. (ed.). *Geography in the Twentieth Century: A Study of Growth, Fields, Techniques, Aims and Trends*. Methuen: London, pp. 463–99.

Wright, J.K. (1925). *The Geographical Lore of the Time of the Crusades: A Study in the History of Medieval Science and Tradition in Western Europe*. American Geographical Society: New York.

Wright, J.K. (1947). '*Terrae incognitae*: the place of the imagination in geography'. *Annals of the Association of American Geographers*, 37, pp. 1–15.

# Taking a cultural turn?

## Struggles over the social in social policy

*John Clarke*

This chapter explores the significance of the 'cultural turn' in relation to social welfare. It argues that focusing on issues of social welfare reveals important intersections between social, cultural and political changes in relation to the welfare state and analytical developments in the study of social policy. It suggests that the 'cultural turn' needs to be understood as embedded in both sets of developments. The most significant change in social welfare in Britain has been the collapse of the series of assumptions, conceptions and beliefs that both supported and were sustained by the post-war welfare state. This collapse has become equated with the retreat from a 'statist' form of welfare provision and is identified with the dissolution of the political consensus (or political-economic settlement) that bound the provision of welfare by the state into Britain's economic development and political institutions. By the mid-1970s, Labour politicians were announcing that Keynesianism was dead and beginning the withdrawal from public spending on welfare. The New Right then spent long years picking over the corpse of Keynesianism and performing the public rituals of nailing it ever more firmly into its box. The playing out of this drama has tended to overshadow the break-up of other founding assumptions of the old welfare state: in particular, those of the 'social settlement' that underpinned its policies and practices (Clarke and Newman, 1997, Chap. 1; Hughes and Lewis, 1998). As Fiona Williams has consistently argued, the development of the welfare state in Britain has been informed by a series of assumptions about the character of British society and the proper purposes of welfare provision within it (e.g. F. Williams, 1989, 1993, 1998). These assumptions have centred on conceptions of Family, Work and Nation in which images of gender and age relations, occupation and class formations and intersections of 'race', ethnicity and nationality have been elaborated. Welfare policies have been shaped by such images and have attempted to reproduce them in social life. I want to come back to some of the contradictory consequences of these founding assumptions later, but first I think it is worth reflecting on what actually counted as 'social' within those conceptions of British society and social welfare. What sorts of conditions, problems or needs were seen as socially produced and capable of being remedied by social welfare?

Until the late 1960s, the conception of the social in social policy mainly referred to patterns of socio-economic inequalities (usually referred to as 'class inequalities'), in which the measurably unequal distribution of income, wealth and other life chances has been the focus of both academic enquiry and policy formation. This conception of social inequality has also been the centre-piece of popular politics, in the development of what T.H. Marshall (1950) referred to as 'social citizenship' through a set of conflicts, alliances and compromises (see also Saville, 1975). It has provided the core of political and academic arguments about 'redistribution': the possibility, desirability and means of realigning access to wealth, income and opportunity. The focal point of social policy in practice and the academic study of social policy has been this conception of the 'social' – a specific classificatory schema of social (or socio-economic) inequality. This concern with socio-economic inequality does not mark the limit of social policy but it does mark the edge of what social policy conceived of as the social circumstances that could be redressed through collective action. Beyond this lay the realm of the 'natural' – the biological or psychological conditions that made people different, deviant, dependent or in need. Such circumstances might be *relieved* (through benefits or services) or *repaired* (through medical intervention) but could not be *redressed*. Belonging to the 'extra-social', such circumstances were eternal, universal and inevitable.

The cultural turn in social policy has centred on the recovery of these circumstances from the realm of Nature and the remaking of them as elements of the Social. In different ways, age, 'race', gender, disability, sexuality have been revalorized as social rather than natural products. They have been reclaimed as the effects of processes of social construction and identified as the site of possible social reconstruction (Saraga, 1998). The social settlement, formed in the intersection of a social imaginary of 'family, work and nation', proved to be no more stable in practice than the political-economic one. The interplay of social changes – reshaping the composition and meaning of households, workers and the people – and new forms of social, cultural and political movements destabilized this social settlement. Forms of social differentiation became a focus for collective action and political conflict. Many of these centred on divisions that had previously been treated as natural categories (such as 'race', gender and sexuality) but which collective action sought to redefine as *socially produced and constructed*. These struggles arose in part from the contradictions inherent in different features of the social settlement of the post-war years. For example, the '1945' settlement had assumed a homogeneously white British citizenry. This conception of the Nation involved an attempt to cast 'race' out into the realms of biology and colonial geography (Lewis, 1998a; see also Cooper, 1998). From this starting point, it was assumed that welfare universalism would extend unproblematically to minority ethnic groups through the mechanisms of social assimilation. But the half-hearted and stumbling attempts at assimilation and accommodation in British social policy highlighted the contradictions of trying to sustain such a racialized conception of citizenship in the face of a multi-ethnic populace. As Britain moved uneasily out of Empire, so the social character of citizenship became increasingly exposed as a contested issue. Attempts to maintain the equivalence constructed between white and British encountered struggles against the racialized status of 'second class' citizenship. Growing evidence of, and struggles for, divergent 'needs' made uneven impacts upon the system of welfare (see, for example, Ahmad, 1993; F. Williams, 1989). At some points, such differences were recognized in the form of limited multicultural approaches or in the recruitment of minority ethnic staff identified as more representative of, and better 'attuned' to, the needs of the populations they were expected to serve (Lewis, 1996). Cultural difference was also construed in ways that exposed minority ethnic groups to the repressive dimensions of state welfare, not merely in the form of discriminatory policing, but to diagnoses of 'maladjustment' and 'deviance' of various kinds in respect of education, social work, and health services which in turn evoked greater intervention or surveillance from the state's agencies (Lewis, 1998b).

At the same time, material shifts in the alignment of family and work and the rise of feminism exposed gendered divisions of labour and power where once God and Biology had happily coincided in the invention of the 'normal family'. These changes included the growing involvement of married women in paid employment (alongside their 'vital role' as housewives and mothers); the rise of divorce and remarriage producing serial families; the rise in lone parent families; the spread of alternatives to the family form – communal living, gay or lesbian households, and so on. Conventional patriarchal assumptions about the inevitability and necessity of the family form became increasingly detached from social experience. At the same time, other assumptions about the interior life of the family that imagined it as an intimate and tranquil 'haven', protecting its members from the rigours of the public world, were undermined. Campaigns led by the women's movement revealed a less protective interior, highlighting economic and power inequalities between men and women and the capacity of the family to both produce and conceal abuse of its members (wife-battering, marital rape and the physical and sexual abuse of children). These developments challenged the assumption of the normality of the family as the focal point of state welfare, in which it functioned both as something that contributed to welfare and as something to be sustained by welfare provision. The normalizing assumptions that were built into – and were in turn sustained by – the post-war welfare state had been embedded in a strong distinction between the social and the natural. The ordering of people, positions, relationships and needs grasped most of these as being based in a realm beyond society – a non-social realm of biology – where they were simply part of the 'natural order of things'. So the normal family – with its wage-earning, bread-winning, patriarchal head of household and its nurturing, domesticating and domesticated housewife/mother and its other dependants – was not merely normal. Its normality was legitimated and underwritten by reference to these patterns being natural: men's nature being to hunt and gather (even if only the wage packet) while women's biology predisposed them to bring up babies and to mop up what babies bring up. The same points can be made about the structuring of age, where the period of mature independence is preceded and followed by the biologically inscribed dependency of childhood and old age; about the delineation of disability, codified as biological conditions under the authority of medical power; and about the categorization and regulation of sexualities, where the intersection of the distinctions between normal and abnormal on the one hand and the natural and the unnatural on the other is at its most dense.

Challenges were made across a range of social differentiations: disability, age, 'race', gender and sexuality, all insisting on the *social* character of such identities. At the same time, they refused the dependent and pathologized statuses attributed by biological or psychological essentialism produced and reproduced in the policies and practices of social welfare (see, for example, Carabine, 1992; Oliver, 1990; Oliver and Barnes, 1991; Saraga, 1998; D. Taylor, 1997). The cumulative effect of such challenges might be seen as the 'return of the repressed' in social policy. Those elements of social relations which had been expelled to the realm of the Natural in the structuring assumptions of welfare provision became the focus for collective social action intended not just to enlarge access to social welfare but to transform its structuring principles of provision. Each of these dimensions of inequality and division had been ideologically consigned to the realm of the extra-social: they were simply the more or less fortunate effects of nature. The struggles to make them 'social' opened up new dimensions of partiality and discrimination to critical attention. Both the formal and informal rules of bureaucracies and the knowledges and practices of professionalism were subjected to challenges about their social biases (Clarke and Newman, 1997, Chap. 1). Policy formation and implementation, employment practices and the characteristic organizational cultures of welfare institutions were all implicated in the production and reproduction of power and inequality (Newman and Williams, 1995). Such challenges bore heavily on the 'front lines' of state welfare agencies, even though they were also directed at the commanding heights of policy making. Welfare workers, by virtue of the fact that they carried the day to day contact between the people and the state, were prone to being captured or co-opted by these 'challenges from the margins' and their demands for greater equity or redress (e.g. G. Taylor, 1993). Welfare professionalism was at least partially open to the attempts by these new social movements to socialize definitions of social problems, and became one of the sites in which issues about 'discrimination', 'empowerment' and inequalities of different kinds were played out (Clarke, 1996).

These struggles over what counts as social in social welfare have been paralleled by and linked with struggles over the content, methods and perspectives of studying social policy, involving a local version of the 'cultural turn' visible elsewhere in the social sciences. I am using the idea of the cultural turn to refer to the diverse sources of interest in the cultural, ideological, discursive or symbolic features of social welfare (see also Clarke, 1998a and 1998b). Social policy had previously explored the topic of ideology, but mainly in the form of ideology conceived of as a relatively systematic body of politically oriented ideas which influenced the development of social welfare (see, for example, George and Wilding, 1976; Clarke, Cochrane and Smart, 1987). By contrast, the cultural turn in social policy has treated the realm of ideology in a more expansive way – as involving the implicit social assumptions as well as explicit ideological conceptions that have informed the social character of welfare. At different times – and using different theoretical means – the class, gendered, familial, racialized, disabling and heterosexualized tendencies of welfare policies and practices have been explored (e.g. Lewis, 1998c; D. Taylor, 1997). What most of these have had in common is the commitment to deconstructing the naturalizing effects of particular ideological formations and the consequent revelation of the social relations that are produced or reproduced through ideologies or discourses (Saraga, 1998).

My understanding of the cultural turn sees it as the confluence of very different theoretical and methodological strands, including structuralist Marxism, particularly in the Althusserian conceptions of ideology and subjects; other varieties of Marxism, for example, the Gramscian concerns with both hegemony and the contested contradictoriness of 'common sense'; poststructuralist approaches to the arbitrary yet conventionalized power of language; the Foucauldian examination of discourses together with the knowledge–power relations embedded in them; and the legacy of phenomenological and symbolic interactionist approaches to social construction, particularly in the sociology of deviance. Each of these developments fed into the attention to the 'cultural' within social policy (and the social sciences more widely), with the effect of systematically deconstructing, denormalizing and denaturalizing welfare policy and practice. It is this ferment that has placed new conceptions of the social in the study as well as the politics of social policy. These have challenged both the dominant structurings of social arrangements and identities within social welfare and in the academic field of social policy (e.g. Hughes, 1998; Hughes and Lewis, 1998; Langan, 1998; Leonard, 1997; Lewis, 1998c; O'Brien and Penna, 1998; Saraga, 1998; D. Taylor, 1997; F. Williams, 1996 and 1998). They have emphasized the constructed – and therefore contingent – character of conceptions of natural and unnatural; normal and abnormal; and have demonstrated how social welfare policy and practice has pathologized patterns of social differentiation. The result is a series of approaches within social policy that are linked by commitments to deconstruct social welfare: opening up not just how a particular client group is defined; not just how a particular policy is conceived and enacted; but examining how all the terms of the field are constructed and articulated. From 'consumer' to 'nation'; from the 'social bath' to the 'welfare state' itself, we can find examples of how the 'cultural turn' has enhanced and enlarged what studying social policy means (Hughes, 1998; Lewis, 1998c; Twigg, 1997).

The 'cultural politics' of the new social movements and their assaults on social welfare have significant links with the 'cultural turn' in the study of social policy in a range of ways: personnel, ideas, political mobilizations, and above all a commitment to the theory and practice of construction/deconstruction/reconstruction understood as 'cultural'. Those links are strong and important but they do not mean that the cultural politics of social welfare and the cultural turn in social policy are identical. The academic and analytic version is necessarily reflexive about the constructions that have been created and deployed by activist movements: recognizing that the argument that 'race', disability or age are socially constructed is itself a social construction being strategically deployed for specific social and political purposes (Saraga, 1998; see also Cooper, 1998, on discursive strategies in forms of governance). As a result, even the claim that social arrangements are socially constructed has to be treated as a provisional 'claim to truth' which may be displaced by alternative formations. The cultural turn highlights the temporary, fragile or contingent quality of specific social constructions. The accomplishment of 'socialized' understandings of social welfare is itself a social construction and as such is vulnerable to erosion, challenge and change. Indeed, it is possible to see a range of ways in which the 'social' character of differences, needs and problems in relation to social welfare are vulnerable to attempts to *desocialize* them. There have been a range of moves that seek to take the 'social' out of social policy by transforming problems, needs, inequalities or differences into 'non-social' issues. Here I want to sketch some desocializing reconstructions that have been developed around individualism, morality, biology and geography.

The most obvious drive to desocialize social welfare has been articulated around revitalized individualism, particularly in its institutionalized forms of neo-classical economics and neo-liberal politics. The process of desocializing 'society' is, in some senses, a global one, partly because of the ways global organizations have provided powerful forms of globalizing (and naturalizing) the discourses of neo-classical economics and neo-liberalism (see Deacon, Hulse and Stubbs, 1997). Nevertheless, these processes have also taken different national forms, particularly in terms of the density of neo-liberalism in domestic politics. The promulgation of economic individualism, the promotion of market-like relations in formerly 'public' services, the reinvention of the citizen as consumer, have all affected the publicly available conceptions of social welfare in profound ways (Hughes, 1998; see also Mackintosh, 1998). Some have suggested that the rise of neo-liberal governmentality marks the 'death of the social' (Rose, 1996; Fitzpatrick, 1998). There is no doubt that, especially in such countries as Britain, the USA and New Zealand, the attempt to reconstruct the alignment of public/private divisions in favour of a variety of forms of privatization (consumerist, contractual, corporate and familial, for example) has gone very deep (Clarke and Newman, 1997). Nevertheless, I think announcing the 'death of the social' is premature. Particularly but not exclusively in social welfare, the 'social' remains an intensely contested focus of cultural and political attention, where forms of identity, responsibility, attachment and belonging are struggled over (Clarke and Newman, 1998; Cooper, 1998; Hughes, 1998; Langan, 1998).

It is worth distinguishing neo-liberalism's individualism from the related, but different, attempts to turn social policy into a subdivision of economics, posing welfare as essentially a question of 'what we can afford'. This formulation has circulated globally, in national politics and within specific welfare providing organizations. It shifts the calculative framework within which issues of social needs, social rights or social justice might be posed. In the process, it attempts to close off the social by subordinating it to the more fundamental 'economic'. This is a recurrent problem for the social – it always appears as secondary or residual – what is left after the big or serious processes have been dealt with. So, the social takes second place to the 'natural' (the variations left over after dealing with the biological essences) or the 'medical' (in the determination of forms and priorities of care: see Twigg, 1997). The significance of economics for social welfare has not been limited to this reductionist reasoning about the costs of citizenship. Economics has contributed to the remaking of social welfare, particularly around the introduction of market-like processes into the co-ordination of welfare provision, and, in the process, has produced some distinctive analytical problems (Mackintosh, 1998). Defining the character of these market-like processes is one of them: are they 'markets', 'internal markets', 'quasi-markets', or merely contracting mechanisms (Le Grand and Bartlett, 1993; Hudson, 1994; Walsh, 1995; Harden, 1992)? Harden's conception of the 'contracting state' is one of the more elegant academic puns – registering the intersection between the legal form of co-ordination and the shrinking of state provision. But even here, how the state is being realigned, and the complex power flows across axes of centralization, decentralization and devolution leave a number of unresolved issues about how the 'public' and its interests are to be served, represented and imagined.

The discursive limitations of the neo-liberal conception of individualization and neo-classical conceptions of markets are reflected in the uneasy relationship that these positions occupy in relation to discursive formulations of welfare and morality (see Clarke and Newman, 1997, Chap. 6; F. Williams, 1998). There have been sustained efforts to redefine social problems or needs as 'moral problems' or 'moral disorders'. Here, it should be clear that 'moral' is not a descriptive category but above all signals the aim of reinstalling traditional morality and the forms of authority in which it is embodied. It links the neo-conservative critics of social welfare (such as Charles Murray), the 'anti-liberal intelligentsia' politics articulated by Melanie Phillips among others and the strange strain of self-proclaimed 'ethical socialism' associated with Norman Dennis. Here we see some of the tensions around the alliance of neo-liberalism and neo-conservatism in the political formation of the New Right (Clarke, 1991, Chap. 5). Neo-conservatism is not individualist – it does not subscribe to the view that 'there is no such thing as society'. Rather, its proponents believe in society – but are extremely upset that it is not what it used to be. More importantly, they are deeply committed to restoring past glories, eternal truths and a way of life in which everyone had their place and knew it. Murray, for example, reminded us of the 'popular wisdom' that US liberals ignored while they were building up welfare spending:

> The popular wisdom is characterised by hostility towards welfare (it makes people lazy), towards lenient judges (they encourage crime), and towards socially conscious schools (too busy bussing kids to teach them how to read). The popular wisdom disapproves of favoritism towards blacks and of too

*many written in rights for minorities of all sorts. It says that the government is meddling far too much in things that are none of its business.* (1984, p. 146)

There are strong intersections between these neo-conservative arguments and the more polite – and possibly more influential – strand of communitarian thinking (Etzioni, 1993; see also Campbell, 1995 and Hughes and Mooney, 1998). Alongside such attempts to restore traditional social and moral orders, we can see a resurgent interest in returning 'social' issues to the realm of nature: the rediscovery of biology in a variety of forms. There has been a revitalization of what we might call the 'old biology' in arguments about the 'racial' distribution of intelligence (Jencks, 1992). This has happened alongside the 'new biology' of genetic investigation and management: the relentless pursuit of 'genes for difference' (criminality, sexuality, poverty). We even have the return (if it had ever really gone away) of Parental Investment Theory whose intended social effect is to define and naturalize the difference between men's restless and active sexuality and women's passive and nurturing behaviour (Barash, 1981; Trivers, 1978). These diverse challenges to putting the social into social policy intersect with others, for example defining social needs and problems as matters of geography. Gail Lewis has argued that 'race' is continually articulated through 'place' in the formulation of social policy, such that racialized differences are continually displaced from the nation to the 'local' concern or to the 'elsewheres' of 'racial origin' (Lewis, 1998b; see also Cooper, 1998). The combined effect of all these divergent challenges to the 'social' has been the creation of a complex and contradictory landscape for the current politics of social welfare. Organizational modernizers jostle with genetic modifiers; efforts to expand civic rights encounter moral fundamentalists; free marketeers find themselves at loggerheads with social authoritarians, while some dream of a third way through which all differences may be reconciled (Clarke and Newman, 1998).

The shift to understanding all of these formations as socially constructed has been an important one – no matter what the specific theoretical route has been that informs such discoveries. It is clear that within social policy – as within the social sciences more generally – these developments have been conducted through a variety of theories, perspectives. Some centre on concepts of ideology, ideological struggle and hegemony (Hall, 1996); others have taken discourse and the articulation of knowledge and power as the focal terms (Burr, 1995; Cooper, 1998); still others have addressed these issues through concepts of culture, signifying practices and cultural formations (Morley and Chen, 1996). What they share, albeit from very different starting points, is a concern with the constructed/constituted and contingent character of social arrangements. They focus attention on the underlying conditions through which even the most apparently solidified and permanent formations or constructions have to be produced and reproduced. The 'naturalness' of such formations requires continual maintenance in a range of forms, precisely because they can change or, more accurately, they can be changed: reconstructed in new forms, with new meanings and possibilities. Not even the 'taken for granted' can be taken for granted – certainly not by generations that have seen the break-up of both the Soviet bloc and the apartheid regime in South Africa. Even those who are attempting to reassert the 'natural order of things' know that they have to negotiate a society where 'naturalness' is no longer a shared assumption.

The socially constructed character of particular sets of norms, relations, differences has become a significant focus for the analysis of welfare policies and their institutional expressions (e.g. Saraga, 1998; Lewis, 1998c). There are good theoretical as well as political reasons for foregrounding the contingent, fragile and fluid quality of social constructions, not least for breaking the universalizing and eternalizing effects of constructions of the normal and natural. However, I think it is important to temper the attention to fluidity with a concern with what might be called 'solidification'. The capacity of words, meanings, constructions and identities to be fluid or polyvalent is a central feature of the 'cultural turn'. Nevertheless, there are important analytical issues about how some meanings or constructions take on socially solidified forms, why some differences rather than others become socially valorized, and why some norms become empowered as truth. The concepts of fluidity and poly-valence identify capacities or potentialities rather than a permanent condition of social or cultural flux. They are abstract terms that set out possibilities. But the analysis of specific cultural formations or discourses has to address the extent to which the capacity for flux or change has been stabilized: fixed around a limited set of meanings. Rather than polyvalence, particular social constructions are characterized by their everyday solidity – their passage into common sense as truth. How to grasp solidification, institutionalization or sedimentation when examining social constructions and how to assess the different 'densities' of different formations or discourses are significant analytic tasks. Between the extremes of absolute polyvalence and absolute solidification are a range of different densities or plasticities that social constructions may achieve (and which may have something to do with the extent to which they are contested). In part, these are issues about the level of analysis at which studies are con-ducted. So, for example, one might argue that, abstractly, the construction of the Nation is changeable and contested. One might also point to some of the specific and concrete struggles that have attempted to fix particular meanings to a Nation. But it is also important to register how some constructions of a Nation become embedded, institutionalized and sedimented in formal and popular understandings (see, for example, Cooper, 1998; Lewis, 1998c; and G. Williams, 1985).

There are some difficulties of terminology here. I confess to a fondness for Gramsci's geological metaphors. I like the conception of searching out the 'deposits', 'sediments' and 'traces' of earlier philosophies that have settled into common sense. For example, there is the range of religious, moral and social deposits that have become embedded in the idea of 'desert' in relation to social welfare (Langan, 1998). This concern emphasizes the analytical task of differ-entiating the more fluid, changeable and unstable cultural formations from the more densely compressed, socially resilient and resistant formations. It requires the examination of social constructions/cultural formations in ways that assess their 'relative density'. One of the attractions of these geological metaphors is that they provide a sense of 'weight' to social construction. There are ways in which the cultural turn and the accompanying language of social construction risk giving an unreal lightness – a weightlessness – to the constructions that we find in social policy. By contrast, the geological metaphors restore a sense of solidity, density or massiveness about some of these constructions that properly reflects the ways in which they do weigh very heavily on us. Put most simply, social constructions kill people. Poverty is certainly an ideological formation. It is also a truth produced by particular discursive strategies. It is a social construction – and people die from it. One could go on and make similar points about racialized identities and their social consequences or about other forms of constructed social differences. These are immensely solidified constructions, with deep roots in political discourse and popular common sense: they do not change easily or lightly. But they do change: in the face of challenges, contestation and conflict even the most heavily sedimented and institutionalized formations can move and new formations can be created.

This seems to me to be the heart of the 'cultural turn'. It is the challenge to conduct difficult analyses, to think complex and contradictory thoughts about complex and contradictory cultural formations. The challenge is, in part, the problem of what we do now that we have discovered culture or discourses or social constructions. One of the significant developments in the study of social policy has been the capacity to reveal the constructed or discursively constituted character of a phenomenon that has hitherto been taken for granted as 'natural'. Certainly, one of the political effects of all the variants of the 'cultural turn' is the practice of revelation: the uncovering and laying bare of the socially constructed form of the need, problem, difference. Revealing the 'social' in these phenomena also opens them up to forms of social contestation. But there is a point where the continued practice of 'revelation' becomes disingenuous – the 'discovery' of a discourse stops being a surprise. If we approach the social world from a perspective of cultural analysis or social constructionism, we are likely to discover cultural formations or social constructions. For me, the questions then become matters of why this discourse or construction (rather than any of its contending alternatives)? Through what means did it come to be accepted as 'truth', 'common sense' or the 'taken for granted'? Who – what social forces or alliances – mobilized it and who has been empowered by its institutionalization? What contending or alternative positions surround it, which have to be negotiated, displaced or subordinated? What contradictions have to be managed or contained within it? The political and intellectual challenge – and promise – of the cultural turn lies precisely here: these are necessarily 'unfinished business'. They resist attempts to close them off by appeals to theoretical certainties or established disciplinary truths. The cultural turn is difficult to sustain because of the temptations and pressures to put a stop to things and return to safer and more comfortable ways of thinking. Taking the cultural turn involves trying to resist those temptations and, instead, continuing to think difficult thoughts.

# References

Ahmad, W.I.U., ed. (1993) *'Race' and Health in Contemporary Britain*. Buckingham, Open University Press.

Barash, D. (1981) *Sociobiology: The Whisperings Within*. New York, Harper and Row.

Burr, V. (1995) *An Introduction to Social Construction*. London, Routledge.

Campbell, B. (1995) 'Old Fogeys and Angry Young Men: a critique of communitarianism.' In *Soundings*, issue 1, Autumn, pp. 47–64.

Carabine, J. (1992) ' "Constructing Women": women's sexuality and social policy.' *Critical Social Policy*, 34: pp. 23–39.

Chaney, D. (1996) *The Cultural Turn*. London, Routledge.

Clarke, J. (1991) *New Times and Old Enemies: Essays on Cultural Studies and America*. London, HarperCollins.

Clarke, J. (1996) 'After Social Work?' In N. Parton (ed.) *Social Theory, Social Change and Social Work*. London, Routledge.

Clarke, J. (1998a) 'Thriving on Chaos: Managerialism and Social Welfare.' In J. Carter, ed., *Postmodernity and the Fragmentation of Social Welfare*. London, Routledge.

Clarke, J. (1998b) 'Coming to Terms with Culture.' Paper presented to the Social Policy Association conference 'Social Policy in Time and Space', Lincoln, July.

Clarke, J. and Cochrane, A. (1998) 'The Social Construction of Social Problems.' In E. Saraga, ed., *Embodying the Social: Constructions of Difference*. London, Routledge.

Clarke, J., Cochrane, A. and Smart, C. (1987) *Ideologies of Welfare*. London, Routledge.

Clarke, J. and Newman, J. (1997) *The Managerial State: Power, Politics and Ideology in the Remaking of Social Welfare*. London, Sage.

Clarke, J. and Newman, J. (1998) 'A Modern British People? New Labour and the reconstruction of social welfare.' Paper presented to the Discourse Analysis and Social Research conference, Denmark, September.

Cooper, D. (1998) *Governing out of Order: Space, Law and the Politics of Belonging*. London and New York, Rivers Oram Press.

Deacon, B., Hulse, M. and Stubbs, P. (1997) *Global Social Policy: International Organizations and the Future of Welfare*. London, Sage.

Dennis, N. (1993) *Rising Crime and the Dismembered Family*. London, IEA Health and Welfare Unit.

Etzioni, A. (1993) *The Spirit of Community: Rights, Responsibilities and the Communitarian Agenda*. New York, Crown.

Fitzpatrick, T. (1998) 'The Rise of Market Collectivism.' In E. Brunsdon, H. Dean and R. Woods, eds, *Social Policy Review 10*. London, Social Policy Association.

George, V. and Wilding, P. (1976) *Ideology and Social Welfare*. London, Routledge and Kegan Paul.

Hall, S. (1996) 'Gramsci's Relevance for the Study of Ethnicities.' In D. Morley and K.-H. Chen, eds, *Stuart Hall: Critical Dialogues in Cultural Studies*. London, Routledge.

Harden, I. (1992) *The Contracting State*. Buckingham, Open University Press.

Hudson, B. (1994) *Making Sense of Markets in Health and Social Care*. Sunderland, Business Education Publishers.

Hughes, G., ed. (1998) *Imagining Welfare Futures*. London, Routledge.

Hughes, G. and Lewis, G., eds (1998) *Unsettling Welfare: The Reconstruction of Social Policy*. London, Routledge.

Hughes, G. and Mooney, G. (1998) 'Community.' In G. Hughes, ed., *Imagining Welfare Futures*. London, Routledge.

Jencks, C. (1992) *Rethinking Social Policy*. Cambridge, Mass., Harvard University Press.

Langan, M., ed. (1998) *Welfare: Needs, Rights and Risks*. London, Routledge.

Le Grand, J. and Bartlett, W., eds (1993) *Quasi-Markets and Social Policy*. Basingstoke, Macmillan.

Leonard, P. (1997) *Postmodern Welfare*. London, Sage.

Lewis, G. (1996) 'Welfare Settlements and Racialising Practices.' *Soundings*, issue 4, pp. 109–19.

Lewis, G. (1998a) 'Welfare and the Social Construction of "Race".' In E. Saraga, ed., *Embodying the Social: Constructions of Difference*. London, Routledge.

Lewis, G. (1998b) 'Same Place, Different Cultures? Thinking welfare through the postcolonial.' Paper presented to the Social Policy Association conference 'Social Policy in Time and Space', Lincoln, July.

Lewis, G., ed. (1998c) *Forming Nation, Framing Welfare*. London, Routledge.

Mackintosh, M. (1998) 'Social Markets'. In A. Trigg *et al.*, *Understanding Economic Behaviour*. Milton Keynes, The Open University.

Marshall, T.H. (1950) *Citizenship and Social Class*. Cambridge, Cambridge University Press.

Morley, D. and Chen, K.-H., eds (1996) *Stuart Hall: Critical Dialogues in Cultural Studies*. London, Routledge.

Murray, C. (1984) *Losing Ground: American Social Policy 1950–1980*. New York, Basic Books.

Newman, J. and Williams, F. (1995) 'Diversity and Change: gender, welfare and organisational relations'. In C. Itzin and J. Newman, eds, *Gender, Culture and Organisational Change: Putting Theory into Practice*. London, Routledge.

O'Brien, M. and Penna, S. (1998) *Theorising Welfare: Enlightenment and Modern Society*. London, Sage.

Oliver, M. (1990) *The Politics of Disablement*. Basingstoke, Macmillan.

Oliver, M. and Barnes, C. (1991) 'Discrimination, Disability and Welfare: from needs to rights.' In I. Bynoe, M. Oliver and C. Barnes, eds, *Equal Rights for Disabled People*. London, Institute for Public Policy Research.

Rose, N. (1989) *Governing the Soul: The Shaping of the Private Self*. London, Routledge.

Rose, N. (1996) 'The Death of the Social? Re-figuring the territory of government.' *Economy and Society*, vol. 25 (3), pp. 327–56.

Saraga, E., ed. (1998) *Embodying the Social: Constructions of Difference*. London, Routledge.

Saville, J. (1975) 'The Welfare State: An historical approach.' In E. Butterworth and R. Holman, eds, *Social Welfare in Modern Britain*. London, Fontana (first published in 1957).

Taylor, D., ed. (1997) *Critical Social Policy: A Reader*. London, Sage.

Taylor, G. (1993) 'Challenges from the Margins.' In J. Clarke, ed., *A Crisis in Care? Challenges to Social Work*. London, Sage.

Trivers, R. (1978) 'Parental Investment and Sexual Selection.' In T.H. Clutter-Brock and P.H. Harvey, eds, *Readings in Sociobiology*. Reading and San Francisco, W.H. Freeman.

Twigg, J. (1997) 'Deconstructing the "Social Bath": help with bathing at home for older and disabled people.' *Journal of Social Policy*, 26 (2), pp 211–32.

Walsh, K. (1995) *Public Services and Market Mechanisms: Competition, Contracting and the New Public Management*. Basingstoke, Macmillan.

Williams, F. (1989) *Social Policy: A Critical Introduction*. Cambridge, Polity Press.

Williams, F. (1993) 'Gender, "Race" and Class in British Welfare Policy.' In A. Cochrane and J. Clarke, eds, *Comparing Welfare States: Britain in International Context*. London, Sage.

Williams, F. (1994) 'Social Relations, Welfare and the post-Fordism debate.' In R. Burrows and B. Loader, eds, *Towards a Post-Fordist Welfare State?* London, Routledge.

Williams, F. (1996) 'Postmodernism, Feminism and the Question of Difference'. In N. Parton, ed., *Social Theory, Social Change and Social Work*. London, Routledge.

Williams, F. (1998) 'New Principles for a Good-Enough Welfare Society in the Millennium.' Paper presented to the World Congress of Sociology, Montreal, July.

Williams, G. (1985) *When was Wales?* Harmondsworth, Penguin.

Wilson, E. (1977) *Women and the Welfare State*. London, Tavistock.

# PART II

# Popular culture and cultural texts

Czechoslovakia. October 1981. Sunday morning underground
market (pop and rock records on sale) in a forest near Prague
© Henri Cartier-Bresson/Magnum Photos

# Introduction

*David Crouch*

Both cultural texts and popular culture hold central positions in geographies of culture. After three decades of geography essentially being informed by cultural studies it is now able to participate itself in the multidisciplinary cultural debate where the spatial dimension of culture is widely acknowledged and this Introduction sketches its developing remit. Popular culture is now recognized to be part of everyday life and space is a pervasive part of that practice. In the absence of a positioning paper in this section the Introduction permits a slightly larger exploration of its themes.

The chapters in Part II demonstrate the diversity of cultural texts and their practice, each engaging these cultural texts in a discussion of a practised popular culture. Cook, Crang and Thorpe's chapter pursues this through a consideration of the imprecise interface between production and 'consumption' in supermarket displays: displays, cultural texts, consumers buying goods producing a different set of meanings and knowledges in their cultural practice. Crang's texts are artefacts of regional identity used by intermediaries in different versions of identity and nationalism and made sense of in a further plurality of ways. Melanie Wall considers rap music as text by young people in Aotearoa and conveys a very different popular culture. 'English music' is the text used by Revill that emerges in a very complex popular culture.

Cultural geography is persistently concerned with the constitution and construction of meaning in human activity and in its encounter with space and this is very present in the four chapters on popular culture and cultural texts in this volume. Cultural geography developed early on in the enculturation of geography and closely along the lines developed in cultural studies. As such, cultural geography has experienced pronounced changes over a quarter-century. These may be summarized as shifts from representation to an interest in practice, from the semiotics of intended landscape to popular culture (Cosgrove and Daniels, 1989; Jackson, 1989). Popular culture was envisaged successively as a potential source of power and empowerment, and then as commercial cultural products rendered free choice (although hardly ever 'free') in the construction of lifestyles. Throughout this period of continuing adjustment there is a persistent interest in meanings held by individual human subjects as well as various levels of collectivities through to nation and internationality, and the currents of power that run through those meanings (Warren, 1993; Miller *et al.*, 1998; Jackson, 1999). Evident in these chapters is the further shift from textual interpretation to an interpretation of the working of those texts in popular cultural practice. These tensions are found working through these chapters.

Melanie Wall is insistent on the need for cultural geography to go further in its attention to practices. Taking on both representations of race and the commercial contexts of kinds of music, Wall explores what happens 'in practice' for Maori youth doing rap. She directs attention to the importance of under-standing popular culture through its embodiment, practice and performance. Thus, making everyday and popular geographies using music technically 'pro-duced' elsewhere, that practice can 'provide shared texts situated in wider organ-izations of power' and provide for the negotiation of group boundaries. This all happens through the practice. The material text (music in this case) is practised, its technically produced meanings are practised in a wider frame of meanings in everyday life, embodied in listening practices. Wall argues for an enlargement of discourse-analysis centred methods to be able to engage and investigate com-ponents of embodiment. Space becomes significant in different ways, not only as local neighbourhoods of shared rap identities but as numerous imagined com-munities with which other everyday practices overlap. This refiguring of texts through their own lives provides a source of empowerment yet does not draw its power from the exact forms of the texts themselves, but through the way they are appropriated into their own lives. This appropriation is a process of embodiment whereby material is engaged in a wider lived practice and becomes lay geography (P. Crang, 1996). Such methods seek to track ways in which the human subject, frequently intersubjectively, grasps the world around her using in the process both metaphors and material gleaned from elsewhere and reach-ing the world expressively (Radley, 1995). Wall uses this interpretation of rap music in its enactment through popular culture to explore the refiguring of ethnic identity, where its enactment is embodied and informs geographical knowledge of youth identities and of the self (as explored further in Part V of this book).

The chapter by Mike Crang on cultural landscapes of Sweden considers contemporary uses of culture, as popular culture and culture-as-heritage, in pro-moting particular ideas. This concerns the use of culture in the representation of national and regional abstracted ideas and ideals of 'culture'. However, popular culture is used also in this complex case in the pursuit of distinctive identities, and in resisting nationalizations of culture. However, this is further complicated in two ways. The 'folk' is abstracted into an organic yet faceless collective. That 'folk' is also denied time and space and negotiation and its own adjustment and flows. In this Crang demands attention to practice and the importance of geographical practice acknowledging the reflexivity of popular culture. Thus the question is not one of authenticity in 'folk' or any other

version of cultural product or practice but one of power and cultural ownership that is negotiated through practice and space. This paper engages the
debates in cultural geography on the role of representations. Here representations are vitalized through practice. That practice can of course be practice by
the nation-state and coercive, or practice amongst people in the making of lay
geographies, or in the relationship between them. Of course representations have
been pervasive, and often persuasive, in the commercialization of space and
images of space and of materiality that pertains to space but is never complete.

This is the concern of Cook, Crang and Thorpe's chapter on food and the
representations of food on shop counters. The presumption that there is a clear
flow between representation and meaning is found difficult to sustain. There is
also an issue of authenticity here, centrally concerned with meaning and essentially cultural. Authenticity further concerns the translation, acknowledgement
or appropriation of meaning in particular processes of refiguring, and is again
a concern of empowerment. When meaning is constituted through practice and
can be appropriated in new ways at several conjunctures of practice, tracing
authenticity becomes a project of interpreting practice, here in terms of human
subjects as food counter operatives and as customers. It is by investigating
practices and their reflexivity that it becomes possible to critique the consumption/
production polarity, as recent work on shopping, once pre-eminently a triumph
for metaphor, has shown (Miller *et al.*, 1998; Gregson and Crewe, 1997). The
processual nature of buying things, shopping and consumption is thus imaginative and productive rather than one of fixing meaning in representation or of
making bounded distinction between meaning across different social groups.
However, as each of these chapters makes clear, the power of social distinction
in relation to cultural meaning remains strong, although the way it works is
recognizably more complex.

Ian Cook edits the third part of this book. This evident cross-over between
one and the other is appropriate as it responds to MacDowell's reference
['Economy, culture, difference and justice'] to culture as 'symbolic order' and
Cook's chapter with Crang and Thorpe in this section discusses the exchange
in lay geography. In this way, Cook *et al.* contribute to the debate on the turn
of culture wrought through economic practice in popular culture, at the literal
shop floor. Wall also considers relations of economic power through the merchandising of music and the refiguring of that merchandise in popular cultural
practice. Of course products such as the spices considered by Cook *et al.* emerge
through notions of nation, and these thread through both Crang and Revill.

The continuing searches for Englishness in musical tradition and its use to
inform new formal music production also remain complex and defy simple
binaries of definition. The pastoral is still a reference point in this quest but is
a contested space, used to convey through notions of nature and naturalness
an ideal of Englishness where 'the rural' is held in a tension between different
knowledges that penetrate contemporary everyday practice (Crouch, 1992).
However, within what may be identified as Englishness are other complex
tendencies between culture and nature that connect with Part IV of this volume.
In his chapter Revill works through different historical moments in recent formal

musical production in order to demonstrate the complicated encountering between ideas of nation, nature, politics and culture and how these are persistently incorporated in, and are incorporated by, everyday life. This again confounds ideas of separation between lived practice and representation, and Revill observes the representations being made through an embodied identity with lived experience that demonstrate the limits to identifying ideas only in the abstract and knowledge based on preconstructed notions of land, nature, country.

Across these different papers and the examples they take there is an engagement with metaphor and materiality. Culture is rendered as metaphor but produced in a mutual encounter with the material, constituted of practice. Of course representation remains important, as stages in the flow of meaning and knowledge can occur in the spaces between one representation and another. However, representation becomes part of a wider flow of culture that may be characterized by embodiment. Representations are themselves practised and are the products of practice (Crouch and Matless, 1996; Crouch and Toogood, 1999). In a turn to focus cultural geography on and through the human subject it is practice that becomes more important and the process of meaning necessarily acknowledges lay geographies, lay knowledge, as it is increasingly acknowledged that it is through practice that meaning is 'lived'. Embodiment has become a feature of research strategies involving observation and interview (P. Crang, 1994; Crouch, 1999). Numerous agencies, interests and structures can produce meaning that speaks of the producing interests. However, this may be of little value to anyone else until it is engaged in practice. There is a large space in geographical understanding in terms of the ways in which texts have been considered replete with semiotics but where investigation into meaning beyond semiotic production is absent. Places, shopping malls, theme parks, landscapes of all sorts and ordinary streets are at once spaces of representation and of practice (Crouch, 1998). Representation may be part of the semiotic content intended in particular structures and their design but there is also the representation of the self in the spaces in which that self practises the reflexive content of everyday life. 'Received representation', as in designed products or spaces, is utilized as resources amongst others in the practice of everyday life and in the constitution of subjective geographies. An approach to embodied practices of identity can make use of the expressive, multi-sensual and multi-dimensional components of embodiment and 'feeling of doing' (Radley, 1995; Harre, 1993). Human subjects make practical knowledge socially, a material semiotics in process (Shotter, 1993; Nielsen, 1995; Game, 1991). Yet all of these events and practices do not happen in a cultural vacuum, and thereby they are each situated, each operates in relation to contexts and socializations that, however, do not themselves remain fixed (Young, 1990, p. 11).

The human subject is rendered consumer, producer, active agent, imaginative, engaged in an encounter with the world that is neither Promethean nor subservient. This encounter is explored increasingly through embodied practice. Cultural texts, pictures, malls, parks are rendered resources in this process (M. Crang, 1997). The idea of the encounter of embodiment and process do not avoid conflict and power and instead make it more complex. People may not figure space in an exercise of resistance but may do so in making what they do and what they use their own (Thrift, 1996, 1997). Meaning is produced in the encounter between human subject and place, and other human subjects, and a range of material artefacts that may be commercially produced or laden with particular symbolism such as those of nation. These meanings may be considered as geographical knowledge constituted through a collision and a negotiation of everyday, lay practices and numerous representations that may be accorded distinctive meaning through practice and that may shape that practice.

The working of representations, whether essentially visual or appealing to other senses, is made much more complex, uncertain, fractured. Cultural texts become especially interesting when they are worked through embodied practice. Popular culture emerges from these chapters as constituted through embodied lay practice, where ideas are felt. 'All ideas and materials are necessarily em-bodied . . . and all matter embodies meaning and derives its place in the human world by virtue of that meaning' (Crossley, 1995, p. 59). Crossley argues from Merleau-Ponty that '[t]here is no [inner] theatre of the mind where "shows" from the outside are projected . . . There are not two tables, one in the world and one in the mind but rather one table which is seen . . . The active body embodies meanings and ideas' (1995, pp. 46, 48). We get to know a place 'with both feet'. In this way, contemplation and reflexivity are achieved through prac-tical and embodied involvement in the world and this becomes a major concern for cultural geographic research. The degree to which every aspect considered in these chapters acquires cultural affects is a further challenge for cultural geo-graphers to unravel.

# References

Cosgrove D. and Daniels S. (1989) *The Iconography of Landscape*, Manchester University Press, Manchester.

Crang M. (1997) 'Picturing Places: research through the tourist gaze', *Progress in Human Geography* 21, 3: 359–73.

Crang P. (1994) 'It's Showtime: on the workplace geographies of display in a restaurant in southeast England', *Environment and Planning D: Society and Space* 12: 675–704.

Crang P. (1996) 'Guest Editorial', *Environment and Planning D: Society and Space*. 14: 631–3.

Crossley N. (1995) 'Merleau-Ponty, the Elusive Body and Carnal Sociology', *Body and Society* 1, 1: 43–63.

Crouch D. (1992) 'Popular Culture and What we Make of the Rural', *Journal of Rural Studies* 8, 3: 229–40.

Crouch D. (1997) ' "Others" in the Rural: leisure practices and geographical know-ledge', in P. Milbourne ed. *Revealing Rural Others*, Cassell, London.

Crouch D. (1998) 'The Street in the Making of Popular Geographical Knowledge', in N. Fyfe (ed.) *Images of the Street*, Routledge, London, pp. 189–216.

Crouch D. (ed.) (1999) *Leisure/Tourism Practices, Knowledges, Geographies*, Routledge, London, pp. 160–75.

Crouch D. and Matless D. (1996) 'Refiguring Geography: the parish maps of common ground', 21, 1: 232–55.

Crouch D. and Toogood (1999) 'Everyday Abstraction: geographical knowledge in the art of Peter Lanyon', *Ecumene* 6, 1: 72–89.

De Certeau (1984) *The Practice of Everyday Life*, University of California Press, Berkeley.

Fiske J. (1989) *Understanding Popular Culture*, Routledge, London.

Game A. (1991) *Undoing the Social: towards a deconstructive sociology*, Open University Press, Buckingham.

Gorz A. (1982) *Paths to Paradise*, Verso, London.

Gregson and Crewe (1997) 'The Bargain, the Knowledge and the Spectacle: making sense of consumption in the space of the car-boot sale', *Environment and Planning D: Society and Space* 15: 87–112.

Harre R. (1993) *The Discursive Mind*, Blackwell, London.

Hebdige D. (1993) *Hiding in the Light*, Routledge, London.

Jackson P. (1989) 'Commodity Cultures: the traffic in things', *Transactions of the Institute of British Geographers* N.S 24, 1: 95–108.

Miller *et al.* (1998) *Shopping, Place and Identity*, Routledge, London.

Nielsen N.K. (1995) 'The Stadium in the City', in J. Bale (ed.) *The Stadium and the City*, Keele University Press, Keele, pp. 21–44.

Radley A. (1995) 'The Elusory Body and Social Construction Theory', *Body and Society* 1, 2: 3–23.

Shotter J. (1993) *Cultural Politics of Everyday Life*, Open University Press, Buckingham.

Thrift N. (1996) *Spatial Transformations*, Sage, London.

Thrift N. (1997) 'The Still Point: resistance, expressive embodiment and dance', in Keith and Pile (eds) *Geographies of Resistance*, Routledge, London, pp. 124–52.

Warren (1993) ' "This Heaven gives me migraines": the problems and promise of landscapes of leisure', in J. Duncan and D. Ley (eds) *Place/Culture/Representation*, Routledge, London, pp. 173–86.

Young I.M. (1990) *Throwing like a Girl and Other Essays in Feminist Philosophy*, Indiana University Press, Indiana.

# The popular and geography

## Music and racialized identities in Aotearoa/New Zealand

*Melanie Wall*

> We're just different on this side of the bridge . . . well it's white isn't it? Sure music's important, like we listen to different things. They're all homies, wannabe G's, and the wiggers, they're the worst. I mean, like, if I was going to ask these questions I'd ask about race . . . (pause) . . . I suppose you can't really. Everyone is so fucking P.C. these days. I don't know why they bother. Who are they kidding? I mean, everyone knows it's colour that's important.
>
> (Taken from an interview with a white female respondent from the North Shore of Auckland)[1]

The above excerpt is taken from a series of interviews that were conducted throughout the Auckland region with 16 and 17 year old high school students. The respondent highlights a number of themes which are integral to my research: place, the racialization of music listening practices, and the politics of researching race in Aotearoa/New Zealand. In the germinative stage of this research project, my primary objective was to explore the lived experiences of racialized identities in Aotearoa/New Zealand. And therein lay my first hurdle. As the above respondent indicated, in terms of asking direct questions about race in the current political climate in Aotearoa/New Zealand, she states 'I suppose you can't really'. In a nation that prides itself on 'doing' as opposed to 'saying' (see Phillips, 1987; Berg, 1994; Bell, 1996), issues surrounding race join the more taboo subjects of religion and politics. Unless you wish to be labelled a 'stirrer' looking for trouble, discussion of these topics must be avoided for fear of dissension, and opinions only aired if you are fairly certain that you are surrounded by like-minded others. The one-nation myth still retains much popular currency, as Aotearoa/New Zealand is, after all, a racially harmonious

Pacific paradise, and while we may have some 'problems', these are *surely* inconsequential compared to elsewhere?

In response to the specificities of the context of racial politics in Aotearoa/New Zealand, I resolved to approach these issues using popular culture as a means to access racialized identities. As race tends to be constructed, articulated and experienced through a complex set of intersections with a number of other dominant discourses (such as gender, nation and sexuality) and practices (such as sport and music), I selected music listening practices as a means to consider race. From my experiences of growing up in Auckland as a Maori woman at a predominantly 'White' school, I was well aware that exploring racialized identities through music listening practices would be a strategic choice as, in my experience, music preferences were colour coded. While many of my respondents were either unwilling or unable to discuss race directly, they were reasonably accustomed to talking about race in relation to music listening practices.

You could question, from this rather oblique introduction, how this chapter fits under the auspices of a section entitled 'popular culture and cultural texts'. I will argue, however, that attention to popular culture (in this instance popular music) stimulated the depth of insight that I have received into the complex construction of racial coding amongst youth identities in Auckland, which would otherwise have remained elusive. In this chapter I discuss the relationship between geography and popular culture, with particular reference to the significance of experiential research and context to my approach to music and racialized identities in Aotearoa/New Zealand. Throughout this chapter I will draw on a number of examples from my research in Aotearoa/New Zealand to highlight how this approach, by accessing racialized discourse through popular music, has provided a unique insight into racialized practices and experiences. In many ways, this chapter represents my efforts to reconcile key issues which have arisen from attempting to discuss race in conjunction with popular culture, rather than the more traditional reified notions of what culture should entail. While a number of these issues manifest themselves globally, I will contend that the popular is also important for imagining the local; while much of popular culture involves globally significant cultural texts, it is how these texts are appropriated and negotiated for the reproduction of localized identity groupings, such as race, that makes the popular significant.

## *Geography does popular culture?*

> *Respondent 1:*  Why are you asking all these questions about music? Do they pay you to do this? Man, I should get a job like that. Easy money . . . (group laughs) . . . But really, who cares about this stuff? It's not important.
> *Melanie:*  Are you saying that music's not important to you?
> *Respondent 1:*  Course not.
> *Melanie:*  Then why not ask questions about music. What questions should I ask then?
> *Respondent 1:*  Ah . . . I dunno . . . I guess it's cos no one else thinks it's important. Um, it . . . it's just no one really cares about what we think.

*Respondent 2:*   Oh, no one cares . . . poor boy (group laughs)
*Respondent 1:*   Na, don't take the piss . . . No, if I think about it, you're right. If you want to find out what's important to people our age, music would be a good way to go about it. Don't you reckon? . . . (group murmurs agreement)

This excerpt is from a group interview at a West Auckland school. The group are trying to contemplate why their opinions and experiences are relevant to a university researcher, as they understand only certain types of knowledge (often that which is outside of their experience) to have value. This devaluation of popular knowledge is reflected not only in wider society but also within geography. Geographers have traditionally paid scant attention to popular culture, preferring to dwell on high cultural artefacts which fell within a retarded definition of what constituted the arts. While I am aware that this is not a new criticism, I feel that it is one that requires reiteration. I contend that popular culture remain stigmatized, a residual category existing as high culture's Other: mere entertainment too ephemeral and superficial to be worthy of scholarly attention. In this section I argue that popular culture is an area in the relationship between culture and geography which, despite notable exceptions (see, for example, Jackson, 1994; Halfacree and Kitchin, 1996; and selected articles contained within Skelton and Valentine, 1998), is neglected. I realize that this is a sweeping and highly contentious statement given a number of recent efforts over the past fifteen years to amend this shortcoming. However, if this cultural turn is to attend to notions of the popular and popular culture, it must confront the continuing reticence in geography to address these issues.

While geography has managed largely to avoid the current critique of cultural relativism which has been levelled at a number of the social sciences, geography only need look at the current so-called crisis of cultural studies which has been produced, in part, by criticism of the attention of cultural studies to popular culture. In a recent book entitled *Back to Reality*, prominent cultural studies scholars have felt compelled to refute this criticism. Meaghan Morris (1997), for instance, goes to considerable length to underline that cultural studies is not a 'theory' of popular culture. This 'crisis' in cultural studies has significant implications particularly for the current cultural turn in geography which, I would allege, has been generated by the increasingly interdisciplinary nature of the social sciences. As such, if cultural studies has been placed on the defensive, the legitimacy of popular culture as a subject for research in geography is bound to be challenged.

To meet this challenge, the recognition of the continuing denigration of the study of popular culture must be stressed. Whether it is popular culture or the emergent notion of 'popular geographies' that is emphasized, what remains significant for geography is the importance of researching the popular. This does not mean that the most favoured popular cultural products, produced by the so-called culture industries, are necessarily more worthy of serious contemplation, as notions of the popular vary between places and across spaces, between localized groupings and more transnational affinities. Again, with notable exceptions though, I would question why there is so little research by geographers into popular cultural products and practices, such as popular music (music that dominates the charts), popular magazines (magazines that have a large circulation), popular films (films that have high box office and video receipts) and so on. This may seem pedantic but the point I wish to make is that the popular is a question of relevance.

To illustrate the relevance of cultural texts, I will use an example which has arisen on a number of occasions during the research process. A key element in my research into racialized identities is the wholesale appropriation by Polynesian youth in Auckland of African American hip hop culture.[2] On a number of occasions when I have raised this issue in academic forums, the common response has been: 'Is there such a thing as Polynesian rap?' Rap music does dominate the singles charts in Aotearoa/New Zealand and while there is the occasional rap song by Polynesian artists, rap music sales are dominated by African American performers.[3] Acknowledging that the emergence of local rap artists is a significant issue (see, for example, Mitchell, 1994, 1996a and 1996b), for the purposes of my research it is a subsidiary one as the vast majority of the popular music that influences local Polynesian music listening practices is not locally produced. The question regarding Polynesian rap is understandable, though, given that the majority of research exploring identity and music, particularly racialized identity and music, is dominated by research into a specific type of music and the locality from which it originated (see, for example, Cross, 1993; Hayward *et al.*, 1994; Rose, 1994; Mitchell, 1996b). I would contend that this emphasis is derived from a complex set of interlocking dualistic narratives, such as local/global and authentic/inauthentic, which privilege the so-called local sound as a more authentic expression and experience.

Alongside this privileging of locally produced music, a considerable amount of discussion on popular music muses over what music listening practices *should be*. In a paper discussing Polynesian music, for example, Tony Mitchell (1994) cites Mahinarangi Tocker, a Maori singer/songwriter, who claims to derive her inspiration from traditional Maori themes. She maintains Maori singers should not pretend to be Black Americans. Instead, Maori should have their own style and rather than identifying with Black Americans, they should seek connections with the indigenous peoples of America. Yet as Mitchell (1994) asserts, musical preferences are not necessarily decided by 'indigenous ideological choices'. While connections between the historical experience of indigenous Americans and Maori may be present, a case could be made that a significant proportion of young Maori, who listen to rap music, identify more with the racialized experience of urbanized African American youth. Music listening preferences are constructed from a multitude of influences and rather than solely being a political choice, music preferences are derived from an often unconscious but shared notion of racialized music listening practices (as explored later in this chapter). The following excerpt is from a group interview with a number of Maori respondents and highlights their understanding of the significance of music listening practices in the articulation of contemporary racialized identities.

*Melanie:*   From that story then, you seem to care a lot about the origin of the music you listen to.
*Respondent 1:*   Yeah course. Like I know it seems stupid not liking a song just cos a White group does it and then if the same song's Black to automatically like it . . .
*Respondent 2:*   but it's not just colour . . .
*Respondent 3:*   like it is though, that's what . . .
*Respondent 2:*   Yeah I know and I'm not saying it's not but it's also about beat . . . and the voices too. Like for me, Black voices just sound better you know. I like the RnB stuff, melodic . . .
*Respondent 3:*   yeah we all do. Can you imagine being Maori and not listening to hip hop. Like I'm not saying there aren't some who listen to White music like Maori metallas . . . (group laughs) . . . but that's really strange aye. It's part of being brown. Like I reckon a White guy can listen to Black stuff too but for us it says who we are. Like I know the traditional stuff's good but man it's not like you can listen to *Poi-E* all the time . . . (group laughs) . . . Like with the American stuff we can relate aye . . . (group murmurs agreement) . . . it's a brown thing. Like they know what goes on for us aye . . .[4]

This excerpt highlights that the respondents make a conscious distinction between Black and White music, and employ these differences to distinguish their own racialized identities. These comments were typical of those given throughout the interviews I conducted. Popular music was perceived to be relevant to the lived experience of race. To this group of urban Maori teenagers, it was not Polynesian music which they saw as particularly relevant to their identity, but rather it was African American hip hop culture that was preferred. As such, the favouring of locally produced music in academic approaches to popular music often privileges the relative significance of local music and the ways in which the local can be accessed. Examining locally produced music has favoured sites of music production as legitimate cultural expressions of the local whilst avoiding the complexities which underpin the negotiation and appropriation of popular music within localized music listening practices. The 'local' is often represented through the value laden terms of authenticity and the 'real', whereby local music is commonly imagined as a cultural expression of resistance set up in opposition to the all-powerful global cultural industries which manufacture the 'popular'.[5] Considering local music, as evidenced by comments from Mahinarangi Tocker, often leads to a deliberation on what popular music should be. While I would readily acknowledge that there are spaces for political activism within writings on popular music, my concern is for the silences which arise from the lack of discussion on those more globalized cultural texts which are most popular.

Returning to the initial premise of the chapter, in terms of geography and the popular, popular geographies are significant in that they are relevant to everyday lived experiences. This emphasis is in no way intended to valorize the 'everyday', but rather to position everyday experiences as a site of cultural production. It is the consideration of the negotiated place of popular culture in the everyday delineation of connections and differences between people which could provide access to the complexities of contemporary cultural practices. Philip Crang has claimed, in a recent editorial entitled 'popular geographies', that 'focusing on the performativities of knowledge allows an examination of the popular (and indeed unpopular) geographies produced and consumed in embodied, interrelational, contextual practices, not as disembodied cognitions, floating commodity-signs, or imaginative projections' (1996, p. 632). Performativity implies action, the embodied action involved in the practice of popular culture, its production and consumption.[6] What defines the popular is a question of relevance, whereby popular culture can provide common ground as shared texts situated in wider differential relations of power, expressed within material contexts through the negotiation of group boundaries.

In the next section I elaborate upon these themes examining the possibilities for experiential research as a means to explore the popular within geography, with reference to a number of issues that have arisen throughout my research. As highlighted in this section, not only have geographers been negligent in their attention to popular culture, when popular culture has been approached it has been dominated by research into the local and localized production of cultural texts. What I hope to illustrate in the next section is how – through attention to the consumption of popular culture – identity is a negotiated process involving the engagement with cultural texts in producing the local. Whilst the production of globally significant cultural texts is situated largely outside of Aotearoa/New Zealand, the negotiation and appropriation of these cultural texts by Auckland youth engenders localized sites of reproduction which, in turn, inform the performance of identity within material contexts.

## How does geography do the popular?

This section is in no way intended to be a definitive statement on how popular culture should be 'done' in geography but rather I will offer a few insights from my research experience on how it could be approached. In light of previous comments on the importance of relevance in determining what constitutes the popular, I explored the relationship between music and racialized identities through an audience-based approach. This enabled those interviewed to define their music preferences and to discuss the significance of popular music to their experience of racialization.[7] As such, the designation of what constitutes the popular is determined by a group of music listeners, not instigated by a politicized desire for what should be or by a romanticization of the local as a site of resistance. In this section I discuss two primary themes essential to my research: experiential research and context.

### EXPERIENTIAL RESEARCH

The main argument which I make in this section is that experiential research has also been neglected in the current cultural turn. This neglect has been to the detriment of research into the lived experience of 'culture', as experiential research has the potential to offer unique perspectives on the relationship between culture and geography. For example, in relation to my research into the performance and experience of contemporary racialized identities, experiential research is an essential component. I would readily acknowledge that there are problems inherent to experiential research; that there is no essential truth out there waiting for me to uncover. However, the alternatives concern me. I remain convinced that cultural geographers need to place greater emphasis on experiential research, to counteract the dominance of the text. With regard to popular culture, textual analysis is the dominant means of research. For example, in academic approaches to rap music, the vast majority of papers construct rap as an expression of Black vernacularism, a 'Black thing'. Paul Gilroy (1997) claims that a number of Black scholars have used the 'purity and power' of the vernacular, constructed through absolutionist definitions of culture, to create a special authority for themselves, an authority which they are reluctant to concede. These papers tend to read rap as a text, drawing from it themes in the lyrics which they then elaborate upon. As rap music has been widely hailed as an outlaw culture, the voice of the oppressed, particular narratives have become associated with writing rap. These range from ethnocentric accounts of the origins of hip hop culture (Rose, 1994 and Perkins, 1996), to discussions of authenticity, nihilism, misogyny, and homophobia in rap lyrics (Craddock-Willis, 1991; Dyson, 1993; Decker, 1994; de Genova, 1995; and White, 1996).

What is generally ignored in these interpretations, however, is the question of cultural practice. I have yet to find any work on rap music that addresses the issue of how music listeners (as participants and performers) experience rap, and how these knowledges intersect with the dominant discourses which underpin writing rap. What would music listeners think of the issues that academics and other commentators deem meaningful? The following excerpt provides a partial answer to this question. This debate took place during a group interview where 8 of the 11 respondents identified themselves as Samoan. During my first interview with this group, they laid claim to a unified voice, maintaining that in being Samoan they only listen to 'Black music'. In a later interview, all of the Samoan respondents forcefully expressed the importance of Christianity in their lives. Following on from this theme, I queried the apparent incompatibility of their music listening practices with their religious beliefs.

> *Melanie:*   Given this, how does your church life fit with the music you like?
> *Respondent 1:*   Oh you mean how can we listen to rap and still be Christians? With all that swearing and violence . . . (laughs) . . . It's like for me I just don't hear it. Like what we were talking about before aye, it's the beat that's important . . .
> *Melanie:*   What do you mean though, that you don't hear it?
> *Respondent 2:*   Well, like, we know what it's about and all that, but that's not what the music's about for us aye. Do you know what we mean? Like only the hardcore homies are into that stuff . . . You know the gangs and that. But that's not for us aye. It's about the music. It's about who we are.

What this excerpt highlights is that these teenagers were consciously participating in a process of negotiation and appropriation of rap music as a popular cultural text. They were aware of the dominant discourses which surround rap music, representing it as a global cancer which has degenerative influence on young Black identities. For this group, however, African American rap music was about being Samoan in Auckland. They had rationalized the more negative aspects of the rap that they listened to, separating the lyrics from the music and choosing to hear only the latter. Textual approaches to rap, whether a celebration of its apparent Blackness or a critique of its debasing lyrical content, ignore the role of the music listener as an active participant in the performance of rap as a popular cultural text. What these approaches disregard is that while rap music is a cultural text, it cannot solely be approached in terms of the written as rap music is also about sound, the sensory experience of listening.[8] A number of other issues are raised in this excerpt which could also be addressed, however my intention for including this example was to emphasize the significance of experiential research, and the potential for this approach to enhance exploration into the popular and geography.

## CONTEXT

Within this discussion, I have included this subsection on approaching the popular in geography in context, as I regard context as essential to my research process. Earlier in the chapter I was critical of much work exploring popular culture as it tended to romanticize the local, particularly local production of cultural texts, reified in opposition to notions of the global. For my research process, context is significant in that it provides a setting for texts, in this case the lived experience and performance of popular music texts. Conceptualizations of the local are significant insofar as they are perceived to be influential to identity processes, constructions which are not necessarily mutually exclusive from globalized cultural products. For example, Paul Gilroy (1993, 1997) claims that while African American hip hop culture aggressively asserts its localism, these so-called vernacular expressions can inform other localized conceptions of Blackness. Gilroy contends this can be achieved as the ethnic rules, which attempt to delimit rap music as an exclusively localized Black experience, can be taught and learned. This process, I would argue, has occurred amongst Polynesian youth, who have drawn on a notion of a shared Black aesthetic to inform and enhance their collective racialized identity. The earlier examples from my interviews testify to the interrogation and appropriation of African American popular cultural texts by Polynesian respondents from an urbanized New Zealand setting.

Locating the popular in the context of racialized politics in Aotearoa/New Zealand provides a space within which a discussion of the lived experience and performance of popular cultural texts can occur. Simon Frith (1996) maintains that while the performance and experience of popular music facilitates the differentiation of music listeners, simultaneously creating an awareness of cultural boundaries and contributing to them, he contends that what makes music special is that it can also define space without boundaries. While I find the notion of space completely absent of boundaries problematic, I would assert that popular cultural texts can facilitate new ways of approaching notions of space and spatialized practices. The process of identity is experienced through constructing boundaries between Self and Other, the conceptualization of groupings: an inclusionary sense of shared connection and affinity held by people, contrived in opposition to exclusionary imaginings of difference. As such, it is unsustainable to represent the local as detached from the global (rather than as part of a mutually constitutive process of realizing space), as defining the local is undoubtedly intrinsic to identity as process. My interviews revealed a variety of responses in relation to understandings of the local and whilst there were common answers from discernible groupings amongst those interviewed, these were in no way limited to physically bounded spaces such as suburb, city and nation. Rather notions of the local were also derived from imagined connections, some of which were transnational in relevance, such as the notion of a shared Black aesthetic. My imagining of context – Auckland as a setting for the performance of text – provides a physical space from which to access both relations within and across spaces, affinities constitutive to the experience of racialized identities.

## And what was that all about? Music and racialized identities in Aotearoa/New Zealand

Simon Frith (1996) argues that the significance of music rests in the emotional response it invokes from the listener, where listening is a means of performance. He alleges that musical appreciation is not only about pleasure and desire, but rather listening is a form of collusion, whereby music informs our subjective and collective identities. In the context of Aotearoa/New Zealand, music listening practices are one of the few socially acceptable ways to overtly discuss racial differences that supposedly delineate subjective and collective youth identities. While avoiding a discussion of the complex reasons that underpin the accepted colour coded nature of music tastes, what is significant for this chapter is that without accessing these racialized discourses through experiential research on popular music, these relations would have remained obscured. Those interviewed indicated that colour affinity, whether it be White or Black, was demonstrated by music listening preferences. These, in turn, influenced membership to peer groupings, clothing styles, type of slang, nature of recreational pursuits and other behaviour. It opened avenues to explicitly discuss racialized practices, something which is generally avoided by the majority of New Zealanders. While spaces for rupture in these racialized discursive strategies were present, and racialized identities were sometimes not as fixed as they were initially articulated, what is significant is that popular musical texts are experienced in context.

Rather than presenting definitive 'findings' from my interview material on racialized identities and music, this chapter has been a more anecdotal account of a few key issues relating to the popular and geography. What I hope to highlight are possible directions for researching the popular and to present a number of associated debates. This discussion was not intended to be prescriptive but rather to question the relative dominance and marginalization of certain research practices in geography. With reference to the popular, if popular culture remains stigmatized, what does the cultural turn in geography mean? Why is a significant proportion of academic attention towards culture and identity dominated by interpretation of discourse through textual analysis? My previous research examining cultural texts employed various techniques, such as discourse analysis and semiotics, to explore racialization in the media (see Wall, 1997). Whilst my readings of these texts were capable, it would be inadequate, in my opinion, to approach my current research without the benefit of experiential interviewing.[9]

To further accentuate this theme I intend to end this chapter with part of a dialogue which ensued at the beginning of a group interview. The following dialogue reinforces that experiential research is not just about the popular. I attempted to approach popular musical texts as a microcosm of lived material contexts of power. Part of that relation is my role as interviewer and as the academic who has final control over whose voices get heard. The following comments directly placed not only my research but also me as the researcher within the racialized relations articulated through music listening practices, evidence perhaps of the importance of experience in context.

> *Melanie:*   First off I want to ask you an easy question about what sort of music you like?
> *Respondent 1:*   Why do you want to know that?
> *Melanie:*   Because it's important to my study which looks at music and youth identity.
> *Respondent 1:*   Why bother, it's not important . . . Why don't you just make it up?
> *Melanie:*   Do you think I could? Or more to the point would I get it right?
> *Respondent 1:*   Hell no . . . (laughs) . . . But would that matter? My brother goes to university and he does geography and he reckons it's all bullshit anyway . . . (group laughs) . . . He says it's all liberal . . . and full of that Maori shit. None of you have got a clue. You're Maori aren't you, I can tell. Well I can tell you now what kind of music you guys listen to. Its all that homie crap. Black shit . . . (laughs) . . . Well I can tell you for a start that I don't listen to that crap.
> *Melanie:*   Well what do you listen to?
> *Respondent 1:*   See, look at that. I insult you and you don't say anything. Doesn't it worry you that I think Black people are all scum and they should just die.
> *Melanie:*   Should it?
> *Respondent 1:*   See you don't really want to know. You won't inlude this. It's too in your face. As if you could study us. You'd have no idea what's important to us.

# Notes

1.  *homies:*   while homie can refer to people who listen to rap music, more often an extended meaning is implied. The usage of the term is a means of avoiding direct reference to race. Used as a replacement term for Maori and Pacific Island youth, homie denotes not only their supposed collective music preference but also their clothing style, language, attitudes, and (particularly anti-social) behaviour.
    *wannabe:*   a particularly important signifier for youth identities, allied to a notion of the 'tryhard'. So-called 'Black' music has been marketed around notions of racial authenticity which have, in turn, been extended to music listeners. To be labelled a wannabe or a tryhard is extremely insulting, as the music listener is seeking authenticity.
    *G:*   short for gangsta, an African American colloquialism which is used in Aotearoa/ New Zealand as a term which is synonymous with homie.
    *wigger:*   short for white nigger. Employed to vilify those 'Whites' perceived to be race traitors; whose music tastes, clothing styles and other practices coincide with those that have been constructed as exclusively Black.
    *P.C.:*   an abbreviation of 'politically correct'. P.C. is a label applied to someone who is perceived to be excessively liberal and is a notion which is generally employed in a pejorative fashion.
2.  For the purposes of this paper, when I refer to 'Polynesian', I use this term to include both Maori and other Pacific Islander respondents. While these groups are culturally distinct, there are a number of connections between these racialized groups, particularly in terms of music listening practices. For example, the colour category 'Brown' was mentioned on numerous occasions, during my research, to naturalize both the assumed racial connection between Polynesian respondents and as an explanation for certain behaviours, such as music preferences.
3.  In this paper, I use the term 'rap music' broadly to infer not only rap but also RnB and other African American music styles loosely associated with hip hop culture. I have extended the traditional meaning of rap as a music category, as respondents tended to group 'rap' and 'RnB' within the category 'hip hop' and would use all three terms interchangeably.
4.  'Metallas' is slang for those music listeners who prefer heavy metal music. Also, Respondent 3 makes reference to the song *Poi-E*, which was released by the Patea Maori Club in the early 1980s and it held the number one position in the New Zealand singles chart for a number of weeks. The use of *Poi-E* in this context is intended as ironic. While the Patea Maori Club is a 'traditional' Maori performance group, *Poi-E* was a mix of traditional Maori music and emerging hip hop styles. This song was viewed with much disparagement by the group, as extremely 'uncool'. The group maintained that while they enjoyed so-called 'traditional' culture, in terms of it fulfilling their music needs, local music left much to be desired.
5.  A discussion of the construction of the local, as bounded by essentialized notions of place, continues in the next section. I contend that many approaches to the local reproduce dominant ways of imagining space, such as national boundaries, which limit understandings of identity as process, which I would argue is strongly influenced by relations across spaces through transnational affinities.
6.  While I would take issue with the rather loose application of the word performativity in this instance, and question whether it could be substituted with performance, the point that Crang (1996) makes remains relevant.

7.  For the purposes of this chapter I am going to avoid a discussion of the specifics of my research methodology other than to mention that I interviewed over 74 teenagers in their final year of secondary school in a series of group interviews at a number of schools throughout the Auckland region over a six month period in 1998. Alongside these interviews, I conducted a series of one-to-one interviews with 16 of the afore-mentioned students.

8.  There are series of significant questions emerging from an allied critique of the representation of sound in academic writings and the dominance of the visual in approaches to cultural texts. While there is not space to explore these debates within this chapter, there is an excellent discussion of a number of these issues in George Revill's paper on soundscape, also contained within this book.

9.  There are a number of methods other than an audience-based approach, which could redress the comparative dominance of the text in studies of popular culture. For example, bell hooks (1994) has repeatedly petitioned for more in-depth analyses of the structures of power which underpin the production of rap music, such as the relationship between the supposedly independent rap labels and their multinational parent companies.

# References

Bell, C. (1996) *Inventing New Zealand: everyday myths of Pakeha identity*, Penguin, Auckland.

Berg, L. (1994) 'Masculinity, place and a binary discourse of "theory" and "empirical investigation" in the human geography of Aotearoa/New Zealand', *Gender, Place and Culture*, 1(2), 245–60.

Caddock-Willis, A. (1991) 'Rap music and the black musical tradition', *Radical America*, 23(4), 29–38.

Crang, P. (1996) 'Guest Editorial: Popular geographies', *Environment and Planning D: Society and Space*, 14, 631–3.

Cross, B. (1993) *It's not about salary: rap, race and resistance in Los Angeles*, Verso, London.

Decker, J. (1994) 'The state of rap: time and place in hip hop nationalism', in Ross, A. and Rose, T. (eds) *Microphone fiends: youth music and youth culture*, Routledge, New York, 99–121.

Dyson, M. (1993) *Reflecting black: African-American cultural criticism*, University of Minnesota Press, Minneapolis.

Frith, S. (1996) 'Music and identity', in Hall, S. and du Gay, P. (eds) *Questions of cultural identity*, Sage Publications, London, 108–27.

de Genova, N. (1995) 'Gangster rap and nihilism in Black America: some questions of life and death', *Social Text*, 43, 89–132.

Gilroy, P. (1993) *The Black Atlantic: modernity and double consciousness*, Verso, London.

Gilroy, P. (1997) ' "After the love has gone": bio-politics and etho-poetics in the black public sphere', in McRobbie, A. (ed.) *Back to reality? Social experience and cultural studies*, Manchester University Press, Manchester, 83–115.

Halfacree, K. and Kitchin, R. (1996) ' "Madchester rave on": placing the fragments of popular music', *Area*, 28(1), 47–55.

Hayward, P., Mitchell, T. and Shuker, R. eds (1994) *North meets south: popular music in Aotearoa/New Zealand*, Perfect Beat Publications, Umina, New South Wales, Australia.

hooks, b. (1994) *Outlaw culture: resisting representations*, Routledge, New York.

Jackson, P. (1994) 'Black male: advertising and the cultural politics of masculinity', *Gender, Place and Culture*, 1(1), 49–59.

Mitchell, T. (1994) 'Flying in the face of fashion: independent music. Part 2 – He Waiata Na Aotearoa: Maori and Polynesian music in New Zealand', in Hayward, P., Mitchell, T. and Shuker, R. (eds) *North meets south: popular music in Aotearoa/New Zealand*, Perfect Beat Publications, Umina, New South Wales, Australia, 53–72.

Mitchell, T. (1996a) 'Once were warriors and new urban Polynesians: Maori and Pacific Islander music in Aotearoa/New Zealand', *Sites*, 33, 102–27.

Mitchell, T. (1996b) *Popular music and local identity: rock, pop, and rap in Europe and Oceania*, Leicester University Press, London.

Morris, M. (1997) 'A question of cultural studies', in McRobbie, A. (ed.) *Back to reality? Social experience and cultural studies*, Manchester University Press, Manchester, 36–57.

Perkins, W. (1996) 'The rap attack: an introduction', in Perkins, W. (ed.) *Dropping science: critical essays on rap music and hip hop culture*, Temple University Press, Philadelphia, 1–45.

Phillips, J. (1987) *A man's country*, Penguin, Auckland.

Rose, T. (1994) *Black noise: rap music and black culture in contemporary America*, Wesleyan University Press, Hanover and London.

Skelton, T. and Valentine, G. eds (1998) *Cool places: geographies of youth cultures*, Routledge, London.

Wall, M. (1997) 'Stereotypical constructions of the Maori "race" in the media', *New Zealand Geographer*, 53(2), 40–5.

White, A. (1996) 'Who wants to see ten niggers play basketball?', in Perkins, W. (ed.) *Dropping science: critical essays on rap music and hip hop culture*, Temple University Press, Philadelphia, 192–208.

# Between academy and popular geographies
## Cartographic imaginations and the cultural landscape of Sweden

*Mike Crang*

## Introduction

This essay tells the story of studying regional cultures in Sweden. It looks at how these cultures were delimited and charted. The occasion for this interest is set in terms of a nationalizing project where a cartographic vision of cultures and regional difference enabled a peculiar version of the Swedish landscape to develop: a version that was not confined to the shelves of the academy but which forms part of popular landscape interpretation. The connections between a sense of regionally defined culture and 'national spirit' forms a background for this essay. The essay looks at how cultural practices were marshalled into territories; how maps were written into the cultural landscape. It takes the county of Dalarna, as an area that has come to be iconic of Swedishness. It is choice, I argue, that owes much to the location of Dalarna in regional and social terms as a distinct and different interior. This essay reflects on the translation and switching within two sets of categories: first, the popular and the academic; second, the historic and the traditional. The second is drawn from the work of Pierre Nora (1989) which follows how living, dynamic traditions become codified, stabilized histories. In the process they become sequestered from popular life, moving into the half-life of museums, archives and other *lieux-de-memoire*.

## National projects and modern times

Contemporary studies have shown that nations are constructed rather than natural and thus need to be created and maintained. The nineteenth century was something of a high water mark for European national makeovers reflecting both the rise of national politics and the imperial expansion. These processes involved twin effects: first, an evolving set of distinctions between Europeans and non-Europeans; ascribing the former with a linear open temporality, enabling them to stand for modernity, while non-Western peoples were consigned to a cyclical temporality. This leads into the second strand, the development and elaboration of national history and/or myths. These include not just textual histories of national development, but the material organization of, say, the Louvre museum that located each stage of an ascending story of art in a different area (Hellenic Greece, Quattroquincento Italy, sixteenth century Netherlands and so on) to conclude in then contemporary French art (Duncan and Wallach, 1980). Museums placed national history on display. At the same time as carcereal and medical institutions were producing knowledge through the individuation and location of subjects within their walls, so technologies of display were orchestrating knowledge through public spectacles (Bennett, 1998). Exhibition and museums intersected in this process as when the Paris exposition of 1878 featured a diorama of a *balastuga* from Halland which went on to form part of the first permanent exhibition of folk culture in Stockholm. Or at the Stockholm World Fair of 1891, amid the wonders of technology, *Gamla Stockholm* offered a romanticized medievalism celebrating a bygone Sweden (Pred, 1993, 1989).

Scandinavian countries defined their modernity against internal 'primitive' others from folk culture. The archetypal folk were pictured as belonging to an eternal or cyclical time which was being torn apart by the open future of modernity. Large scale emigration and urbanization created fertile ground for this in nineteenth century Sweden. However, the discovery and invention of folklore was more widespread. An insecure national identity was often buttressed with the 'finding' of national folk myths (Dundes, 1985; cf. Davies, 1989; Herzfeld, 1982; Wilson, 1976). In the case of Sweden, this mobilization of folk life was not only through verbal forms but through material culture and concepts of the local cultural landscape. The nationalization of folk symbols coincided with the turmoil around the redefinition of the nation at a time that saw the secession of Norway (see also Engman, 1995). It also saw the rise of the National Romantic art movement (Facos, 1998) and the birth of the 'Homeland' movement of local history preservation. This essay thus looks at how this came to inscribe a regional culture on the province of Dalarna and overwrite this with a national ethos.

## *Shattered worlds:* Lanskap *to landscape*

Inherent in the emergence of folk culture as national symbol is a translation from lived reality to spectacle. It is an inevitable 'genre error' when the quotidian is staged and exhibited (Kirshenblatt-Gimblett, 1991). Living worlds of activity and experience were converted into discrete, (primarily) visual sequences of displays. For instance, the Skansen open air ethnographic park in Stockholm, founded by Arthur Hazelius, was the first to gather together exhibits from around the country in order to illustrate the diversity of regional cultures. A Sweden in miniature, made accessible through displays of regional variation; each region being represented through collections of typical buildings.

> This exhibition of old Swedish life affected me like a dream, a great popular poem, set in reality, which after its fashion set in motion all imaginative powers . . . It is the Swedish people's differences, their varied composition, that makes such a vivid impression when one goes for a walk through this project. (Gustaf af Geijerstam (1892) in Facos, 1998, p. 72)

The park represented regional cultures through assemblages of artefacts presented as holistic displays. However, this unity is achieved only through creating discrete instances of local cultures unified through a spectacular gaze. A similar poetics then links the displays of world exhibitions and emerging photographic archives and slide-shows (cf. Mitchell, 1991; Ryan, 1994). We need to develop the suggestion of Heidegger that the world becomes enframed and made pictorial, by adding the sequentiality of these representations. There is a twin process of detachment and sequence.

The history of detachment follows the shifts from high cultural disgust at the customs and manners of Swedish folk life, through its exoticization to its romanticization. Accounts of repulsion emphasize the horrors of proximity, of being contaminated by the folk. The Baedecker guide of 1889 observed 'lake Siljan [in Dalarna] owes much of its interest to the inhabitants of its banks, who have preserved many of their primitive characteristics . . . In their ideas of cleanliness, they are somewhat behind the age.' This distaste is overcome both by the translation through aesthetic distance and by the valorization of the natural and primitive. Thus in the first strategy, the 26 *lanskap* or medieval counties of Sweden are superseded as a vehicle for knowing the country by a visual landscape idea. We can find the rise of this in landscape postcards, bourgeois photographic albums and promotional materials of the turn of the century (Löfgren, 1985). There is a shift from folk knowledge of an area, where regions are formed through being immersed in different ontologies, to a survey knowledge based around visual detachment and disciplining artefacts (Löfgren, 1976). However, this is not to say all the resonances of *lanskap* were lost, and there remained a strong sense of region perhaps akin to that currently articulated by the heritage term of 'country' (as in 'Constable country' *et al.*; Olwig, 1996).

We might take two symptomatic moments from the turn of the century to illuminate the rise of surveillant knowledge. Selma Lagerlof's book *Nils Holgersson's Wonderful Journey through Sweden* (first published in 1906) blends the old sense of *lanskap* as province and the sense of a panorama. As an educational device for children to learn about their nation it seats the eponymous hero upon a magical goose's back so he can behold each region in turn. The visual grasp of the diversity of Swedish regions was thus activated in the service of a nationalizing project. However, it is too easy to take this detached gaze at face value. No one attains this vision so easily. As Matless (1999) notes, no matter how much the vision from the summit of the hill has been valorized in geographical practice, so too has there been bodily disciplining in terms of the practices and efforts to ascend and descend from the summit. The interpretation of the survey as a disembodied eye misses the often celebrated, embodied enculturation of knowledge in the field. Thus the second moment is the Swedish Touring Club's regional guides which used opening panoramic pictures with the motto of 'Know Your Country', an injunction that was also a criticism of the conventions of metropolitan bourgeois society – to really know Sweden meant to get out there amongst it. Distance and involvement are more tightly bound together than is sometimes suggested. A romantic version of nature, to be participated in through outdoor pursuits and simple lifestyles, opened the interior of Sweden as a rural wilderness. It celebrated a romantic version of rural culture for its contact with nature rather than despised it for unsophistication. We can compare these attitudes in the ethnographic observation of paintings like Kilian Zoll's *Midsummer Dance at Rattvik* (1852), with a more participative and involved style in Anders Zorn's 1897 portrait of *Midsummer Dances* in Mora some thirty miles away (Facos, 1998, p. 53; Sandström, 1996). Unlike an aesthetically detached view of landscape we might suggest there is an ethical attachment brought in with an imagination of diverse holistic countries. Rural life was experienced through both representation and travel. Bicycle touring (as promoted in the 1931 *Cykelturer I Dalarna* or the motorcycling account in 1923 *Med bil och motorcykel genom Dalrna*; see Hamrin and Norling, 1997) and rail travel increased the mobility of people, opening new regions, but also changed the mode in which they saw the landscape – the latter conveying them as parcels, on a smooth track separated from the world by a window almost like a picture frame, unable to focus on the nearby and looking further away (Schivelbusch, 1977) but also allowing the totalization of a region (Vernon, 1998; Fryckman and Löfgren, 1987). The possibility of the tour, collecting instants, taking places and reassembling them in a collection meant events along the route became objectified and detached from their context. An example from the Albert Kahn archive – an unfulfilled project to survey the diversity of human culture and preserve it in a photographic archive – is the 1910 collection *Med Hyrbil och kamera till Dalarna 1910* (By automobile through Dalarna with camera) (Kruse *et al.*, 1994). Here we find the key moments of folk culture in the region being snapped and frozen in an ethnological lens. The region is rendered available as a series of instants.

If we invert this development of going out to tour the country we can read Skansen Museum as an attempt to bring all Sweden into one place. This open air museum assembled not just artefacts but complexes of farms or buildings representative of different regional types.[1] Rather than 'distributing the nation's cultural heritage without attention to regional specificity – the idea of the classical museum – the ecomuseum pits its own concept of the refraction of museum culture in discrete environments' (Poulot, 1994, p. 73). Overegging the case, these open air museums are museums of spaces and regions rather than temporal series and progression. Instead of assembling artefacts by type or abstract schema, they gather them into their functioning units. This portrays a two pronged patrimony, on the one hand, descent over time and, on the other, an occupation of space provides a coherence to the idea of culture (de Certeau, 1986, p. 124). This linkage of environment and region, of climate and geography, with national characteristics was one of the hallmarks of the National Romantic movement which, along with connected writings by utopian thinkers like Karl Erik Forsslund, spoke to a mystical union of people and place, a cosmic syncretism, through the perseverance of a primitive connection of folk and nature.

## Patrimony and time apart

A treasure chest model – where traditions are in some sense 'contents' that are passed from generation to generation – was often invoked in order to develop a sense of identity through temporal belonging. Its attenuated version haunts accounts that lingeringly describe the development of a region over time, with the landscape building up and evolving. But more bluntly the idea of a set of customs and habits that guide the spirit of a people and set them apart is a standard claim of nationalism. In the case of Sweden this 'spirit' came to be seen to reside in the folk and the peasantry. Thus Erik Gustaf Geijer, who followed the gothic revival across Europe, turned to the transcription of oral traditions of the peasantry and pioneered the publishing of Swedish folk tales (*Svenska folksagor*) in 1814–15, setting the themes for the subsequent National Romantic movement:

> Every folk lives not only in the present, but also in its memories: and it lives through them. Every generation propagates itself both physically and morally, bequeathing to the next generation its customs and concepts. This tradition from one time to another unifies a people, fostering their unbroken consciousness of themselves as a nation; it transmits to them, so to speak, their personality. (Geijer (cited in Facos, 1998, pp. 33–4))

The folk became the container of a national essence. This was not a solely Scandinavian passion; England saw a similar interest in folk song and dance around the turn of the century (Boyes, 1993; Judge, 1993; Stradling, 1998). Influenced by romantic thought to look at affinities binding nations together, the folk and peasants formed a crucial ethical vision of the natural and national character. Thus if we follow the correspondence of Karl Larsson to August Strindberg from a visit to Halland:

> Kindly, powerful, blond people, mostly with honest, innocent faces. You know I was so happy, felt so free in the midst of these, my own folks, as they say. They were a simple, pure breed, not the jumble of human tramps one finds in the cities. And how they 'got into nature', as we painters say. They were like flowers of the field, the result or product of the earth. (Larsson (cited in Facos, 1998, p. 60))

This vision of the folk becomes particularized around key regions – especially that of (Upper) Dalarna. The connection of the physical environment to the spirit of the people makes National Romantic art create landscapes that seek to make the viewer engage with the spirit of the nation. This was an idea expressed explicitly in the Danish Open Air Museum (founded 1901), which in the words of its founder, Bernard Olsen, aimed to 'give a picturesque and understandable image of our folk and provincial peculiarities' and thus show 'how our national characteristics have been formed by climate and the nature of our country' (in Skougaard, 1995, p. 23). The folk were the bearers of an unbroken tradition that was seen as threatened by modern conditions. The National Romantic painters were dominated by *émigrés*, whose relationship with their homeland formed the ambivalent heart of the movement.

> Homesickness or, in Swedish, home-longing (*hemlängtan*), had a visceral dimension for Swedes. It evidenced the belief, central to their culture, in a biomystical link between themselves and the landscape. The romantic metaphor of the return home and the biological metaphor of rootedness in an environment were crucial to National Romantic ideology.
> (Facos, 1998, p. 25)

Sweden was valued for its putative 'primitivism'[2] and naturalness, yet upon returning many artists, such as Anders Zorn, were shocked by the rapidity of social change among the peasantry. The folk life was seen as disappearing. Various versions of songs or tales were sifted to find common elements that might show an original antecedent behind them. The model thus presupposed an authentic original that in some sense became corrupted or lost in transmission through time – all change was degenerative. Thus different versions were treated as showing degeneration not as, say, blossoming variation. We might note the implications of this are, first, a singular originary folk culture was posited behind the diverse forms encountered (see Dundes, 1985), second, it emphasizes the replication and repetition rather than innovation and performance of folk culture. Both these ideas become important in the way regional cultures were imagined. The folk culture is seen as only remaining as relics. It becomes the object of a desperate salvage anthropology – buying artefacts and houses that mark its retreat and passing from the everyday. It also valorized these everyday artefacts – providing a greater attention to material culture as we shall see.

## Charting differences

Interest in regional cultures in Sweden can be connected with linguistic surveys in Germany or the collection of popular stories by Nisard in France (de Certeau, 1986), and the 1893 Ethnographic Survey of Britain (Vernon, 1998). The moment of ethnological inscription was also the moment where folk cultures ceased to be living entities and became curios, aestheticized and museumified. They were portrayed as unconnected to and uncontaminated by high culture (de Certeau, 1986, p. 125). The vision is of the popular cut off from the modern. The folk were confined through pristine orality and Othered from modern culture. This temporal bounding is amplified by the spatiality ascribed to the folk.

The National Romantic concern with the natural and environmental links of cultures fed into academic studies through the influence of figures like Karl Erik Forsslund whose 1914 *hembygdsvård* manual on preserving local natural and cultural environments ran to nine editions (Fryckman and Löfgren, 1987). The journal of the Local Heritage Federation was *Bygd och Natur* and in 1919 a new subject appeared in the school curriculum: *hembygdskunskap* or 'local geography and history'. The materiality of folk ethnology stressed a regional unity through co-adaptation of human and physical landscapes. The approach can be traced to the work of Oscar Montelius who held the first major Swedish conference on ethnological regions in 1874 (Erixon, 1945, p. 7) and the first chair was held by Nils Lithberg from 1918. Perhaps the most influential exponent of regional typologies was Sigurd Erixon (1888–1968) whose empirically detailed analysis of material culture reconstructed cultural regions (Erixon, 1945; Campbell *et al.*, 1957; Helmfrid *et al.*, 1994; Löfgren, 1976; see Figure 1). This is an approach that persists in popular learned works, such as *Swedish Landscapes* (Sporrong *et al.*, 1995). The folk culture of Sweden was split along a range of fault lines where initial commentaries, such as Sunbärg's in 1910, located an east–west split between cultural hearths in Denmark, via Skäne, and Mälardalen (the Stockholm region) – the Flodstrom line (Helmfrid *et al.*, 1994, p. 60). North–south divisions were later added, where Erixon outlined four overlapping (though not identical) boundaries between cultivation patterns. Together these offered the southern limit of Nordsvensk culture – colloquially the shieling line (Erixon, 1945, p. 10; see Figure 1). The Nordsvensk system is one of winter and summer *fabodar*, or shieling, pastures, whereby the main farm has a subsidiary farmstead where animals are taken for the summer months, generally located higher in the mountains and up to 40 km from the main village.[3]

Studies looked at the diffusion of practices and then regional constellations of forms. So Montelius traced the spread of ploughculture and 'havre' cultivation patterns across Scandinavia and the northern British Isles up to the Shetlands and Orkneys (Erixon, 1945, p. 23). But generally less attention went to development than to the particular patterns then fixed in regions.[4] The effect is to look at the particular combinations of artefacts as indicators of diverse regional cultures. Detailed overlays of diverse patterns of artefacts, looking at the creation of boundaries related to the distribution of particular forms, and overlaying each with another produced fine grained regional analyses combining ranges of elements – generally material culture, sometimes natural environmental indicators, sometimes ethnolinguistic. Researchers were drawn to material artefacts, not simply as part of the link of cultures and environments, but as durable relics susceptible to quantitative and statistical mapping. They produced enormously detailed archives both of objects in museums and records of their use and location, and symbolically Lithberg's chair in ethnology was shared between the univer-

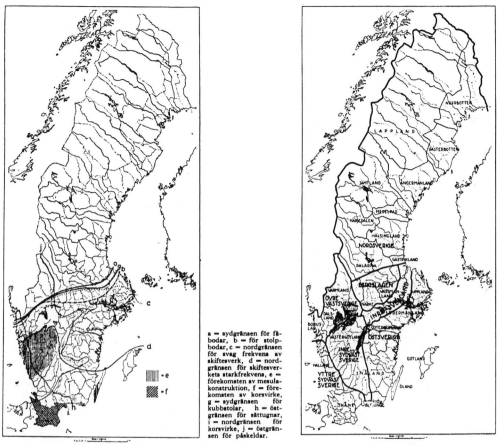

a — sydgränsen för fä-
bodar, b — för stolp-
bodar, c — nordgränsen
för svag frekvens av
skiftesverk, d — nord-
gränsen för skiftesver-
kets starkfrekvens, e —
förekomsten av mesula-
konstruktion, f — före-
komsten av korsvirke,
g — sydgränsen för
kubbstolar, h — öst-
gränsen för sättugnar,
i — nordgränsen för
korsvirke, j — östgrän-
sen för påskeldar.

*De skiljande kulturgränserna mellan övre och nedre Sverige (a, g, b, c)*
*och utbredningen av några sydvästsvenska kulturformer.*

*De svenska kulturområdena.*

**Figure 1A and B** Map of cultural boundaries (left, A) and cultural areas (right, B)
from Erixon (1945, figures 2 and 17 respectively). The conjunction of a, b, c, g on the
left hand map forms the shieling line. The detailed regional pattern of many divisions
in artefacts is shown on the right.

sity and the Nordisk Museet (Löfgren, 1997). For example the relatively common
*Loftbodar* ('loft house'), where the main building has two levels with an external
staircase, could be divided into at least five subcategories according to the
number of rooms, stairs and their arrangement. Löfgren (1997, p. 96) notes the
way this vast effort followed an almost Linnaean ambition to name and classify everything setting itself up as the guardian of mundane objects of peasant
life (see Figure 2).

This delimiting of cultural regions proceeded via the most comprehensive
and detailed accumulation of minute factual detail as part of a grand project to

**Figure 2** The interior of Bergkarlås barn, a threshing barn, preserved at the Zorn gammelgård, which contains an array of preserved artefacts specific to the locality. Assiduous documenting of artefacts like these underpinned typologies used to define cultural regions in Figure 1.

produce the Folk Atlas. However, we should not lose sight of the passions behind it and the shared spirit of the National Romantics. We can find traces of this in the most scholarly treatises. One work by Erixon reflected on the dispersion of villages, such as he grew up in, into individual farms. His account provides elegiac photographs of villages that no longer exist, and accounts of those who remember vanished ways of life in villages where people were happier. We need to keep in view the imaginative landscape being created through these folk studies:

> It is naturally wrong to dismiss Erixon and his generation of scholars as sentimental dreamers who, behind all their maps and tables, longed for the bygone, deep-rooted village communalism, far from the industrialised world. On the other hand, it is true to say that longing for closely knit villages and uncomplicated community in beautiful houses inspired them to a restless study of the rich historical deposits in the Swedish landscape of the twentieth century. (Hellspong and Klein, 1994, p. 33)

However, this powerful imagination could only illuminate declining cultures incapable of modernization. It could not admit change to peasant life that was not destructive. It leaves a picture of unchanging regional cultures suddenly destroyed. While academics set about producing scholarly works, on the ground people were mobilizing behind popular institutions of regional culture. These we may find reflected in art but also in terms of local preservation movements inflected by ideas of regional folk culture.

Here I wish to focus on the way the region of Dalarna and Darlecarlian culture were constructed in a popular historical geographic imaginary. The main parallel is the construction of Dalarna as a bounded regional culture in spatial terms and as belonging to a separate temporality, one that is unchanging, cyclical and naturalized. However, it is also clear that these temporalities mapped on to an imagined geography of primitiveness, which produced contradictory spaces. Dalarna came to occupy such a paradoxical space as antique and also as a national symbol. Its regional position as *Nordsverige* located it sufficiently far from Danish influence to claim to represent distinctive Swedish attributes. Its symbolic, geographic centrality is evident when the opening of a regional account is:

> Dalarna lies almost entirely between the latitudes 60° and 62° North, thus occupying a position transitional between the exclusively forested regions of northern Sweden and the more highly cultivated and industrial parts to the south. (Edwards, 1940, p. 5)

The position is oddly then also one of instability – a regional idea forged out of a point of transition. As we shall see, this centrality had to cope with how Dalarna was, at the same moment, peculiarly set aside from the Swedish mainstream.

The regionalized vision of local cultures and the role of Dalarna can be traced through the Mora farmstead, that was installed as one of the first exhibits at Skansen (bought 1885 moved to Skansen in 1891 when it opened), and comprises several buildings dating from the sixteenth century (with one building dating from 1320) forming the distinctive square enclosed yard. The guidebook has this to say:

> Despite the living quarters and cowhouse being of relatively recent date, the farmstead in its entirety provides a good picture of what a farm in the province of Dalarna looked like during the Vasa period and how they survived in these tracts until the 20th century. (1991, pp. 22–3)

The Mora farmstead is placed next to a *fabodar* pasture represented by Älvdalsbodarna, a distant summer farm which, although the oldest parts, the firehouses, are dated the eighteenth and seventeenth centuries, 'represent very ancient living habits which in some summer pasture farms survived even into the 20th century' (p. 23). The impression is of a continuity that was suddenly shattered, a timeless and unchanging past. The ethnological vision of an enclosed static society echoes one in anthropology where the primitive is set up as untouched and unchanging. In reality though the situation is complicated by how Dalarna folk life coexisted with modernity (Erixon, 1945, p. 14).

The vision of regional cultures is also offered by photographs from the turn of the century through which the local was recorded. These locally produced pictures were taken by resident country photographers who are defined less by style or theme than by their attachment to localities. 'What they had in common was a loyalty to their district, a sort of chronicler's inclination that allowed what was initially occasional annotation to grow into self-assumed mission' (Ågren, 1994, p. 114). From these pictures, filled with incidental as well as deliberate local content, archives have been reassembled and can be consulted now to trace material culture and people. The reconstructive project of preserving a disappearing (or now vanished) culture is strong. Where the landscape itself is now marked by absence and loss, the pictures offer a vast wealth of detail to refill the memorial landscape: 'In the work of a country photographer, devotion to a limited district and its people is reflected in a stubborn will to fix visible changes in pictures that one by one build up a frozen world of the negative archive' (Ågren, 1994, p. 127). Just as the artefacts of material culture made the lost everyday world tangible, here it was visible. However, this was not purely an 'insider' practice:[5]

> It was common for photographers to set up businesses in places where tourism had an early hold. The photographer sought to meet the tourists' demand for pictures of beautiful scenery and picturesque folklife. The motifs favoured by photographers coincided with the genre pictures cherished by the National

Romantic painters (Severin Nielson) and the local traditions that ethnographers were keen to document (A.C. Hultgren and Karl Lärka). (Ågren, 1994, p. 113)

These artefacts are not records of an isolated culture but the juncture of reflexive practices between locality, art, academy and tourism. Couples depicted in folk dress were often posing for special photographs in the dress they knew you should wear for such records. Our idea of Dalarna's regional culture comes through such mediated moments. The relics of 'folk culture' were often already double coded with their own significance. Historical geographical writings on the region thus tend to be both aware of Dalarna's idealized status and also perpetuate it, describing the countryside's appeal as romanticized yet also reiterating the continuance of folk traditions:

> The open cultural landscape with its traditional architecture, often well preserved, appeals to our romantic concepts of country people's way of living in olden days in Sweden. In particular Upper Dalarna has often been seen as a district where old traditions have survived. (Hansen, 1994, p. 122)

## Atypicality – Other within

The relationship of the regional to the national is complex in the case of Dalarna. The national tourist symbol is the cheery, carved wooden Dala Horse from the area,[6] and as early as 1937 the columnist Gustaf Näström published his book *Dalarna som Svenskt ideal* (Dalarna as the Swedish ideal). 'Today to have a chalet or house painted with the red paint of Falun [the copper mining town of Dalarna] is an absolute incarnation of everything Swedish' (Rosander, 1988, p. 121). In adverts throughout the century female figures in folk costume have been used to symbolize Sweden – especially unmarried girls from the Siljan village of Rattvik (*Rattvikskulla*) – to sell all sorts of unrelated commodities like tobacco and coffee, among other items (Rosander, 1988, pp. 107, 120).

In this sense we need to complicate three ideas of local belonging often used to explain the regionalist tendencies in Sweden. First, no simple hierarchical model of attachments – where one first knows the local, then subsumes this into regional and then national scales – is adequate since Dalarna functions at a national level as well (cf. Ekman, 1991; Löfgren, 1989). Second, Darlecarlian culture cannot be seen as typically Swedish, since what makes it remarkable is the sense of difference marked by its retention of folk culture. Equally, the continuance of Darlecarlian culture complicates the idea of it being primitive – since it is used in the national imaginary as a vehicle for a political, ethical vision based on an egalitarian peasantry (Facos, 1998, p. 56). We might reconcile these two by suggesting that Dalarna is seen as an anachronism in preserving or continuing what was once common throughout Sweden but vanished elsewhere. However, thirdly, Dalarna carries the sense of both an inherited or lost patrimony – perhaps a patrimony only accessible through its privileged territory –

*and* the exotic. A similarity thus exists with the status of Karelia for Finland, which as the valorized territory of the *Kalevala*, plays a crucial role in national folklore. It is also atypical in its Eastern Orthodox traditions and monuments. The physical loss of most of the area in wars with the Soviet Union provides a significant emotional and national charge to celebrate this most atypical province (Raivo, 1997). We might also draw out similarities with the Cornish position in England. Here too a raft of institutions surveyed and mapped the region, from its geology and agriculture to its people.

In Cornwall, like Dalarna, the spiritual and natural imagination and land-scapes played a crucial ethical role in the Cornish 'renaissance', which was really the rediscovery of a Celtic essence. Like Dalarna, this was connected to artistic movements – and it is worth noting the artist colonies set up at New Lynn in Cornwall and at Leksand in Dalarna.[7] In the case of painters like Stanhope Forbes, the Cornish landscape and people who fell 'naturally into their place and harmonise[d] with their surroundings', played nature to English artists' nur-ture (Vernon, 1998, p. 160). However, this exoticized Cornwall was imbued with anti-English sentiment. It was the location of mythical Celtic essence held up as one of the outposts of a romantic heritage that had resisted successive invasions. At a time when modernity was being questioned, this mythical region offered symbolic resources for a critique. Dalarna, by contrast, was centrally bound to a story of national self-determination:

> The region has played an illustrious part in the history of the country as a whole and is rich in memories of Gustav Vasa, the hero of Swedish liberty. The peasantry too, of this part of Dalarna shared conspicuously in the struggle for their country's freedom and the inhabitants of today inherit from their sturdy ancestors a traditional mode of life and culture which continue to be steadfastly fostered. (Edwards, 1940, p. 24)

This nationalized past is thus connected to an inherited peasant tradition which is seen as 'unaffected' by large scale industrial development in Southern Dalarna, with the other parts of the province remaining 'true' to its 'agricultural tradition' (Edwards, 1940, p. 7). Conventional accounts thus separate 'Lower' Dalarna as the industrial area below the East and West Dal confluence, and either calling all the remainder Upper or distinguishing the mountainous tract towards the Norwegian border (Upper) and the most symbolically important area around Lake Siljan (Middle).

> The lowlands which surround the Lake [Siljan] support numerous peasant communities and other thriving settlements where the forest has been cleared. The mode of life evolved by the inhabitants during many centuries of occupation, while being typical of Sweden, also exhibits distinctive features of its own. It is indeed the adaptation of the life and activity of these people to the particular features of the local environment that the real distinctiveness of Middle Dalarna is to be found. (Edwards, 1940, p. 5)

The distinctiveness comes both in the preservation of folk life and the Nord-svensk pattern of summer and winter farmsteads, but also an almost unique pattern of landholdings. The land tenure was fragmented through a system of multiple inheritance – including women (Sporong *et al.*, 1995, p. 112). The result was small holdings being continually dissassembled and reassembled through deaths and marriages – indeed the *parstuga* farmhouse, typical of the area, was able itself to be split in two and moved. The inheritance system survived several phases of land reform through the eighteenth and nineteenth centuries and continued up till the 1960s in some cases. The result is that the social system of Dalarna was possibly, and certainly could be presented as, egalitarian and marked by relatively little social differentiation.[8] The peculiar-ity of Darlecarlian agriculture was thus well suited to adoption by groups try-ing to promote a Swedishness founded in an ethics of equality. It also harks back to the mobilization of the idea of *lanskap* as collective common law herit-age, as opposed to elite rules (Olwig, 1996).

However, an alternative story might invoke the one-time largest copper mine in Europe in Lower Dalarna. It is possible to tell different stories of even Upper Dalarna – in terms of histories of peasant revolts against the Crown as well as support for it. The very idea of the region preserving a discrete folk culture also needs some challenging, since Dalarna has also long been marked by cyclical patterns of migration to find work in other provinces and cities. The painter Zorn's father was a foreman in the bottling factory in Stockholm where his mother worked. If we can thus question some of the lurking notions of organic unchang-ing spatial and cultural units behind the idea of a regional folk culture, we also need to look at how the idea of such a folk was institutionalized in Dalarna.

## Institutionalizing culture

The preservation of folk culture in Dalarna needs to be interpreted in light of this inscribing of the region into social memory. The first form addressed here is folk costume, second are local open air museums (*hembygdsgård*), run by local heritage movements (*hembygdsföreningen*), dedicated to preserving the cultural landscape. Studies suggest that throughout history the upper classes have played a significant role in shaping and influencing peasant costume (Hellspong and Klein, 1994, p. 39). The idea that prior to the end of the nineteenth century there was an autonomous sphere of peasant culture is not realistic. However, the idea of pitting the innovative temporality of fashion against a timeless tradition is a powerful myth of modernity. The start of the twentieth century marks a point where active intervention began to shape the development of folk aesthetics and culture in particular ways. For instance, the *Kvinnliga Allmänna Nationaldräktsföreningen* (Women's General National Costume Society founded 1902) campaigned for women to be freed from domina-tion by foreign fashion through more general use of folk costume, which fits an interpretation of women as bearers of national cultural symbols, but which also tied in with an early feminist struggle against the restricted movement allowed by fashionable long skirts and corset. Meanwhile, a national costume was designed by Ankarcrona and Larsson based on Dalarna. Not to be outdone, in Norbotten a committee drew up a standardized 'local' design in 1912, and Värmland developed a standard based around a Västerfärnebo costume.

The sight of people in folk dress going to church in Leksand reputedly prompted Arthur Hazelius to begin collecting ethnological artefacts (Hamrin and Norling, 1997). In this light the early acquisition of the Mora farmstead has renewed significance. Indeed the first summers at Skansen saw the employment of traditionally clad Dalarna women as costumed guides – women who had, as was common, seasonally migrated to Stockholm to work. However, within Dalarna costume was already being promoted through tourism. So in the 1910 trip in the Kahn collection (Kruse *et al.*, 1994), we find a tourist hotel with maids in 'folk dress'. Already then some locals were tourees, bridging insider and outsider. Models of a discrete region tend towards a purified image of a folk that denies outside and middle- and upper-class influences both in the collecting and the folk material itself. Thus Anders Zorn was local, yet also cosmopolitan. He put energy and resources into refounding events and practices – such as folk music contests. The act of collecting and recording was itself part of transforming folk culture where, for instance, Carl Gudmussen stimulated a revival of Dalarna long and short horn blowing by collecting songs and melodies. Today the area is the heartland of folk musicians – the so-called 'Polska Belt' (Figure 3). Zorn helped re-establish midsummer pole raising rituals – which are now popular events nationally, forming an almost mandatory cover picture to introductions to Swedish customs. The largest such event is held at the Mora farmstead in Skansen. In Dalarna the poles, sited in the *hembygds-gård*, not only differ from village to village, but the dates of raising them are staggered so as not to clash.

The preservation of material artefacts was seen as part of trying to shape modern development (Björkroth, 1995) through involvement by *hembygds-föreningen* in regulating buildings and architectural controls. The shape of Swedish modernism was contested between the functionalist schools and those drawing on arts and crafts. These involvements have receded and now they remain custodians of the local outdoor museums.[9] Dalarna has around fifty such open air museums, mostly established in the interwar years, as collections of local buildings and artefacts from each parish (Anderson, 1978) with enormous care and concern for detail. It is an artefactual authenticity that tends to overwhelm any account of the artifice in creating these sites. They also depict a stable landscape rather than its evolution over time, thus the buildings are either restored to a specific period or kept as they were last used – emphasizing the sense of a coherent world that abruptly ended.[10] The effect is compounded by the location of many open air museums where they work to minimize visual intrusion of modern elements into sites (Ehrentraut, 1996). The open air museums in Dalarna tend to immobilize an idea of an organic, holistic folk identity prior to the present.

**Figure 3** Music festivals and revivals of traditional instruments and costumes are often held in hembygdsgård.

The reconstruction of a local cultural landscape thus downplays spatial and temporal complexities. However, I do not mean to criticize this version of regional culture for being inauthentic. Indeed the critique of inauthenticity often presupposes a notion of unchanging and unselfconscious culture opposed to external influences. However, if cultures are more fluid and evolving in different spatial and temporal ways, then interventions to reshape them have to be seen as part of the course. In that sense the intervention of folk collectors and the nationalizing of certain symbols are part of a continuing story of local culture, not its end. Too often conservatives and radicals both produce bounded cultures – the conservatives announce they have ended and the authentic scraps need saving, the radicals that the scraps are thus inventions not the authentic original. Instead let us think of the ambivalent, pluri-spatial, pluri-temporal story of Dalarna where the act of preservation has become almost the defining local characteristic; not only what is preserved but the practices and places of preservation themselves define the local culture. So a careful cultural geography needs to see the process of preservation as located within popular culture not only operating upon it. Dalarna is one of the places where, as Handler and

Linnekin (1984) remark of Quebec, the process of folklore popularization is as traditional as the folklore itself. So if we return to Edwards' (1940) expedition to map the region of Dalarna, already by then:

> During the height of the tourist season Dalarna gives the impression of being a province in which local patriotism is so emphasised that it overlooks the greater stream of Swedish history. This is perhaps because the stress laid on local antiquities, crafts and folklore has become the basis of a lucrative commerce . . . Yet underlying this ostentatious expression of the Dalarna cult the inhabitants have a deep regard for their beautiful region and a strong desire that its customs should not be forgotten. (p. 50)

The ambivalence is already there that this is clearly an act, a restaging and redeployment of folk culture (cf. Brenneis, 1993). As such it is not the continuance of an unconscious and unreflexive tradition, but the knowing and reflexive engagement with that tradition. To return to where this essay began, this means an awareness of the role of *lieux-de-memoire*, that is places where senses of the past and regional identity are produced and living custom, often taken as relatively unreflexive, moves into the half-life of museumified artefacts. From customary and dynamic use to codified history certainly describes part of the story in Dalarna, yet equally these codified versions then come out to animate renewed practice. The popular is not divorced from the formal version – certainly not in terms of an opposition of official and popular cultures. That is official culture's own myth of the popular as a separate and discrete entity. Or indeed, vice versa.

Instead we need to see the official inscription of regional culture as helping to shape and produce practices that act to support that regional identity. Thus the *hembygdsgård* are generally quiet and relatively unvisited. They are places to stroll or picnic not the hyper-real vistas of Disneyland. However, they fill and overflow for special events such as midsummer pole raising, festivals and folk performances. Thus take the annual staging of the *Himlaspalet* play in Leksand. Clearly the large stage, the audience of outsiders and so forth all change this from a myth and ritual told within a culture. It is not that then. Yet we know forms like this do change, and the folk culture they were based upon evolved and changed. Instead perhaps we need a reflexive understanding of these forms that sees the tradition of re-enacting folk culture as part of modern popular culture. The fluid practices of producing the apparently fixed regional culture are very much what defines the contemporary region. In this sense we could, to follow Kirshenblatt-Gimblett (1998), see these displays as not only showing an image of the region but as doing. The practices of imagination do not simply represent an image of a past region but put this into service of a current performance.

## Acknowledgements

Thanks to audiences who have commented on these ideas at the Annual Conference of the Art History Association, the European Urban and Regional Studies Conference, and the Crossroads in Cultural Studies Conference. The fieldwork was supported by the generous help of the Dalarna Research Institute (Dalarnas Forskingråd).

# Notes

1. A notably different sensibility from the classical museum – such as Hazelius's other creation, the Nordic Museum.
2. One can also find quests for primitive Swedish cultures cut adrift by changing national boundaries, see Rausing, 1998.
3. The concurrence of these lines also demarcates the division of Upper Dalarna around Lake Siljan and into the mountains from the Bergslagen copper mining area.
4. Thus prominent theories suggested Dalarna adopted many new practices in the sixteenth century, at a time of relative prosperity, but which then became fixed with declining incomes – becoming associated with a golden era. This oddly suggests a regional nostalgic fixation inside a national nostalgia.
5. The country photographer was an anomalous social position in the countryside, often attracting outsiders and providing one of the few professional avenues open to women (such as Anna Larsson (1900–65) and Gerda Söderlund (1874–1949) in Dalarna).
6. The horse was popularized by being used as a symbol at the New York World Exhibition in 1939. This raises two points. First, that in such exhibitions the layout of national pavilions echoes the Skansen layout of regions. Second, the horse might equally be a symbol of the acumen of the handicraft association at Nusnäs near Mora that sent 10,000 to be offered as souvenirs (Brück, 1988, p. 82; Greverus, 1976) We might also note how this is mapped into a peasant craft idiom in near contemporary accounts such as that by Edwards (1940, p. 34): 'The men of Leksand have long specialised in carpentry . . . Rättvik on the other hand, are noted for their skill in painting. Around Mora wood-carving is widely practiced and it is estimated that during the winter of 1938–9 some 30,000 small wooden horses were made at Vattnäs mainly for export to the USA.'
7. Though too much should not be read into it, it is worth looking at Zorn's paintings of the English West Country, and Cornish coastal scenes in the evolution of his work.
8. Edwards (1940, p. 24–5), using 1932 data, recorded 98% of holdings less than 10 ha, with 48% of holdings smaller than 2 ha and 24% under 1 ha.
9. These act as stages for a wide variety of 'folk' events such as music festivals (for instance the *Visfest* in Borlange) or for midsummer pole raising and so on.
10. Compare for example the Amuri Museum of Worker's Housing in Tampere, Finland, where the working-class accommodation is shown evolving through time, or St Fagan's in Wales.

# References

Ågren, P.-U. (1994), 'Country Photographers', in Klein, B. and Widbom, M. (eds) *Swedish Folk Art: all tradition is change*, Harry N. Abrams, Stockholm.
Anderson R. (1978), *Hembygdsgårdar Dalarna*, Dalarnas Museum, Falun.
Bennett, T. (1988), 'The Exhibitionary Complex', *New Formations* 4, 73–102.

Björkroth, M. (1995), 'Hembygd – a Concept and its Ambiguities', *Nordisk Museologi* 2, 33–40.

Boyes, G. (1993), *The Imagined Village: culture, ideology and the English folk revival*, Manchester University Press, Manchester.

Brenneis, D. (1993), 'Some Contributions to Social Theory: aesthetics and politics in a translocal world', *Western Folklore* 52, Apr., 291–303.

Brück, U. (1988), 'Identity, Local Community and Local Identity', in Honko, L. (ed.) *Tradition and Cultural Identity*, Nordic Institute of Folklore, Turku, p. 82.

Campbell, Å., Erixon, S., Linqvist, N. and Sahlgren, J. (1957), *Atlas över Svensk Folkkultur*. Utgiven Kungl. Gustav Adolfs Akademien. Bokförlaget Niloé, Uddevalla.

Certeau, M. de (1986), *Heterologies: Discourses on the Other*, Manchester University Press, Manchester.

Clive, J. (1985/6), 'The Uses of the Past in Victorian England', *Salmagundi* 68–9, 48–65.

Davies, G. (1989), 'Polish National Mythologies', in Hosking, G. and Schöpflin, G. (eds) *Myths and Nationhood*, Hurst & Company, London, pp. 141–57.

Duncan, C. and Wallach A. (1980), 'The Universal Survey Museum', *Art History* 3 (4), 448–69.

Dundes, A. (1985), 'Nationalistic Inferiority Complexes and the Fabrication of Fakelore: a reconciliation of Ossian, the Kinder-und Hausmurchen, the Kalevala, and Paul Bunyan', *Journal of Folklore Research* 22 (1), 5–18.

Edwards, K. (1940), *Sweden: Dalarna Studies*, Le Play Society (Student Group), London.

Ehrentraut, A. (1996), 'Globalisation and the Representation of Rurality: Alpine open air museums in advanced industrial societies', *Sociologica Ruralis* 36 (1), 4–26.

Ekman, A.-K. (1991), *Community Carnival and Campaign: expressions of belonging in a Swedish region*, Stockholm Studies in Social Anthropology 25, Department of Social Anthropology, Stockholm University.

Engman, M. (1995), 'Finns and Swedes in Finland', in Tägil, S. (ed.) *Ethnicity and Nation Building in the Nordic World*, Hurst & Company, London.

Erixon, S. (1945), *Svenska kulturgränser och kulturprovinser*. K. Gustav Adolfs Akademiens småskrifter 1. Lantbrukförbundets Tidskrifts AB, Stockholm.

Facos, M. (1998), *Nationalism and the Nordic Imagination: Swedish art of the 1890s*, University of California Press, Berkeley.

Fryckman, J. and Löfgren, O. (1987), *Culture Builders: an historical anthropology of middle class life*, Rutgers University Press, New Brunswick.

Greverus, I.-M. (1976), 'Nothing but a little Dala-horse, or How to De-code a "Folk" Symbol', *Folklore Today*.

Hamrin, Ö. and Norling, O. (1997), 'I Turisternas lanskap', *Svenska Turistföreningens Årsbok*, 26–41, Stockholm.

Handler, R. and Linnekin, J. (1984), 'Tradition, Genuine or Spurious', *Jnl of American Folklore* 97, 385, 273–90.

Hansen, B. (1994), 'The Siljan District', in Helmfrid, S. (ed.) *Landscapes and Settlements: national atlas of Sweden*, SNA Publishing, Stockholm.

Hellspong M. and Klein, B. (1994), 'Folk Art and Folklife Studies in Sweden', in Klein, B. and Widbom, M. (eds) *Swedish Folk Art: all tradition is change*, Harry N. Abrams, Stockholm.

Helmfrid, S. (ed.) (1994), *Landscapes and Settlements: national atlas of Sweden*, SNA Publishing, Stockholm.

Helmfrid, S., Sporrong, U., Tolin, C. and Widgren, M. (1994), 'Sweden's Cultural Landscape: a regional description', in Helmfrid, S. (ed.) *Landscapes and Settlements: national atlas of Sweden*, SNA Publishing, Stockholm.

Herfeld, M. (1982), *Ours Once More: folklore, ideology and the making of modern Greece*, University of Texas Press, Austin.

Hooper-Greenhill, E. (1989), 'The Museum in a Disciplinary Society', in Pearce, S. (ed.) *Museum Studies in Material Culture*, Leicester University Press, Leicester, Chap. 6.

Hooper-Greenhill, E. (1992), *Museums and the Shaping of Knowledge*, Routledge, London.

Judge, R. (1993), 'Merrie England and the Morris, 1881–1910', *Folklore*, 104, 1–2, 124–43.

Kirshenblatt-Gimblett, B. (1991), 'Objects of Ethnography', in Karp, I. and Lavine, S. (eds) *Exhibiting Cultures: the poetics and politics of museum display*, Smithsonian Institute Press, Washington DC.

Kirshenblatt-Gimblett, B. (1998), *Destination Culture: tourism, museums, and heritage*, University of California Press, Berkeley.

Kruse, H., Jobs-Björklöf, K. and Andersson, R. (1994), *Med Hyrbil och kamera till Dalarna 1910*, Leksands Kulturnämd, Dalarna.

Löfgren, O. (1976), 'Peasant Ecotypes: problems in the comparative study of ecological adaptation', *Ethnologia Scandinavica*, 100–15.

Löfgren, O. (1985), 'Wish You were here! Holiday images and picture postcards', *Ethnologia Scandinavica*, 90–107.

Löfgren, O. (1989), 'The Nationalization of Culture', *Ethnologia Europaea*, XIX, 2–23.

Löfgren, O. (1997), 'Scenes from a Troubled Marriage: Swedish ethnology and material culture studies', *Journal of Material Culture* 2 (1), 95–113.

Matless, D. (1999), 'The Uses of Cartographic Literacy: mapping, surveying and citizenship in twentieth century Britain', in Cosgrove, D. (ed.) *Mappings*, Reaktion, London.

Mitchell, W. (1994), *Landscape and Power*, University of Chicago Press, Chicago, Illinois.

Nora, P. (1989), 'Between Memory and History: les *lieux-de-memoires*', *Representations* 26, 7–25.

Olwig, K. (1996), 'Recovering the Substantive Nature of Landscape', *Annals of the Assoc. of Amer. Geogr.* 86 (4), 630–53.

Poulot, D. (1994), 'Identity as Self-Discovery: the Ecomuseum in France', in Sherman, D. and Rogoff, I. (eds) *Museum Culture: histories, discourses, spectacles*, Routledge, London.

Pred, A. (1994), *Recognising Contemporary European Modernities*, Routledge, London.

Pred, A. (1991), 'Spectacular Articulations of Modernity: the Stockholm Exhibition of 1897', *Geografiska Annaler* 73 B (1), 45–84.

Raivo, P. (1997), 'The limits of tolerance: the Orthodox milieu as an element in the Finnish Landscape', *Jnl of Historical Geography* 23 (3), 327–39.

Rausing, S. (1998), 'Signs of the New Nation', in Miller, D. (ed.) *Material Cultures: why some things matter*, UCL Press, London, 189–214.

Rosander (1988), 'The "nationalisation" of Darlecarlia', in Honko, L. (ed.) *Tradition and Cultural Identity*, Nordic Institute of Folklore, Turku, p. 82.

Ryan, J. (1994), 'Visualising Imperial Geography', *Ecumenic*, 1, 157–76.

Sandström, B. (1996), *Anders Zorn: an introduction to his life and achievements*, Zorn-samlingarna, Mora.

Schivelbusch, W. (1977), *The Railway Journey: trains and travel in the nineteenth century*, Blackwell, Oxford.

Skougaard, M. (1995), 'The Ostenfeld Farm at the Open-Air Museum: aspects of the role of folk museums in conflicts of national heritage', *Nordisk Museologi* 2, 23–32.

Sporrong, U., Ekstam, U. and Samuelsson, K. (1995), *Swedish Landscapes*, Swedish Environmental Protection Agency, Stockholm.

Stafford, W. (1989), ' "This Once Happy Country": nostalgia for a pre-modern society', in Chase, M. and Shaw, C. (eds) *The Imagined Past: history and nostalgia*, Manchester University Press, Manchester.

Stradling, R. (1998), 'England's Glory: sensibilities of pace in English Music, 1900–50', in Leyshon, A., Matless, D. and Revill, G. (eds) *The Place of Music*, The Guilford Press, New York, pp. 176–96.

Urry, J. (1992), 'The Tourist Gaze and the "Environment" ', *Theory, Culture and Society* 9, 1–26.

Vernon, J. (1998), 'Border-Crossings: Cornwall and the English (imagi)nation', in Cubitt, G. (ed.) *Imagining Nations*, Manchester University Press, Manchester, pp. 153–72.

Wilson, W. (1976), *Folklore and Nationalism in Modern Finland*, Indiana University Press, Bloomington.

# Regions to be cheerful
## Culinary authenticity and its geographies

*Ian Cook, Philip Crang and Mark Thorpe*[1]

## Introduction

This chapter has a rather mundane, some would say dull, start. We would like to you to imagine a trip around your local supermarket and to think, in particular, about those sections where you might buy your bread and, perhaps, a cooking sauce and some pasta. If you live in the UK, you might have in mind a part of the store where you could buy your standard sliced white or brown loaves, 'continental' or 'speciality' breads (e.g. ciabatta or naan), 'morning goods' (e.g. croissants or crumpets) or 'oven-fresh' bread (finished off in the instore bakery). Further along your journey, you might come across the cooking sauce fixture whose products are arranged in a different way. Higher up, you might find the shelves divided into blocks of 'Italian', 'Indian', 'Chinese', 'Mexican' and 'traditional' (i.e. 'British') sauces (see Figure 1). And, lower down, you might find noodles, rice and pasta (see Figure 2). You might pick some of these things off the shelves. You might have a good look at them. You might check out the instructions, the ingredients and/or the price of particular items. You might buy some. You might not. You might walk straight past them. Maybe you're not the one who does the food shopping. But we have been studying bread, cooking sauces and pasta regardless, as three case study foods in a larger project on the internationalization of the 'British' diet.[2]

A great deal of popular and academic attention has been paid to this internationalization process. Great significance has been attached to the now 'mainstream' popularity of 'ethnic' cuisines in the UK. It is said that there is now an Indian and Chinese restaurant or take-away in almost every town. A kebab or a curry is now the stereotypical end to a beery 'lads' night out'. 'Curry and chips' is said to have displaced 'fish and chips' as the nation's favourite dish. The British Tourist Board advertises the diversity of 'ethnic' foods to attract overseas visitors to the UK.[3] And commentators have argued that the mainstream popularity of these foods reflects, and perhaps promotes, changing understandings

**Figure 1** Masterfoods' suggested arrangement of regional/ethnic foods along the supermarket aisle. Source: Masterfoods' *Wet Cooking Sauce Market file*, no date.

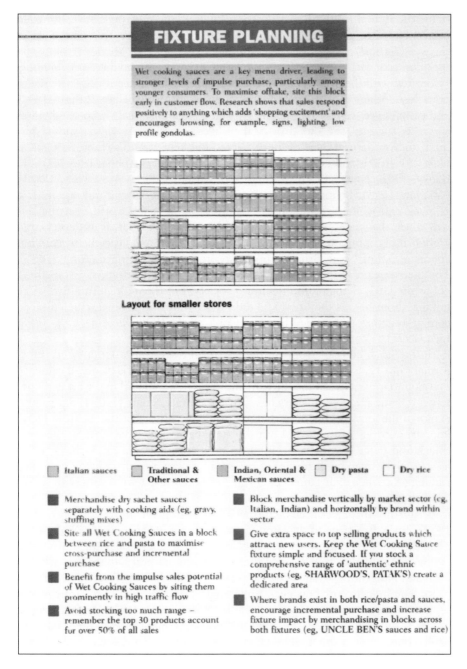

## FIXTURE PLANNING

Wet cooking sauces are a key menu driver, leading to stronger levels of impulse purchase, particularly among younger consumers. To maximise offtake, site this block early in customer flow. Research shows that sales respond positively to anything which adds 'shopping excitement' and encourages browsing, for example, signs, lighting, low profile gondolas.

**Layout for smaller stores**

☐ Italian sauces    ☐ Traditional & Other sauces    ☐ Indian, Oriental & Mexican sauces    ☐ Dry pasta    ☐ Dry rice

- Merchandise dry sachet sauces separately with cooking aids (eg. gravy, stuffing mixes)
- Site all Wet Cooking Sauces in a block between rice and pasta to maximise cross-purchase and incremental purchase
- Benefit from the impulse sales potential of Wet Cooking Sauces by siting them prominently in high traffic flow
- Avoid stocking too much range – remember the top 30 products account for over 50% of all sales

- Block merchandise vertically by market sector (eg. Italian, Indian) and horizontally by brand within sector
- Give extra space to top selling products which attract new users. Keep the Wet Cooking Sauce fixture simple and focused. If you stock a comprehensive range of 'authentic' ethnic products (eg. SHARWOOD'S, PATAK'S) create a dedicated area
- Where brands exist in both rice/pasta and sauces, encourage incremental purchase and increase fixture impact by merchandising in blocks across both fixtures (eg, UNCLE BEN'S sauces and rice)

**Figure 2** A closer view: Masterfoods' suggested arrangement of regional/ethnic foods on the shelf fixture. Source: Masterfoods' *Wet Cooking Sauce Market file*, no date.

of the UK as 'multicultural' society. These kinds of interpretations have, however, been subject to a great deal of criticism. Much of this has centred on the use of the word 'ethnic' to describe these 'Indian', 'Chinese' and 'Mexican' foods and the first, second, third, etc. generation 'minority' migrant populations commonly associated with them. Here, 'ethnic' has been seen as a category constructed to house the 'others' of the white mainstream of 'British' society. 'Ethnic' foods and peoples are seen as 'exoticized' by this white mainstream who see them as a way to 'spice up' the 'dull dish' of their lives, as a window on to a more 'spiritual' and/or 'natural' way of life, accepted but kept at arm's reach, as signs to play with in the politics of white middle-class identity (Alibhai-Brown, 1998; hooks, 1992; James, 1996; Martens and Warde, 1998; May, 1996a, 1996b; Narayan, 1995). To most, this relationship is out and out cultural racism. It is a 'black and white', 'them and us' way of looking at the world.[4] Seen in this way, then, the rising popularity of 'ethnic' cuisines in the UK is not necessarily a good thing if, as on the cooking sauce shelf, the boundaries between what counts as 'ethnic' and 'British' appear secure and unchallenged (Attar, 1985). The 'mainstream' popularity of 'ethnic' foods may, therefore, say something about the UK's changing multicultural society, but the multiculturalism we're talking about here is the 'mosaic' kind: of separate and distinct 'true cultures' coming together to live side by side (Cook *et al.*, 1999; Kaplan *et al.*, 1998; Lien, 1997).

In the large body of work on the historical travels, translations and transformations of cuisines and their ingredients, however, there is plenty of evidence that boundaries between the 'British' and the 'ethnic' should make little, if any, sense. A wide range of authors have, to paraphrase Stuart Hall (1992, p. 49), explored the 'outside' histories that are inside the history of the 'British'. Here, imperial histories – Roman and British – have been seen as the most far-reaching (Black and Bain, 1993; Riley, 1993). In both eras, imperial expansion was (at least partly) driven by the desire of European powers to gain control over spice trading (Alcock, 1993; Laurioux, 1985). In both eras, a wide variety of ingredients became equally 'indigenous' to different 'national' cuisines through this trade and also through the changing geographies of plant cultivation (Agyeman, 1991, 1995). Referring to the later imperial era, for instance, James Walvin has argued that 'The migrations of peoples, animals, flora and fauna, from one region to another so transformed the human and natural habitat that what is local and indigenous, what is alien and exotic, is now barely discernible' (1997, p. x). The influence of European imperialisms on the 'discovery', transportation, cultivation, trade and use of chillies, for example, links the hot pepper sauces of the Caribbean, to the 'English' Worcestershire sauce, to the 'Mexican' Chilli con Carne, to the 'Indian' curry, to the 'Hungarian' Goulash[5] and so on (Achaya, 1998; Hudgins, 1993; Walker, 1993; Wells, 1997). And, of course, the recipes and other culinary knowledges regarding such foods have had their own travels, translations and transformations. The 'curry', for instance, has been part of 'British' culinary culture since at least the early eighteenth century (Black and Bain, 1993; Burton, 1993). In the nineteenth century, what began as informal trading by 'curry loving colonels' returning to Britain from the Raj grew into quite substantial businesses importing chutneys, pickles, pastes and ketchups to an increasingly wider public (Allen, 1993; Chaudhuri, 1992). But this was an 'Indian' food *invented*, rather than simply transported, *between* India and Britain (Burton, 1993; Narayan, 1995). To this way of thinking, then, a 'curry' is a hybrid 'British Indian' food – it is not one *or* the other – and has been for over 200 years.

To this latter way of thinking, then, the UK has had a changing multicultural society from as far back as anyone can see (Fryer, 1984, 1990) – as Mayerline Frow has put it 'Little is known about the first people who inhabited the land we now call Britain, except that they came from somewhere else' (1996, p. 7) – and the histories of food are an excellent way to think through and to illustrate this. Botanical, culinary and migratory histories can make a mockery of distinctions between 'ethnic' and 'British' foods, and of 'mosaic multiculturalism' in general.[6] They can show how cuisines are continually hybridizing processes rather than fixed things. They can show that the components mixed together in any hybridizing process are themselves always hybrid. And they can show that ingredients, knowledges, technologies and practices – culinary and otherwise – cannot have any straightforward 'origins' (MacKenzie and Wajcman, 1985). In such a 'polycentric' multicultural world (Shohat and Stam, 1995; Stam and Shohat, 1994), 'cultures' might have centres but they don't have sharp edges (Massey, 1991, 1995). Elsewhere, we have thought through the various 'multicultural imaginaries' in/of the world of food by drawing, in particular, on academic debates about commodification, ethnicity and hybridity (Cook *et al.*, 1999). Here, though, we want to take a different tack, but to do much the same thing. We want to delve into the world of the UK food trade and its press. And our reason for doing this is rooted in one of the most quoted articles in this academic debate: a chapter in bell hooks' (1992) book *Black Looks* entitled 'Eating the other'. Here, she has argued that, in the mainstream consumer cultures of advanced capitalist societies, histories of imperial relationships between white and black people have been 'eradicated, via exchange, by a consumer cannibalism that not only displaces the Other, but denies the significance of that other's history through a process of decontextualization' (*ibid.*, p. 31). The reason we want to examine the UK food trade and its press in the following pages, though, is that great efforts seem to be being made here to contextualize. People working in the UK food trade (press) regularly make claims to 'know their history' (or, more precisely, their historical geographies) and attempt to capitalize on this. In the process, their pronouncements regularly bring together the two supposedly contradictory ways of thinking about the internationalization of the British diet just outlined (i.e. the neat boundary making on the supermarket shelf and the messy boundary crossing of ingredients, cuisines, people and their histories). And these come together most notably in claims and counter-claims about the 'authenticity' of different companies' products. Below, we have attempted to characterize and to think through these trade debates, to use them to reflect back on these more political/academic concerns, and to explain how there is more to culinary categorizations than meets the eye on the supermarket shelf.

## Let's twist again, on a roll, in a hot market

By all accounts, whatever 'authenticity' means, it has been at the centre of recent developments in UK food manufacturing and retailing. Market reports and the trade press continually stress that consumers are becoming more 'sophisticated', trying a wider variety of international cuisines at home, and wanting to go beyond the 'anglicized' versions they have been used to in the past (Barnard and McLintock, 1994; Matthews, 1997; Mintel, 1996a). And, in the pun-infested world of the food trade press, this means that pasta sales are 'twisting again', that bread sales are 'on a roll', and that cooking sauces constitute a 'hot market' (Anon, 1995a; Deeprose, 1996; Atkinson, 1995). In the otherwise saturated and stagnant bread market, for instance, 'continental and speciality bread' is the fastest growing sector (Mintel, 1996c, 1997a). Reports suggest that this growth will continue as 'French' baguettes, 'Italian' ciabatta, 'Indian' naan, 'Mexican' tortillas, 'Greek' pittas, and UK regional breads become more mainstream purchases (Anon, 1996d; Deeprose, 1996; Walsh, 1995). In the 'Italian' food market, research has shown that pasta-based meals such as spaghetti bolognese and lasagne have been around for such a long time that many UK consumers don't consider them to be part of a 'foreign' cuisine (Anon, 1994c). Yet, in this expanding market, the fastest growing sectors are in 'genuine Italian' products, recipes and ingredients such as pizzas baked in stone ovens, pesto-based sauces, sundried tomatoes, 'Italian' cheeses, fresh pasta, and a wider variety of dry pasta shapes as available on the Italian market (Mintel, 1996d). Finally, ambient[7] cooking sauces have been at the centre of the retail boom in international cuisines. Here, reports stress that sales in the three main 'ethnic' food areas – 'Indian', 'Chinese' and 'Tex-Mex' – have continued to grow. However, within these areas, regional dishes – e.g. 'Goan', 'Cantonese' and 'Cajun-Creole' – are becoming increasingly popular (Figure 3); beyond these areas, other 'ethnic' cuisines – e.g. 'Caribbean', 'Japanese', and 'Thai' – are making inroads; and, for all of these cuisines, sales of the 'authentic' ingredients needed to prepare meals from scratch – e.g. cumin, soy sauce, tortillas, hot pepper sauces, rice vinegar, and coriander – are also growing significantly (Anon, 1994a, 1995b, 1995d, 1996a, 1996b; Cull, 1995a; Mintel, 1996a, 1996b).

Market reports stress that these new developments have been encouraged by the popularity of (a) dining out in 'Italian', 'Chinese' and 'Indian' restaurants, along with a growing number of restaurants devoted to their regional variants and the 'new ethnic' cuisines mentioned, and of (b) long-haul holidays to North America, the Caribbean, the Far East and elsewhere. In both cases, the boom in retail sales for these foods is said to be driven by consumers wanting to re-create the tastes enjoyed on these visits in their own homes. To help them to do this, these reports also stress the importance of two other influences: (c) the rise of the 'celebrity chef' both on TV and in the recipe book market,

---

**Figure 3** Regionalising Chinese cuisine: the opening pages of Amoy's *Chinese Recipes* booklet (1989).

# Introduction

## THE PRINCIPLES OF CHINESE COOKING

The main distinguishing feature of Chinese Cooking is the emphasis on the harmonious blending of colour, aroma, flavour and texture both in a single dish and in a combination of dishes for a meal.

Balance and contrast are the key words here, based on the ancient Chinese philosophy of Yin & Yang. Consciously or unconsciously, a Chinese cook from the housewife to the professional chef, all work to this yin-yang principle – i.e. harmonious balance and contrast in conspicuous juxtaposition of different colours, aromas, flavours and textures by varying the ingredients, cutting techniques, seasonings and cooking methods.

Perhaps one of the best examples of the yin-yang principle is found in the way we blend different seasonings in complementary pairs: sugar (yin) and vinegar (yang) salt (yin) and Sichuan pepper (yang), spring onion (yin) and ginger root (yang), soy sauce (yin) and rice wine (yang) and so on.

### THE NORTHERN SCHOOL

The noted cuisines of Peking and Shandong form the main cooking style of the Northern School, with Tianjin and Henan, as it were, in the second division. Although the northern school may not be the best, it is certainly the oldest of Chinese culinary art: as its origin lies in the basin of the Yellow River – the cradle of Chinese civilisation. Because Peking has been the capital of China for many centuries, it has accumulated the best dishes from all the regional cooking to become the culinary centre of China and has, at the same time, developing a distinct style of its own.

### THE WESTERN SCHOOL

The Western School of Chinese cuisine is represented by the richly flavoured food of Sichuan (or Szechuan as it used to be known). The use of hot chilli in everyday cooking is not supposed to paralyse your taste buds, but rather to stimulate your palate. The neighbouring province of Hunan is also renowned for hot and spicy cooking, but has a distinctive style of its own.

### THE EASTERN SCHOOL

The Eastern School is represented by the sophisticated cooking style from the Yangtze River delta known as Huaiyang, which also includes China's largest city Shanghai to the east, and the scenic city Hangzhou to the south. The characteristics of this region can best be summarized as exquisite in appearance, rich in flavour, and sweet in taste.

### THE SOUTHERN SCHOOL

The Southern School is represented by China's most diverse school of cuisine from Canton in the Pearl River delta. Because Canton was the first Chinese port opened for trade, foreign influences are particularly strong in its cooking. Together with its neighbouring province of Fujian (Fukien) Canton is the place of origin of many Chinese emigrants overseas, therefore the southern school is also the best known style of Chinese cooking abroad.

# China

China is a vast country, with varying climates and different natural products from region to region, therefore the cooking style reflects these divisions. Yet the fundamental character of Chinese cooking remains the same throughout the land – the Peking cooks in the north and the Cantonese cooks in the south all prepare, cook and serve their food in accordance with the principles mentioned later: some of the cooking methods may vary a little from one region to another, and the emphasis on seasonings may differ, but basically food from the four main regions are all 'Chinese'.

Yellow River

Beijing

Tianjin

Shandong

Henan

Shanghai

Yangtze River

Hunan

Amoy

Sichuan

and (d) the increasing availability of fresh, ambient, chilled and frozen ingredients and processed international foods on the supermarket shelf (Anon, 1996e; Mintel, 1996b, 1996c, 1997b). However, although the popularity of some of these cuisines is acknowledged to be widespread, only a small proportion of consumers are at the 'sharp end' of this market. The most 'adventurous', for instance, are said to be primarily 'ABC1s and working managers' living in London, the South East of England and parts of Scotland (Matthews, 1997; Mehta, 1996; Mintel, 1996a). As well as being those with the widest variety of international cuisines on their doorsteps (particularly in the form of restaurants), these are the people who are supposedly willing to pay higher prices for goods which are the 'real thing' (Anon, 1995c, 1996d, 1997; Murphy and Bray, 1996; Walsh, 1995).

In the wider scheme of things, this increasing concern with 'authenticity' in the UK food business is unimportant. It doesn't appear to be important in the shopping and eating decisions of the vast majority of British people and, although growing rapidly, the business serving these sectors is not worth a particularly significant amount of money.[8] However, the most significant thing that it has done (in business terms at least) has been to open up spaces for 'smaller regional or product specific suppliers to develop a considerable influence on market development' (Mintel, 1996a, p. 4). These influential and rapidly growing companies include Le Pain Croustillant, Pasta Reale, Patak's and Discovery Foods (leading UK manufacturers of frozen part-baked 'continental' breads, fresh pasta, 'Indian' foods and 'Tex-Mex' foods); WT Foods, BE International, G. Costa and Unimerchants (leading UK-based importers and distributors, each marketing ranges of 'Indian', 'Oriental', 'Tex-Mex' and other regional foods); and Sacla and Busha Browne (leading Italy- and Jamaica-based manufacturers supplying these expanding UK markets). As well as having each other to worry about, these companies also have to compete with the (usually more 'anglicized') branded products made by subsidiaries of much larger TNCs like the British Tomkins Group (whose Rank Hovis McDougall subsidiary owns British Bakeries whose brands include Hovis & Mother's Pride); the Dutch Unilever (whose Van den Bergh Foods' Ragu and Chicken Tonight brands are market leaders in cooking sauces); and the Swiss Nestlé (whose Buitoni subsidiary/brand is a market leader in dry pasta).[9] In terms of opening up new markets through imaginative and somewhat risky new product developments (NPDs), the smaller manufacturers and importers claim to lead the way (Cook *et al.*, forthcoming). What these larger TNCs have tended to do, then, is to wait until these openings are of a sufficient size to enter them themselves through introducing their own NPDs and/or, in true Victor Kiam style, buying the smaller companies which did all the hard work (as the Tomkins Group did in May 1998 with its purchase of Le Pain Croustillant: Hemmington Scott Ltd, 1998; Watson, 1998).

To reiterate perhaps the most obvious point, an ability to claim that products are, in some way, 'authentic' is an essential part of the marketing pitch of companies in this line of business. Yet, through reading the trade press and talking with our interviewees, the easiest conclusion to come to was that the word 'authentic' (along with its close relations 'ethnic', 'exotic' and 'traditional') had so many definitions that it didn't describe anything at all. Some trade commentators have pointed out that these words are, more often than not, simply 'used by reflex' (e.g. Anon, 1997, p. 31) and those who have pondered their precise meanings have often ended up unravelling them completely. One recent discussion of 'ethnic' foods in the UK, for instance, concluded that 'British is still best, just so long as you broaden out the term "British" to include everything we like, regardless of origin. There is a logic here, I'm just not quite sure where' (Wise, 1996, p. 106; see also Paulson-Box and Williamson, 1990). Thus, it is perhaps not surprising that the managing directors of some importers and manufacturers – like Tim Barlow of G. Costa and James Beaton of Discovery Foods – are said to be 'trying to break down various barriers, [and] encouraging cross-cuisine, mix and match ideas' (Anon, 1997, p. 32; Anon, 1994b).

Under these circumstances, most of our interviewees appeared to have *working definitions* of 'authenticity' which they used in strategic ways, even though they knew that these would ultimately fall apart under closer examination. Perhaps the most interesting, here, were the ones provided by Jim, a supermarket cooking sauces buyer:

> The way I would look at it as a buyer here would be to say that there are actually two ways of looking at this: there's 'authenticity' and there's 'genuineness'. Um, 'authenticity' does not necessarily mean that it's produced in the point of origin that it should be, that you would expect it to. 'Genuineness' is. If it's 'genuine' it's actually coming from India or coming from the Far East. Um, you could have a 'genuine' product that's not necessarily 'authentic' because it's been changed according to what we in the UK would view as acceptable. *Yeah, like watering down jerk sauces, that kind of thing.* Exactly, Yeah, but it could be produced in Jamaica or somewhere. Um, so what you actually have very rarely is something that is fully 'authentic' and fully 'genuine'. For example, our range of classic Italian sauces all come from Italy. Now, it doesn't shout out they are Italian, but we just call them classic Italian. They are 'genuine' but they're not necessarily 'authentic' because 'authentic' pasta sauces are made, they're not just eaten.
> (Jim, cooking sauces buyer, Big-4 supermarket chain. All words in italics are those of the interviewer.)

Jim's lexicon of 'authenticity' and 'genuineness' was by no means common to all of our interviewees. But there was an architecture in it which was common across the trade (press): i.e. that it was the relationships between specific locations and particular kinds of connections which 'authenticated' different foods. However, where he identified one kind of geographical 'origin' here – i.e. where a product was manufactured – we found this to be complicated by three others – where it was sold, where it was prepared, and where it was eaten. To manufacture and market foods which were 'authentic' to these locations was a matter of bringing together the right ingredients, the right recipes, the right technologies, the right personnel and/or the right 'stories'. The 'geographies of culinary authenticity' that we found always *combined* claims about the 'right' locations and the 'right' connections. And this was by no means in the sanitized 'geography of authenticity' identified in critiques of 'mosaic multiculturalism'.

## A passage from India

Having said this, debates about 'authenticity' were almost always rooted in places of manufacture. Companies making/importing 'Indian' foods in/from India, 'Italian' foods in/from Italy and so on used this fact as a major part of their 'authenticating' strategies. As the MD of UK importer Unimerchants has put it, UK consumers want 'their Indian dishes made in Madras, not in Manchester' (in Anon, 1995b, p. 42), and the Hong Kong-based 'Oriental' food manufacturer Amoy boasts 'Authenticity – all products are the same formula used in the Orient with most imported from Hong Kong, China, Thailand and Sri Lanka.'[10] Most companies adopting this strategy emphasized their use of locally sourced ingredients, to make 'traditional' local products, using 'traditional' local recipes and expertise, and to make products sold to and used by local people (Anon, 1996b; Cook and Harrison, 1998). Yet, these 'traditions' were rarely static, bounded or completely localized. For a start, authenticating accounts often acknowledged that whatever 'traditional' foods might be in each part of the world, they have ended up being mass produced in factories for mass consumption in local markets. Thus, for instance, the MD of Rank Hovis McDougall's 'ethnic' food subsidiary Sharwood's has explained that the lifestyles of Asia's three billion people 'are changing just as ours are. They are living in a dynamic and contemporary environment and have needs similar to ours. They too are looking to serve quick and high quality products' (in Matthews, 1997, p. 40). Unimerchants has stressed that it imports foods from a *modern* India, that the Ashoka products which it imports are made there 'in a modern food factory of course' (Peattie, 1995, p. 40: Figure 4). Similarly, the publicity for the UK market produced by Pastificio Rana – a fresh pasta manufacturer based in Italy – has explained how Italian people have experienced the same kinds of social changes as British people in recent years. In both countries, more women have taken jobs outside the home, fewer meals are being produced from scratch at home, and therefore factory-made foods are increasingly becoming part of 'authentically' Italian home cooking (Anon, 1994d). The 'authenticity' which all of these companies attached to their foods *was* based on their place of manufacture, then, but this was modern manufacturing which kept pace with changing consumer trends in the 'home' market.

This did not mean, however, that history and geography had no place in such companies' authenticating strategies. Many acknowledged, and even promoted, the fact that the products they made had had their travels over time and space. Publicity materials produced by the UK industry-funded Pasta Information Centre, for instance, outline a history of pasta which has no definite beginning but which passes through Egypt 7,000 years ago, China 6,000 years ago, Italy 700 years ago, and the UK 100 years ago. The historical geographies of pasta production and consumption have therefore meant that 'In one form or another, it is a component of many cuisines, from Asia to Europe; and where it isn't part of a cooking tradition, it is still universally available, from Bogota to Boulogne and from Singapore to Saskatchewan' (Botsford, 1996, p. 2). Unimerchants has used these kinds of arguments to introduce the Peloponnese

**Figure 4** Blending the traditional and modern: Noon's 'Indian' food factory in Southall, West London. Photo from company publicity.

and Mediterranean Sunripe brands of 'authentic' Greek pasta accompaniments to the UK market. According to the company's marketing director, 'Italy is by no means the sole source of pasta-based cuisine and with this new Greek range we are introducing the best of two interrelated Mediterranean cultures' (in Anon, 1994e, p. 55). Perhaps the country containing the most 'authentic' foods usually associated with other countries, however, is the USA (and the state of California in particular: Botsford, 1996). As well as the well known examples of 'Italian' foods which have also become 'authentically American' like spaghetti bolognese and pizza (Levenstein, 1985; Cull, 1995b), the 'authenticity' of the 'Mexican' foods now becoming increasingly popular in the UK has also been associated with the USA. Recipes for salsa, for instance, are part of any 'Mexican' recipe book. Yet Discovery Foods, one of the main manufacturers of salsa for the UK market, states outright that:

> There is no such thing as a definitive Mexican salsa – as well as regional variations, almost every home has its favourite recipe . . . Salsas have been adopted by Americans in a big way and we have gone to Texas as well as Mexico for inspiration for the new *Discovery* range.[11]

The company has gone further, however, by broadening out its regional identity from a manufacturer of 'Tex-Mex' foods to foods of 'The Americas'. In the process, it has enthusiastically embraced and promoted the hybridity of Californian cuisine, and passed on the latest developments to its UK cookery club members. Take the following excerpt from the Spring 1996 edition of *Discovery News*:

> Picked up in the San Francisco Examiner . . . BURRITO BOOM. A report on the latest fashion in exotic, upmarket Burritos. Multi-ethnic, gourmet fillings like Thai-spiced chicken, jasmine rice, Peking duck and goat's cheese have taken the trendy young Bay Area professionals by storm. 'A burrito made with such chichi fillings would have been a gastronomical faux pas 10 years ago', says reporter Wendy Tamaka, 'but these days the formerly humble fare is getting the red-carpet treatment. It's so 90s.'

Promoting this hybrid cuisine further, the company has just introduced king-size flour tortillas to the UK market to enclose these and other combinations of fillings. These tortillas are referred to as 'Californian Wraps', and their packaging describes the process of using them as 'Wrap and Roll'.

Thus, there is often more to these historical geographies than simple stories of one kind of food developing and changing as it (was) moved across national borders. In concert, the travels, translations and transformations of numerous individual ingredients, recipes and practices have produced hybrid cuisines all over the place which are centuries old. In the trade (press), these wholesale changes were often placed in wider world historical contexts. The histories of European imperialism have not necessarily been brushed over, for instance, by companies marketing food 'originating' in the UK's former imperial territories. Marketing materials produced by those supplying the growing Caribbean food market in the UK, for instance, can include potted imperial histories of the region in which successive waves of Arawak, Carib, Spanish, African, Dutch, English, French, Indian and Chinese invaders and (forced) migrants brought their culinary traditions and ingredients to the region (e.g. chilli peppers, okra, gungo peas, yams, curries and chutneys). It is therefore the Caribbean's thorny history that has produced its distinctive and unfolding 'cosmopolitan' culinary scene.[12] While the colonial histories attached to contemporary manifestations of such foods are sometimes highly decontextualized,[13] some companies outline brand histories that go into perhaps unexpected depth. The Jamaica-based hot pepper sauce manufacturer Busha Browne,[14] for instance, has adopted a branding strategy which draws on 'old colonial imagery' (Figure 5). However, the 'story' the company provides to explain this strategy is far from decontextualized:

> In 1836, Howe Peter Browne, 2nd Marquess of Sligo returned to Westport, Ireland, taking with him memories of the spicy and exotic dishes he had been served during his governorship of Jamaica. Known as 'The Emancipator of slaves', the noble Marquess had earned himself a renowned place in Jamaican history for his two year term of office. His had been the unenviable task of supervising the first stage of Emancipation which was unpopular with the reactionary planters for whom the abolition of slavery assuredly meant

**Figure 5** Old colonial imagery: the Busha Browne logo.

financial ruin. In desperation, many of these planters subsequently sold their vast estates to the local managers who were known as 'Bushas'. Sligo himself was among the first to free his slaves on his Jamaican estate – Kelly's and Cocoa Walks estates – which he had inherited from his ancestors, the Kellys and Brownes of Ireland; Jamaican settlers from the late 1600's. The Brownes became wealthy and were numbered among the respected members of the plantocracy who were famous for their entertaining and the variety of food they served. In Jamaica their name has always been pronounced 'Browney'.

A descendant of the family Sligo, Charles Adolphus Thorburn Browne – familiarly known as Charlie Browne – has spent most of his 75 years 'cooking up a storm'. He whiled away many hours of his boyhood days in the kitchen of his family home, Tryall Estate in Hanover, western Jamaica. Later, cooking became a hobby and he retrieved from his family archives recipes for authentic Jamaican sauces, jams, pickles and condiments made from the great variety of exotic Jamaican fruits and vegetables; all unique, spicy and delicious. Specially selected recipes from his treasure trove are now being prepared and bottled for Busha Browne's Company in Jamaica to be enjoyed today as much as they were over 200 years ago.[15]

In a number of cases, UK manufacturers and importers have also acknowledged their roots in such imperial histories. According to BE International's company history, for example, the 'BE' stands for the 'Bombay Emporium' store. This was opened by a former British Indian Army officer in London in 1931 and, at the time, was allegedly 'the country's only shop specialising in the sale of Indian spices, pickles and dry foods', a place where 'the Memsahibs could recapture the true taste of India' (Mintel, 1996a).[16] Sharwood's company history, on the other hand, is similarly rooted in the Raj, but the company is currently trying to 'throw this off' (Anon, 1997). Other overseas manufacturers have felt it necessary to locate themselves in the context of more recent political economic history, however. Nestlé, for instance, has owned the Italy-based pasta manufacturer Buitoni since 1988. However, as this acquisition was explained in the trade journal *European Supermarkets*: 'Buitoni's history goes back a long way, right back to 1827 in fact. A family run business for many years, the company was given a much needed injection of capital in 1988 when it was bought out by the omnipresent Nestle' (Anon, 1996c, p. 55).[17] The Hong Kong-based 'Oriental' food manufacturer Amoy, for instance, has recently produced publicity materials charting the company's 'Dynasty' from its beginnings in the Chinese city of Amoy in 1907, through its 'loss when the communists took over' and subsequent move to Hong Kong in 1949, the spread of its markets throughout the Far East in the 1950s, the 'shar[ing of its] success with a number of companies, including the American Pillsbury Group' between 1972 and 1987, its eventual purchase by the Danone Group in 1991, and to its current sales to 'over 20 countries world-wide'.

## The right ingredient

What seems to be accepted in the UK food trade (press), then, is (a) that 'ethnic'/regional cuisines are part of the 'modern world', (b) that they have always developed across national/regional borders, (c) that these developments have been shaped by larger global processes such as imperialism and transnational business practice, and (d) that UK interests have been involved in this trade for some time. Thus, many UK-based manufacturers argue that, if the world of food is and always has been thoroughly interconnected, 'authentic' 'Italian', 'Indian', 'Chinese', 'Mexican' or any other 'ethnic'/regional foods can be made in the UK. So, for instance, Patak (Spices) Ltd manufactures its 'Indian' cooking sauces, pastes, chutneys and breads in factories in the North of England and in Scotland; Abel Eastern International Ltd, the leading supplier of 'Indian' breads to the UK market, manufactures them in Scotland; Pasta Reale manufactures its fresh pasta and pasta sauces in a factory in the South of England; Discovery Foods manufactures its cooking sauces, ingredient sauces, and flour tortillas in a factory in Central England; and Le Pain Croustillant manufactures its frozen part-baked continental breads in West London. All of these companies claim, in some way, that their products are 'authentic' to the cuisines concerned. So, what authenticating strategies do they use in the trade press and in their own publicity to make this case? In short, most claim that the right combinations of key inputs – factory equipment, manufacturing techniques, ingredients, recipes and/or personnel – have been imported from the right parts of the world. In its *Guide to Indian Cuisine*, Patak's claims that its success, for instance, can be attributed 'to our faithful adherence to only the very best traditional recipes and intimate knowledge of exotic spices, fruits and vegetables plus our unrivalled manufacturing standards'. There is no 'magic formula' for 'authentication' across these companies, though. Some have different combinations of these inputs, and some make a bigger deal of some inputs than others. But it is usually the case that they agree that, just because a product is made in one part of the world, it is not necessarily 'authentic' to the cuisine there. For example, just because 90% of dry pasta sold in the UK is made in Italy, it does not necessarily follow that this constitutes 90% of the 'authentic' dry pasta sold in the UK as, 'So seriously do the Italians take their pasta that it is illegal to market inferior products. Inferior means not made from 100 per cent durum wheat. This does not stop the Italians from making products from mixed wheat (soft and hard) for the UK market, [though]' (Roper, 1996, pp. 39–40).

To illustrate how UK manufacturers' authenticating strategies can work in detail, we can take the case of the 'French' breads manufactured by Le Pain Croustillant. Its publicity materials state that it produces 'the best French bread in the world' in its factories in Southall, West London. This claim is 'authenticated' through stories of its bakers and engineers travelling to France to learn about 'traditional' French baking techniques to ensure that these could be replicated in its modern, high-tech factories (in which they have recently installed the largest stone ovens in the world); of its researchers' search for 'ancient recipes for delicious French breads that were overlooked in the rush for mass production'; and of their search for 'exactly which strains of wheat, from which particular regions, made the finest French bread'. Thus, the company declares, their 'bread's journey to the dinner table starts in the wheat fields of France'[18] via a factory in Southall where this research is put into practice to manufacture breads with 'true French character in terms of the appearance, taste and aroma' (Anon, 1996h, p. 27). Unfortunately, its publicity materials conclude, while these principles are still kept to by *craft* bakers in France, mass producers supplying the French market do not take anything like so much care to produce 'authentic' French bread. Hence a British-owned, British-based company can claim to produce the 'best French bread in the world'.[19]

While the authenticating strategies of Le Pain Croustillant are based largely on knowledge gleaned from historical and technical *research*,[20] the strategies of other UK-based companies (and overseas companies with UK subsidiaries) are based more on 'insider' knowledges and assertions of the 'ethnic authenticity' of key personnel. We questioned one supermarket buyer about the continental and speciality breads that she sourced from UK manufacturers, for instance:

> How could you say 'This is genuine. This is the proper stuff.' But it's made in Solihull? It's very very difficult. Something like a Naan bread is made in Scotland, but the MD and his wife are from Pakistan and they know their product and they know how to make it and it's made reasonably traditionally. Our Caribbean baker is Caribbean. They are people who do have the skills to reproduce the product. (Louise, bread buyer, Big-4 supermarket chain)

Thus, it is not only the movements of ingredients, recipes and cuisines from Italy, India, China, the Caribbean and so on which bolster the 'authenticity' of UK-made 'Italian', 'Indian', 'Chinese', 'Caribbean' and other products, but also the movements of people between these parts of the world and the UK. Three companies which play heavily on this kind of 'ethnic authenticity' are S&A

Foods, Patak's and Pasta Reale (Figure 6). In a newspaper article charting the success of Derby-based S&A Foods,[21] the life of its managing director Perween Warsi is outlined in some detail. Growing up in the Northern Indian province of Bihar, she had her culinary skills and fascinations encouraged by her mother and grandmother from the age of 6. And, after moving to live in the North of England, this local knowledge became invaluable as:

> 10 years ago, she turned the cramped kitchen of their Derby home over to experimentation, and spent every waking hour devising and refining recipes. On her arrival in England she had been struck by the poor quality of Indian foods on sale in supermarkets, and had conceived the idea of preparing dishes as flavourful, rich and varied as those she had enjoyed back in Bihar. Surely, she reasoned, there was a market for *authentic* Indian cuisine here? In a matter of months, she won her first contract, supplying Indian finger foods to a local Greek restaurant, and *S&A Foods* – named after her sons Sadiq and Abid – was up and running. (Weir, 1996, p. 5: emphasis added)

Similarly, *Patak's Guide to Indian Cuisine* states that the prime reason to buy its products is their '*authenticity: authentic* recipes, handed down from generation to generation' (emphasis added), and then paints a picture of how these recipes moved from India to the UK in the person of Meena Patak. She reveals that, 'As a young girl growing up in Bombay, I was introduced to the finest traditions of Indian culinary art at a very early age. Over the years, I have perfected this art, the fruits of which I offer you now in my range of Patak's Authentic Indian Foods.' Similarly, Pasta Reale's publicity claims that its products are made to 'the most *authentic* recipes' (emphasis added), the principles of which were learned by one of its joint founders – 'development chef' Virginia Lopalco – as she grew up in Italy before moving to the UK. As described in one trade press report, she 'grew up in poverty as a tenant farmer's daughter in Treviso, Italy, leaving school at the age of 11' (Roper, 1997, p. 26) and it was there, she has said, that 'Mama taught me to cook and make pasta. I was brought up from the old Italian background of large families. My mother was an excellent cook and taught me everything I know' (in Atkinson, 1995, p. 73).[22] In all three cases, company publicity states that these women have always prepared these dishes in their own kitchens for 'you' to savour in the comfort of 'your' own home. Pasta Reale puts this most boldly, stating that its NPDs 'are cooked up and tested in a homely kitchen, not a laboratory. They are supervised by Virginia Lopalco, who sees them through the eyes of a wife and mother, not a food scientist.'[23] In all three cases, then, the authentication strategies of these first generation migrant family firms attempt to capitalize on the gender and 'ethnic authenticity' of their female co-founders to produce personalized NPDs for the UK market.

---

**Figure 6** Virginia Lopalco and Roberto Santi celebrating Pasta Reale's success in 'true Italian Style'.
**Figure 7** Patak's setting out the authenticity of their foods.

# Pasta Reale now UK's Numero Uno in fresh pasta!

The company that pioneered the commercial fresh pasta market in the UK has emerged as leader of this dynamic market, now worth £55 million and growing at 22% year on year.

Pasta Reale's growth has been achieved through an aggressive product development programme working closely with its customers. Further major capital investment in equipment and processes makes this the most efficient pasta business in Northern Europe.

In true Italian fashion, Virginia Lopalco, Product Development Director and her brother Roberto Santi, Chairman, celebrate the company's success.

**MAJOR ADVERTISING CAMPAIGN**

For 1997, Pasta Reale are investing in a major consumer advertising and promotion campaign (see story overleaf) that will consolidate its position as "Numero Uno" in the U.K. pasta market.

Having come from Italy and set up in business originally as the owner of a trattoria outside Croydon, Virginia Lopalco still develops all her recipes in her kitchen in the new purpose-built Pasta Reale factory – except that now she supervises not 500 meals a week

but *one million!* The founding family are looking ahead to an exciting future, with a constant stream of innovative fresh pasta products pouring off the Crawley production lines.

*It's a success story with a Reale Tasty flavour.*

---

## Duetto is top UK brand in fresh pasta

The startling success of Pasta Reale's two-colour Duetto is confirmed by the latest figures which show that

this popular brand is now the biggest seller in the UK fresh pasta market.

Originally launched with two varieties – White & Green Spinach and Ricotta Cappelletti and White & Red Cheese and Sundried Tomato Cappelletti – the Duetto line-up is now extended to include Duetto Tagliatelle with red and green peppers.

Launched in November 1995,

Duetto now accounts for over 5% of the entire U.K. fresh pasta market and is stocked by most retailers. The two-colour product proves a powerful attraction on the chilled shelf, and at a typical price of £1.69, Duetto represents outstanding originality – and value for money. Duetto has created an entirely new market segment in fresh pasta, and features boldly in Pasta Reale's consumer advertising for 1997.

# The True Taste of India

## Authentic Products

For three generations the Patak family have produced some of the finest Indian Cuisine in the world, thus earning an International reputation for quality and taste.

This success we attribute to our faithful adherence to only the very best traditional recipes and intimate knowledge of exotic spices, fruits and vegetables plus our unrivalled manufacturing standards.

"One of the most difficult achievements, when preparing authentic Indian food, is to create the true taste, aroma and texture of traditional dishes. Now, I've made it possible for you to achieve this special Indian experience with my exciting recipe book Patak's Guide to Indian Cuisine . . . which uses Patak's unique range of sauces, pastes, chutneys, pickles, pappadums and breads.

When you enjoy a meal prepared the Patak's way, you can easily close your eyes and imagine the crowded bustle of a Bombay restaurant . . . the scented breeze of the spice fields, or the towering majesty of the magnificent Taj Mahal.

My specially prepared blends are used all over the world, by everyone from professional chefs to the more adventurous home cook . . . who wishes to experience the delights of real Indian cooking.

As a young girl growing-up in Bombay, I was introduced to the finest traditions of Indian culinary art at a very early age. Over the years, I have perfected this art, the fruits of which I offer to you now in my range of Patak's Authentic Indian Foods. If you wish for the true taste of India, my promise is that you will never experience anything more delicious and authentic than the recipes I've prepared for you here."

*Meena Patak*

## Patak's Pastes

The Patak family have perfected the art of preserving the original taste of spices by encapsulating them in a vegetable oil. Fresh spices and fragrant herbs are freshly ground, to bring out the subtle and varied flavours which are then further enhanced by sophisticated roasting and blending and finally sealed in the jar to make a meal preparation — whilst still retaining the essential flavours and aromas of the original spices.

## Patak's Sauces

The most convenient of all food preparation methods, this range of regional sauces will allow speedy meal preparation.

## Patak's Pickles

The average Indian household in the U.K. does not make its own pickles because the right quality ingredients are not readily available. Patak's Pickle range is extensive, reflecting the varying tastes within India's complex region-based cuisine and is used by most Indian families in the UK. They are precisely the same whether used in Gujerat, or the World's leading restaurants or by the Patak family at home.

## Patak's Pappadums

The perfect partner for almost any Indian meal. Traditionally topped with different chutneys and pickles. Patak's plain or spiced Pappadums make the art of producing a truly authentic meal simplicity itself.

## Patak's Chutneys

The art of making chutney dates back hundreds of years. Indian chutneys are preparations of a delicate balance of exotic fruits or vegetable, blended with spices and preserved by cooking with sugar.

Chutney, although traditionally served as an Indian meal accompaniment, can also simply be added to curries whilst cooking.

## Patak's Breads

Lightly-spiced Naan bread, traditionally cooked in a clay tandoor and served with a cool cucumber raita and fresh vegetables, is a common sight on many Indian dinner tables. Chapatti (or Chapatty), a thinner and plainer version of traditionally unleavened Indian bread, served with Lime or Mango Chutney, is equally delicious with all main courses.

# The Spice Regions of India

## Hyderabad

The capital of the state of Andhra Pradesh, Hyderabad is one of the richest Muslim states – most famous for its Korma and Biryani dishes. A wealth of cultural tastes emanates from this region, from mild dishes to incredibly spicy curries!

## Bengal

Calcutta – Bengal's capital – is world famous for its concentration of all that is most exotic and interesting in Indian cuisine. Fish courses are a passion in Bengal and no meal is complete without one, as are Rasgoola, Sandish and Rasmalai dishes.

## Rajasthan

From this region, comes the most popular and adventurous Indian cuisine – Tikka and Tandooris – invented by the Rajput warriors centuries ago and still enjoyed by millions of people all over the world! Meat and game are the traditional base ingredients, with roasted vegetables a speciality, usually served with yoghurt and puris.

## Kashmir

Kashmir is situated at the foot of the Himalayas. Snow-capped peaks surround lush valleys where valuable crops of the crocus are grown, from which is produced the most expensive and rare spice Saffron.

## Gujerat

Mahatma Gandhi's birthplace, Gujerat is the only Hindu vegetarian state in the whole of India. A typical Gujerati meal will consist of traditionally dry ingredients, such as dhall, rice, chapattis, puri, pickle and Bhajias – all eaten from a 'Thali' – a stainless steel plate. The world famous red chilli pepper is grown in this area, as are the exquisite spices Coriander and Fennel.

## Kerala

Most popularly known as the main spice state of India. Kerala is famous for its black pepper plants and mango trees, which form the main ingredients for many Indian dishes. Very hot and spicy recipes abound in this region, spiced with Cinnamon, Cloves, Cardamom, Ginger and Turmeric – usually laced with rice and coconut. Traditionally, meals are still eaten from a fresh banana leaf.

The NPDs of these and other UK-based companies have also been personally 'authenticated' through the bought-in backgrounds of celebrity chefs. For instance, S&A Foods has publicized the bought-in expertise of 'top Bombay chefs' (Cull, 1995a, p. 24) and, for its new 'Oriental' range, of celebrity Chinese-American chef Ken Hom who was employed to develop and promote some of his favourite recipes for the company (Anon, 1996a, 1996l; Weir, 1996). Discovery Foods' publicity makes much of the fact that many of its recipes have been developed in association with New Orleans TV personality, author and celebrity chef Paul Prudhomme whose 'famous New Orleans restaurant, K-Paul's, is an inspiration to all who seek the authentic tastes of America'.[24] But celebrity chefs with this kind of background are not the only ones to have been employed to authenticate regional foods. British Bakeries, for instance, employed English TV chef Keith Floyd in 1994 to launch its 'Floyd on . . .' range of continental breads. With his reputation in the UK as someone who has 'covered a multitude of products and countries' in his 'Floyd on . . .' TV series, this range was not limited to the breads of one country or region (Jenkinson, 1996, p. 8). At launch, it included 'Italian' Garlic & Herb Focaccia and 'Scottish' Selkirk Bannock breads (Barnard, 1995; Mintel, 1996c).

## Home and dried

As we have shown, the authentication strategies of 'ethnic'/regional food manufacturers within and beyond the UK very often concentrate on where cuisines, recipes, ingredients, foods, and key personnel come from. But what about where they go to, where they are consumed? As we said at the beginning of this chapter, much of the drive towards more 'authentic' and less 'anglicized' foods in the UK has been encouraged by consumers tasting the 'real thing' in particular places and wanting to re-create these tastes at home. Thus, for manufacturers and importers, a key authenticating strategy is one that connects their products to the 'benchmark originals' found in these places: to the sandwich shops where new continental and speciality breads are usually first introduced to the mainstream UK market (Mehta, 1996; Walsh, 1995); to the sites where overseas manufacturers attempt to capture the interest of the UK consumer who 'spends a fortnight each year lying on the beach outside [their] door' (Whitworth, 1995, p. 49); to the 'ethnic'/regional restaurants which have so successfully captured the interest of the UK consumers on holiday and at home; to the stores that have supplied the UK's ethnic minority 'communities' with specialist regional foods; and to the homes where consumers take the products they have bought and make something out of them.

Some companies, like Amoy, Enco and Patak's authenticate their products through claims to be leading suppliers to UK Chinese, Caribbean and Indian communities respectively (Anon, 1995d, 1997). Some overseas manufacturers can authenticate their products if they are on sale in tourist shops for UK holiday-makers and in UK supermarkets on their return (Cook and Harrison, 1998; Whitworth, 1995). But, as we indicated in the introduction, the prime locations in which British people encounter 'ethnic'/regional foods are restaurants. Thus, when planning their 'Indian' NPDs, for instance, Sharwood's claims to aim 'for the look, taste and aroma associated with an English Indian Restaurant' (Cull, 1995a, p. 24) and to work towards products which allow consumers at home to put 'restaurant "authentic" food on the table in no more than 20 minutes from the time of preparation' (Anon, 1995a, p. 40). Other companies do not have to try so hard as, they claim, many of these restaurant meals have been made using their products anyway (Anon, 1995c). Amoy, for instance, claims that the popularity of its products in the UK developed over a thirty year period when it supplied thousands of Chinese restaurants and take-aways and, in the process, became the 'favourite brand of famous Oriental chefs'.[25] Other companies recount how they *began* in the early stages of the boom in the UK 'ethnic' restaurant trade. We have already discussed the early history of S&A Foods in this respect,[26] but this is not an exceptional story. Pasta Reale started in 1973 when Salvatore and Virginia Lopalco set up the first Italian restaurant in Croydon, Surrey from which they began to sell their fresh pasta to local delis as a sideline, before moving to produce it on a factory scale in 1979 (Anon, 1996c; Roper, 1997). And Patak's publicity tells how the company started in 1957 after 'L.G. Patak arrived in the UK and began making and selling Indian entrees for his home to family and friends'; two years later, he was able to buy a shop near London's Euston Station supplying the local Indian community with 'Indian' pickles, chutneys, pastes and more; and, after the company had expanded to involve his son Kirit and Kirit's wife Meena, it went into larger scale production in 1978 after purchasing its first factory in North West England.[27] Over this time, Patak's had supplied the UK's growing 'Indian' restaurant trade so that, today, it claims to supply 90% of these restaurants and therefore has its finger firmly on the pulse of changing tastes and trends (Anon, 1995b, 1995d; Cull, 1995a; Matthews, 1997).

It is interesting to note what kinds of 'ethnic'/regional foods have developed in the UK restaurant trade, then, particularly as the success of small 'ethnic'/regional food manufacturers is said to have resulted primarily from their constant innovations in NPD. In no way do NPDs in these areas simply involve replicating 'traditional' dishes from other parts of the world. For one thing, they have to pay heed to the kinds of cuisines and people that were caught up in the restaurant trade that initially popularized such foods. Those brought together in the 'Indian' restaurant sector are particularly complicated, for instance (Anon, 1995b; Jamal, 1996). As Ted, the marketing director of a big UK 'ethnic' food manufacturer/importer, told us:

> What we have here, if you talk about Indian food, is a peculiar perception of Indian food. It is a menu that was really devised quite a number of years ago. Maybe just after the war by Punjabis who were the first people to come over to Britain just after the war and open restaurants. It's kind of a Punjabi menu, inasmuch as Rogan Josh the Punjabi dish, Korma is a bit Punjabi, Madras is a bit Punjabi inasmuch as, in Madras they eat quite different food because they're principally vegetarian. . . . (And) there's another paradox in Britain inasmuch as, at the moment, there are something like 8,000 Indian

restaurants. It depends on where you are regionally . . . but, by and large, the
restaurants tend to be run by Bangladeshis. . . . Bangladeshi cuisine is *nothing*
to do with what's on the menu. Bangladeshi cuisine is very very simple fish
dishes which wouldn't be palatable to Brits at all. So you have an interesting
situation where the Bangladeshis are serving a hybrid, uh, restaurant menu
which is based on, um Punjabi food. (his emphasis)

Another complicating factor needs to be added at this point: 'British taste'.
What restaurateurs want to put on their menus and what their customers will
eat are not necessarily the same thing. As Virginia Lopalco has explained, 'I like
to retain the tradition of pasta and Italian food' (in Atkinson, 1995, p. 73) but,
'after 40 years in the country, I've learned something about British tastes, too.
Now I combine this experience with traditional Mediterranean flavours to make
my exclusive recipes.'[28] What people in her position are said to have learned is
that, for instance, 'the British' like their main meals to have sauces. In one trade
press article about her company, it is explained that

Fresh filled pasta is designed to be eaten without a sauce. But by tradition,
Britons add large quantities of sauce to all pasta, unlike the Italians. 'That's
why we produce a range of sauces, so all one needs to do for a meal is mix
the two together' says [Pasta Reale's marketing director]. (Anon, 1996i,
p. 44; Anon, 1996g)

Similarly, Ted told us that the most popular 'Indian' dish in the UK,

Chicken tikka masala, is a dish they say doesn't exist in India, although it
probably does now. And it was a fusion of chicken tikka, which is a Punjabi
dish, and a sauce because Brits love sauce. So, um, it's a new dish that's
emerging. New dishes are emerging all the time.

And Mr M, the general manager of a UK-based 'ethnic' bread manufacturer,
explained the meaning of the new 'Middle Eastern' breads which his company
was developing to suit 'British' tastes:

We are the melting pot for this internationalisation. Like, for example, pitta
bread. It's not British but now, if you come up with a pitta bread with a crust,
it might be that is what you call British in the long run. *But it's British and
Middle Eastern at the same time.* We're the second generation of bakery
product. The third generation.

Such hybridities produced out of compromises between production and consumption seem even more pronounced when it comes to what consumers do
with 'restaurant (and other) authentic' foods when they get them home. Here,
the trade press contains highly contradictory notions of 'the consumer'. S/he
is described, on the one hand, as becoming increasingly sophisticated, know-
ledgeable and critical of the (in)authenticity of foods on the supermarket shelf
and, on the other, as not paying very much attention to these knowledges when

shopping and not necessarily using products 'properly' once s/he gets them home (Cook *et al.*, 1998, forthcoming; Jamal, 1996). As Jim the cooking sauce buyer told us:

> I think, in this business, you have to recognise that the usage of products is very much up to the consumer. You can dictate so far. I know for a fact that my pasta sauces are being used as pizza toppings. I know they're being used as casseroles. If we honestly believe the customers are just using our products in the way we expect them to be using them, we're being naïve.

Faced with this situation, some companies have paid great attention to producing all the right products, technologies and knowledges in kit form for consumers to make 'truly authentic' meals at home. For example, under its 'Indian' Curry Club brand, G. Costa 'launched the Balti Cooking Kit, an authentic karahi (the Balti cooking vessel) containing a Balti spice mix; Balti curry paste and Balti cooking sauce' (Anon, 1995a, p. 41; Murphy and Bray, 1996). But most companies seem to acknowledge, and some even play on, this 'mix and match' consumer creativity (Matthews, 1997). With cooking sauces, for instance, recipe tips provided with Uncle Ben's and Discovery products often suggest how they can be used to make an 'authentic' dish, but also ways that they can be mixed with sour cream to make dips or added to casseroles to spice them up a bit (Murphy and Bray, 1996). Similarly, Virginia Lopalco has explained her meal suggestions in the following way:

> I have used readily available ingredients, which you need not follow precisely. You can change the added ingredients, using whatever you have to hand as a substitute – don't be afraid to improvise. Your meal will be your individual creation and will be perfect every time because the main ingredient remains our delicious and fresh pasta – ready to eat in minutes![29]

Under these circumstances, many companies do not even try to authenticate their products by their material 'origins' as, among other things, consumers are often intimidated by having to know so much about them (Cook *et al.*, 1998; Graimes, 1988). One UK-based dry pasta manufacturer (Rank Hovis McDougall's subsidiary Pasta Foods) has stated that its consumer research found that the 'country of origin of pasta is not at all important to consumers, but quality and "cookability" is' (Anon, 1996j, p. 200). Thus, one marketing consultant has recommended that such companies should market their products in terms of 'a motivating position that is interesting in its own right, such as fun and wild or a Latin Lover promise' (Jenkinson, 1997, p. 34). In Ireland, this advice has been taken up as Van den Bergh Foods has marketed its Ragu pasta sauces with the slogan 'It brings out the Italian in you' (Brophy, 1994, p. 32), and one Caribbean rum manufacturer has said that 'The more we can fit in with the British desire for escapism the better. When it's your fortieth rainy day in a row, a little bit of sunshine – even if it's manufactured – is a good thing' (Whitworth, 1995, p. 49). There were plenty of other examples in the trade press and in our interviews of products being developed to meet and stretch 'British' tastes, knowledges and prejudices,[30] but perhaps our most interesting discovery was that the production and consumption of some of these hybrid products were being exported to the places where the 'originals' were supposed to come from. Louise the supermarket bread buyer, for instance, told us that the style of 'Italian' ciabatta that was most popular on the UK market was one based on a recipe developed for customers of the up-market food retailer Marks & Spencer. This recipe, she told us, contained olive oil and therefore meant that:

> People's perception in this country is that a ciabatta should have an olive oil type flavour. If you go to Italy it isn't like that at all, and Italian people would not recognise it as Italian ciabatta. They would think they were getting something completely different. The product we take from Italy, we have to try and make them make it more like an English ciabatta to make it acceptable. So it's quite a strange one.

Here and elsewhere, then, authentication by place of consumption and place of production were confused as 'English style' ciabatta was authenticated through being manufactured in Italy, even though it would not be recognized as ciabatta *in* Italy. These confusions, however, were not always things to worry about. They could be taken advantage of. The business development manager at Masterfoods and the MD of Pasta Reale, for instance, agreed that the 'freedom' of not having to keep to Italian culinary conventions in the UK means that their NPDs can take 'Italian' food in new and exciting directions: towards the 'flavoured pastas, such as garlic and herb tagliatelle' (the former in Anon, 1994f, p. 23) and the 'new colours, new shapes and new tastes, [that] the British . . . readily accept' (the latter in Anon, 1996i, p. 44). And these kinds of developments have potentially interesting futures as, for instance:

> The force fortissimo behind Pasta Reale, Virginia Lopalco, is reputed to have a long term dream to export her company's products to Italy. Marketing manager Chris Redman does not refute this. 'She thinks it would be sweet irony. And we have just started to go into Continental Europe now. . . . We've just come back from . . . a big food fair in Italy. We were looking for innovations for fresh pasta and I have to say, we are the real innovators', says Redman. 'The products on sale in Italy are straightforward varieties. No colour, no flair. When we talked to manufacturers out there, there was real respect for what we're doing.' (Anon, 1996k, p. 40)

UK manufacturers such as Patak's, Discovery Foods, and Pasta Reale are already exporting second-, third-, fourth-, whatever-generation British-Indian, British-Tex-Mex-American, and British-Italian foods to the rest of the world.[31] Sacla's recent Privilegio Club publicity has stated that traditional Italian consumers have to wake up to the ways that products such as pesto are becoming part of overseas cuisines. The might 'make an old fashioned Genoese shake his head in disapproval. But cooking must develop and change.'[32] New culinary forms and the business which created them were hybrid and hybridizing processes, then, and were continuing to make a nonsense of 'national' culinary boundaries, even though the cooking sauce shelves were still divided up so neatly by 'ethnic'/ regional cuisines.

## Spaghetti Junction

Our main conclusions in this chapter therefore concern the production and consumption of 'ethnic'/regional foods, the place of culinary and other historical geographies in this process, and the implications of all of this for our understandings of popular 'multicultural imaginaries' in the UK. The first point that we want to make is a political/academic point: that, when assessing the significance of the 'ethnic' food boom in a multicultural Britain, it is important not to draw conclusions from isolated elements of the complex and contradictory processes we have outlined above. It would be easy, for instance, to say that the way a range of cooking sauces is set out on the shelves of UK supermarkets reflects wider understandings of (culinary) cultural difference in British society. However, an examination of the processes through which these products, and knowledges about these products, are produced and consumed leads to much messier conclusions.[33] It would also be easy to concentrate on the authenticating strategies which link particular company's products to roots in isolated, unchanging, 'traditional' culinary cultures around the world. But this would have to ignore the fact that many companies using such accounts also promote others which undermine and contradict them. What we want to argue, then, is that the organization of cooking sauces on the supermarket shelf is a small but important point of clarity on an otherwise thoroughly mixed up journey from sites of production to sites of consumption. And, if we compare the cooking sauce shelves with those where breads are set out, we can see that arranging foods by 'ethnic'/regional 'origins' is not the only way that this is done. Here, products are set out by 'meal occasion', i.e. not the places from which they have supposedly come from but the times and places in which they will be consumed.

The second point we want to emphasize is that mixed up, messed up, boundary defying culinary histories and practices were everywhere we looked: in the trade press, in our interviews with trade personnel and, even though we haven't had the space to go into it here, in our research with consumers (see Cook et al., 1998). While many different authenticating strategies were used in the trade to cut through this, no single version of 'authenticity' seemed to be dominant and, indeed, many seemingly contradictory versions were used by the same companies (Patak's being the most notable case, here). A third point we want to emphasize is that these authenticating strategies almost always drew on specific and detailed understandings of the (historical) geographies of food, of the origins, movements and mixings of culinary ingredients, recipes, technologies and practitioners. In the process, detailed and complex understandings of wider social, cultural, political, economic, and other processes were drawn into food trade debates. A fourth point we want to make is that the materialities of foods mean that different kinds of historical geographical narratives were likely to be constructed for different kinds of foods. The geographies of manufacturing are affected by the need for fresh, chilled and frozen foods to have minimal journey times to stores for food safety reasons and because, for example, fresh pasta 'only has a shelf life of three weeks and any delay in delivery cuts into the retailers' profit margins' (Brophy, 1994, p. 33). So, for instance, 'Indian' breads and chilled ready meals need to be made in the UK, but appropriate business strategies and historical geographical narratives *can* be constructed to represent them 'authentically Indian'. Our fifth point is one that we want to reiterate: that concerns about the 'authenticity' of 'ethnic'/regional foods have made a big difference to what kinds of companies make what kinds of products in what kinds of locations. As market reports have stated, in many cases, these concerns have provided opportunities for small manufacturers in the UK, in continental Europe and in some of the poorest countries in the world to grow into larger and more influential companies (Mintel, 1996a; Cook and Harrison, 1998).

The messy, hybrid, multicultural historical geographies set out in these debates are obviously constructed for competitive business purposes, and not for wider pedagogic reasons. The debates we have summarized here are debates primarily within the UK food trade, the majority of the press and publicity materials quoted being produced by the food industry for the food industry. As we have argued elsewhere, these and other materials are often constructed to persuade supermarket buyers to stock that company's products (see Cook *et al.*, forthcoming). In a marketplace where 'authenticity' sells, manufacturers and importers have to convince buyers that what they have to offer *really* is the 'real thing'. If a company's NPDs can be authenticated through a plausible historical narrative – whatever its (multi)cultural implications – then this is likely to happen. As Ted told us, 'you know what history is like: people can write it to suit themselves'.[34] But these competitive pressures have brought a number of progressive ideas about culture and nationality to the surface in materials produced for buyers and consumers alike (e.g. companies' cookery club newsletters). This is the reason that we have refused to simply dismiss their claims to 'authenticity' but, rather, have tried to examine their strategies of 'authentication' to see where they come from, where they go to, and what they throw up in the process.[35] We believe that something interesting and potentially important is happening here.

# Notes

1. School of Geography, University of Birmingham, Department of Geography, University College London, and Simons Priest and Associates, London, respectively. Thanks go to staff at the Information Unit at the Institute for Grocery Distribution for helping us through the trade press, to the audience at the Cultural Turns/Geographical Turns conference, where this paper was first presented, and to David Crouch for his helpful editorial comments.
2. This chapter has come from a larger project entitled 'Eating places: the provision and consumption of geographical food differentiations' which was funded by the ESRC (project number R000236404). We thank them for their support.
3. See its web site @ *http://www.visitbritain.com/*, especially the 'Brit pop' materials for young American tourists.
4. It is interesting to note here that Mintel market reports describe foods as 'ethnic' only if they originate 'from countries outside Europe. Consequently Italian and French food, for example, is excluded' (1996a, p. 1).
5. Here, the key ingredient is paprika: powdered red chillies minus the seeds.

6. This is not necessarily always the case, though, as mosaic culinary histories are promoted in some popular cookery publications (see Cook, Crang and Thorpe, 1999; James, 1996).

7. 'Ambient' foods are those tinned, bottled, bagged and packeted products which can be stored at room temperature for long periods of time before opening.

8. For instance, in 1996 the total retail sales of bread in the UK amounted to £1,750 million, £73 million of which was accounted for by 'continental and speciality breads' (Mintel, 1997a).

9. Supermarket own label products complicate this picture as these are made by both smaller and larger manufacturers.

10. This quotation is taken from the company's 1996 promotional materials.

11. This quotation is taken from an article called 'Seminar on Salsa' in the Spring 1996 edition of its cookery club newsletter *Discovery News*.

12. This is a condensed version of the potted history which introduces Enco's 1996 *Caribbean Cookery* recipe booklet, which has been produced for free distribution to its cookery club members and other interested parties. Enco is a subsidiary of WT Foods and claims to be at the forefront of the increasing popularity of Caribbean foods in the UK. Similar arguments have been made by manufacturers introducing foods 'authentic' to the 'Cajun-Creole' cuisines of Louisiana. Discovery Foods' account of the history of these cuisines in its Winter 1996 edition of *Discovery News* goes into the same depth.

13. Grace Kennedy's failed repackaging of its range of hot pepper sauces to crack the 'mainstream' market in the UK under the Lady Nugent brand is a case in point, here. One trade press writer summed up the depth of this branding strategy in the following way: 'The *Lady Nugent* brand name harks back to (Empire), derived from the wife of the Governor of Jamaica during the Napoleonic wars'. That is it.

14. Busha Browne is a subsidiary of the Pan Jamaican Investment Trust (Whitworth, 1995).

15. This 'Busha Browne story' is at *http://www.bushabrownes.com.jm/company/index.html*.

16. BE International is a subsidiary of the Groupe Danone. These quotations are taken from a newsletter produced to announce the implications of Danone's merger of BE International and HP Foods in 1996 (*The Grocer*, 1997a, 1997b).

17. A similar argument is made to authenticate Van den Bergh's market leading Ragu pasta sauces, whose recipe 'is claimed to be based on the authentic Italian recipes of Giovanni and Assunta Cantisano, who established the brand in 1937' (Mintel, 1997b, p. 25).

18. All quotations so far are taken from the company's publicity materials.

19. The arguments, again, are taken from the company's publicity materials.

20. Le Pain Croustillant is not alone, here. Ted, the marketing director of a large UK-based 'ethnic' food manufacturer/importer, authenticated his company's products in a similar way. As he told us, 'it's perfectly possible for [our company] to go out and discover things about other countries that the inhabitants of the countries concerned often just don't know. So, in other words, if you want to know about London, don't ask a Londoner because they don't know what's going on in London.'

21. This company is a leading manufacturer of 'Indian' and 'Oriental' ready meals under the Shahi and supermarket own label brands.

22. In an undated edition of the company's cookery club newsletter *Numero Uno News*, Lopalco makes similar claims to have 'learned to make pasta and sauces in Italy at my mother's knee'.

23. This quotation is taken from an undated edition of the company's cookery club newsletter *Numero Uno*.
24. This quotation is taken from the Spring 1996 edition of *Discovery News*.
25. This quotation is taken from its 1996 promotional materials.
26. See Alleyne (1994) and Anon (1994a) for more details.
27. All of this information comes from company publicity.
28. This second quotation is taken from an edition of the company's *Numero Uno News*.
29. This quotation is taken from an edition of *Numero Uno News*.
30. While these 'ethnic' and regional stereotypes can 'work' for some parts of the world, and promote sales of goods made in certain parts of the world, they can be disastrous for others. Take the effects of Jim's understanding of British consumers' knowledge of India:

> From all our, um, research, just to stick that it comes from Bombay or, um, Delhi frankly isn't actually a good marketing point and frightens off a lot of people. The perception of India um is very different to the perception of the Far East in terms of consumers' confidence in products. Um, and that may very well come from, sort of, you know, all the, sort of, mass media, I suppose, of what India is all about. *So why would that affect food being made there being . . . ?* Because the perception, you know, would be cleanliness, and hygiene and all those sorts of things. I mean, I suppose, um, really, Third World production does frighten people in terms of, you know – is that a safe product? – or what have you. And, therefore, to market it as, very much, 'this is fully authentic: made in Bombay', whilst it could be actually a positive, from the consumer point of view is less of a positive than it would be for some of the other areas. (Jim, Big-4 supermarket cooking sauce buyer)

For further discussion of these interviews, see Cook, Crang and Thorpe (forthcoming).

31. Patak's exports to 40 countries, this trade accounting for 30% of its turnover (Anon, 1995d), and it aims to become 'the world's leading supplier of authentic Indian food products' (Department of Trade & Industry, 1999); Discovery exports to 10 European countries (Anon, 1994c); Pasta Reale exports to Germany and Malta (Anon, 1996c); and there are stories of British bread manufacturers exporting 'French' bread to Belgium, Spain and France (Anon, 1996f).
32. Sacla produces a red pesto for the UK market which is not (yet) part of the Italian diet.
33. See Burgess (1990), Johnson (1986) and Miller (1998, this volume) for criticisms of the overly 'textual' approach in social and cultural studies.
34. For instance, it is worth pointing out that the historical geography of pasta which stated that it had just 'passed through' Italy on a longer journey was promoted by a UK industry body funded by companies manufacturing pasta in the UK, not in Italy!
35. Elsewhere, we have described this kind of research process as 'getting with the fetish' (Cook and Crang, 1996; Cook and Harrison, 1998).

# References

Achaya, K.T. (1998) *Indian Food: An historical companion.* Oxford: Oxford University Press.
Agyeman, J. (1991) 'The multicultural city ecosystem' *Streetwise* 7, pp. 21–4.
Agyeman, J. (1995) *People, Plants and Places.* Crediton: Southgate Publishers.

Alcock, J. (1993) 'Flavouring in Roman culinary taste with some references to the province of Britain' pp. 11–22 in Harlan Walker (ed.) *Spicing up the palate: studies in flavourings, ancient and modern*. London: Prospect Books.

Alibhai-Brown, Y. (1998) 'Whose food is it anyway?' *The Guardian, G2 Supplement*, August 25, pp. 2–3.

Allen, B. (1993) 'Foreign flavours: the Italian warehouse and its near relations in England, 1720–1880' pp. 23–7 in Harlan Walker (ed.) *Spicing up the palate: studies in flavourings, ancient and modern*. London: Prospect Books.

Alleyne, R. (1994) 'A growth industry' *Derby Evening Telegraph* 2 February.

Anon (1994a) 'Balti joins the curry club' *The Grocer* 6 August, p. 68.

Anon (1994b) 'Discovery's resistance has paid dividends' *The Grocer* 6 August, p. 83.

Anon (1994c) 'Pasta la vista' *Checkout* September, p. 72.

Anon (1994d) 'Sponsored feature: Rana passionate about pasta' *Good Food Retailing* July, pp. 12–13.

Anon (1994e) 'Brands and own label from Mediterranean' *The Grocer* 16 July, p. 55.

Anon (1994f) 'Fresh facings' *Supermarketing* 15 April, p. 23.

Anon (1995a) 'Ethnic food: a hot market' *Ulster Grocer* March, pp. 40–2.

Anon (1995b) 'Exploring new territory' *Checkout* February, p. 42.

Anon (1995c) 'Ethnically sound' *Independent Retail News* 21 July, pp. 32–4.

Anon (1995d) 'Pace of expansion hots up at Patak's Spices' *Eurofood and Drink* March, pp. 25–7.

Anon (1996a) 'A passage from India' *The Grocer* 3 August, p. 42.

Anon (1996b) 'Championing a wondersauce' *The Grocer* 3 August, p. 38.

Anon (1996c) 'Pasta: the shape of exciting things to come' *European Supermarkets* September/October, pp. 54–5.

Anon (1996d) 'White bread takes thickest slice of the nation's bakery market' *Brand Strategy* 21 June, pp. 12–13.

Anon (1996e) 'The staff of life' *European Frozen Food Buyer* January/February, pp. 35–40.

Anon (1996f) 'British baker uses his loaf' *The Sunday Times* 28 January, p. 9.

Anon (1996g) 'Boom in UK pasta market fuelled by fresh pasta' *Eurofood* 29 August, p. 4.

Anon (1996h) 'Le Pain Croustillant opens new plant' *Frozen and Chilled Foods* May, p. 27.

Anon (1996i) 'New colours, new shapes, new tastes' *The Grocer* 27 July, p. 44.

Anon (1996j) 'Focus on pasta and pasta sauces' *The Grocer* 27 July p. 48.

Anon (1996k) 'The Reale thing' *Value Retailing* July, p. 40.

Anon (1996l) 'S&A Foods develops authentic recipe for success' *Food and Drink Business* January.

Anon (1997) 'Latent Orient' *Value Retailing* March, pp. 31–2.

Atkinson, D. (1995) 'Let's twist again' *Checkout* September, pp. 72–3.

Attar, D. (1985) 'Filthy foreign food' *Camerawork* 31, pp. 13–14.

Barnard, S. (1995) 'Grape and grain' *The Grocer* (10 June).

Barnard, S. and McLintock, L. (1994) 'Prospects over next five years are bright' *The grocer* 6 August, pp. 52–4.

Black, M. and Bain, P. (1993) 'The price of three passions: pepper, sugar and chocolate' pp. 40–7 in Harlan Walker (ed.) *Spicing up the palate: studies in flavourings, ancient and modern*. London: Prospect Books.

Botsford, K. (1996) 'The world is just a great big tortellone' *The Independent* (section 2) 5 September, pp. 2–3.

Brophy, M. (1994) 'Profits from pasta' *Checkout Ireland* January/February, pp. 30–3.

Burch, S. (1996) 'On top of the world: entrepreneur wins award of a lifetime' *Derby Evening Telegraph* 1 November.

Burgess, J. (1990) 'The production and consumption of environmental meanings in the mass media: a research agenda for the 1990s' *Transactions of the Institute of British Geographers* 15, pp. 139–61.

Burton, D. (1993) *The Raj at Table: A culinary history of the British in India.* London: Faber & Faber.

Chaudhuri, N. (1992) 'Shawls, jewellery, curry and rice in Victorian Britain' pp. 231–46 in Nupur Chaudhuri and Margaret Strobel (eds) *Western Women and Imperialism: Complicity and resistance.* Bloomington: Indianapolis University Press.

Cook, I. and Crang, P. (1996) 'The world on a plate: culinary culture, displacement and geographical knowledges' *Journal of Material Culture* 1(2), pp. 131–53.

Cook, I., Crang, P. and Thorpe, M. (1998) 'Biographies and geographies: consumer understandings of the origins of foods' *British Food Journal* 100(3), pp. 162–7.

Cook, I., Crang, P. and Thorpe, M. (1999) 'Eating into Britishness: multicultural imaginaries and the identity politics of food' in Sasha Roseneil and Julie Seymour (eds) *Practising Identities: Power and Resistance.* Basingstoke: Macmillan.

Cook, I., Crang, P. and Thorpe, M. (forthcoming) 'Constructing the consumer: category management and circuits of knowledge in the UK food business' *Geoforum.*

Cook, I. and Harrison, M. (1998) *Getting with the fetish? Manufacturing Jamaican hot pepper sauces for the UK market.* Paper presented at the IBG/RGS Economic Geography Research Group conference 'Geographies of commodities', Manchester University.

Cull, C. (1995a) 'Pukka tucker' *Value Retailing* month?, pp. 21–4.

Cull, C. (1995b) 'Home and dried' *Value Retailing* July, pp. 16–18.

Deeprose, J. (1996) 'On a roll' *Frozen and Chilled Foods* May, p. 23.

Department of Trade & Industry (1999) *http://www.dti.gov.uk/exportwinners/case-study/patak-spices-ltd.html.*

Frow, M. (1996) *Roots of the Future: Ethnic diversity in the making of Britain.* London: CRE.

Fryer, P. (1984) *Staying Power: The history of black people in Britain.* London: Pluto Press.

Fryer, P. (1990) *Aspects of British Black History.* London: Index Books.

Graimes, N. (1988) 'Spaghetti junction' *Good Food Retailing* October, pp. 29–33.

The Grocer (1997a) *Food and Drink Directory.* William Reed Directories.

The Grocer (1997b) *The Grocer Guide to the UK's top food and drink suppliers.* William Reed Directories.

Hall, S. (1992) 'Old and new identities, old and new ethnicities' pp. 41–68 in Anthony King (ed.) *Culture, Globalisation and the World-system: Contemporary conditions for the representation of identity.* Basingstoke: Macmillan.

Hemmington Scott Ltd (1998) *Press release: Tomkins acquires Le Pain Croustillant* (5 May) @ *http://www.hemscott.com*

Hendrickson, C. (1996) 'Selling Guatemala: Maya export products in US mail-order catalogues' pp. 106–21 in David Howes (ed.) *Cross Cultural Consumption: Global markets, local realities.* London: Routledge.

bell hooks (1992) 'Eating the other: desire and resistance' pp. 21–39 in her *Black looks: race and representation.* London: Turnaround.

Hudgins, S. (1993) 'Red dust: powdered chiles and chili powders' pp. 107–20 in Harlan Walker (ed.) *Spicing up the palate: studies in flavourings, ancient and modern.* London: Prospect Books.

Jamal, A. (1996) 'Acculturation: the symbolism of ethnic eating among contemporary customers' *British Food Journal* 98(10), pp. 12–26.

James, A. (1996) 'Cooking the books: global or local identities in contemporary British food cultures' pp. 77–92 in David Howes (ed.) *Cross Cultural Consumption: Global markets, local realities*. London: Routledge.

Jenkinson, E. (1996) 'Floyd cashes in on bread' *Checkout* April, p. 8.

Jenkinson, E. (1997) 'Land of shape and glory' *Checkout* January, pp. 31–4.

Johnson, R. (1986) 'The story so far: and further transformations?' pp. 277–313 in D. Punter (ed.) *Introduction to Contemporary Cultural Studies*. Harlow: Longmans.

Kaplan, A., Hoover, M. and Moore, W. (1998) 'Introduction: on ethnic foodways' pp. 121–37 in Barbara Shortridge and James Shortridge (eds) *The Taste of American Place*. Oxford: Roman & Littlefield.

Laurioux, B. (1985) 'Spices in the Medieval Diet: a new approach.' *Food and Foodways* 1, pp. 43–76.

Levenstein, H. (1985) 'The American response to Italian food, 1880–1930' *Food and Foodways* 1, pp. 1–24.

Lien, M. (1997) *Marketing and Modernity*. Oxford: Berg.

MacKenzie, D. and Wajcman, J. (1985) 'Introductory essay' pp. 2–25 in their (eds) *The Social Shaping of Technology*. Milton Keynes: Open University Press.

Martens, L. and Warde, A. (1998) 'The social and symbolic significance of ethnic cuisine in England: new cosmopolitanism and old xenophobia' *Sociologisk årbok* 1, pp. 111–46.

Massey, D. (1991) 'A progressive sense of place' *Marxism Today* June, pp. 24–9

Massey, D. (1995) 'Places and their pasts' *History Workshop Journal* 39, pp. 182–92.

Matthews, V. (1997) 'Regions to be cheerful' *Checkout* February, pp. 35–40.

May, J. (1996a) ' "A little taste of something more exotic": the imaginative geographies of everyday life' *Geography* 81(1), pp. 57–64.

May, J. (1996b) 'Globalisation and the politics of place: place and identity in an inner London neighbourhood' *Transactions of the IBG* 21, pp. 194–215.

Metha, S. (1996) 'Bake, rattle and rolls' *Checkout* December, pp. 80–2.

Miller, D. (1995) 'Consumption as the vanguard of history: a polemic by way of an introduction' pp. 1–57 in his (ed.) *Acknowledging Consumption: A review of new studies*. London: Routledge.

Miller, D. (1998) 'A theory of virtualism' pp. 187–215 in James Carrier and Daniel Miller (eds) *Virtualism: A new political economy*. Oxford: Berg.

Mintel (1996a) *Ethnic Food Report* March. London: Mintel Market Intelligence.

Mintel (1996b) *Seasonings Report* August. London: Mintel Market Intelligence.

Mintel (1996c) *Bread and Morning Goods Report* May. London: Mintel Market Intelligence.

Mintel (1996d) *Italian Food Report* September. London: Mintel Market Intelligence.

Mintel (1997a) *Bread Report* September. London: Mintel Market intelligence.

Mintel (1997b) *Ethnic Food Report* May. London: Mintel Market Intelligence.

Murphy, Y. and Bray, L. (1996) 'A touch of spice' *The Grocer* 3 August, pp. 35–41.

Narayan, U. (1995) 'Eating cultures: incorporation, identity and Indian food' *Social Identities* 1(1), pp. 63–86.

Paulson-Box, E. and Williamson, P. (1990) 'The development of the ethnic food market in the UK' *British Food Journal* 92(2), pp. 10–15.

Peattie, K. (1995) 'Tastes of the world' *Scottish Grocer* October, pp. 38–43.

Riley, G. (1993) 'Tainted meat: an attempt to investigate the origins of a commonly held opinion about the uses of spices in the cooking of the Middle Ages and the

Renaissance' pp. 1–6 in Harlan Walker (ed.) *Spicing up the palate: studies in flavourings, ancient and modern*. London: Prospect Books.

Roper, J. (1996) 'Penne from heaven' *Value Retailing* July, pp. 39–40.

Roper, J. (1997) 'What you Reale want' *Value Retailing* July, p. 26.

Shohat, E. and Stam, R. (1995) *Unthinking Eurocentrism: Multiculturalism and the media*. London: Routledge.

Stam, R. and Shohat, E. (1994) 'Contested histories: eurocentrism, multiculturalism and the media' pp. 296–324 in David Theo Goldberg (ed.) *Multiculturalism: A critical reader*. Oxford: Blackwell.

Walker, H. (ed.) (1993) *Spicing up the palate: studies in flavourings, ancient and modern*. London: Prospect Books.

Walsh, N. (1995) 'Upper crusts' *Checkout* November, pp. 59–60.

Walvin, J. (1997) *Fruits of Empire: Exotic produce and British taste, 1660–1800*. Basingstoke: Macmillan.

Watson, B. (1998) 'Speciality bread company bought' *British Food Industry News* 58, 19 May @ *http://www.london.press.net/issues*.

Weir, P. (1996) 'Perween and the curry factory' *The Independent* (section 2) 4 October, pp. 4, 5.

Wells, T. (1997) *The Spices of Life*. London: New Internationalist.

Whitworth, M. (1995) 'Treasure islands' *The Grocer* 8 April, pp. 45–58.

Wise, I. (1996) 'Fortune cooking' *Frozen and Chilled Food* November, p. 106.

# English pastoral
## Music, landscape, history and politics

*George Revill*

## Introduction

A glance through the classical CD shelves in any music shop will demonstrate the extent to which British music of the period 1880–1940, and English music in particular, is sold on its associations with countryside imagery (see for example Burke, 1983). From Sir Hubert Parry to Edward Elgar, Ralph Vaughan Williams and beyond, British music of the late nineteenth and twentieth centuries has been routinely interpreted by both professional critics and audiences alike as firmly rooted in time and place. It is viewed as a quaint and parochial sideshow in the history of western art music. However, simply to dismiss this music as an over-confident celebration of Edwardian pomp, or the nostalgia for a lost England of rural content, is both to grossly oversimplify and miss the complex engagement of this music with the rural and its representation.

The period in British musical history between 1880 and 1940 has come to be known as the English musical renaissance. Whilst it is true that there was most certainly a burgeoning of musical talent the label 'English' itself is most contentious. The term 'British' was used as well as 'English' but in parallel with other metropolitan-based intellectual movements before and since, Celtic art and English regional diversity were appropriated and represented as a prop to the dominant culture. Many of the composers involved were not English and their music was often inspired either by parts of the British isles well outside 'deep England', or by continental Europe and even the USA. Certainly the attempt to actively cultivate a school of nationalist composition took place largely in reaction to the various European nationalist schools of composition which had developed across both established and emerging nations from Russia in the 1830s. In Britain the major group of nationalist composers was loosely known as the pastoral school. The work of Ralph Vaughan Williams, Gustav Holst and their followers, or the more isolated figure of Frederick Delius, is illustrative of this. Music by these composers has been called pastoral both because it quoted or imitated folk music and also because of the frequent setting to music of poetry redolent with rustic imagery, by Thomas Hardy and A.E. Houseman for example. Yet, rather than suggesting merely the conservative inward looking qualities of British music, an engagement with the pastoral is itself indicative of an intricate complex relationship in British art, land and landscape, between European ideas and conceptions of Englishness. The nature of the pastoral as a cultural form, its structures, symbolism and historical currency are important to an understanding of this complex cultural geography. This paper will first examine the importance of the pastoral to a geography of music. It will trace some of the ways in which the pastoral in music constitutes an important medium in cultural politics negotiating the relationship between landscape, nature, society and politics. It concludes by examining implications of the pastoral for British music in the period 1880–1940. It uses the example of Elgar's cantanta *Caractacus* to examine some of these issues.

## Geography, music and the pastoral

For geography, the pastoral raises a number of parallels between the study of music and the study of landscape. The pastoral as a genre based in classical poetry has informed conceptions of landscape and rural life in Europe since the Italian renaissance of the fourteenth century. In his essay on the 'duplicity' of landscape, Daniels (1987, p. 196) argues that landscape is a concept of 'high tension' because it stands at the intersection of a variety of concepts such as 'institution', 'product', 'process' and 'ideology' which are key to both social interaction and material outcomes. Consequently landscape is at once both cause and effect, process and product, material and ideological. As Matless has recently suggested, 'Landscape can be considered a term which, of necessity migrates through regimes of value sometimes held apart' (Matless, 1998, p. 12). As a set of codes and conventions linking the physical landscape with agricultural treatises, architecture, poetry, painting and music, the pastoral is central to our understanding of the ways in which landscape negotiates a multiplicity of ideological and material realms.

Like landscape, music has played its part in policing the nature–culture divide sorting order from chaos and civilization from barbarism in western thought since classical antiquity. The classical pastoral is of central importance to this process. As Glacken (1967, p. 17) demonstrates, musical theories of cosmic, material, moral and social order are central to classical geography, its imagination and its science (see also James, 1995, pp. 38, 53–4). Throughout the renaissance and into the enlightenment music and mathematical science fuse theories of cosmic and social order within practical treatises on harmony and counterpoint. As Wilfred Mellers (1987) has shown, the pastoral in music is perhaps most directly linked to classical and renaissance science through the Greek legend of Orpheus, an enduring theme in musical history. This story is itself closely linked to the Greek myth of Er which tells of a human world governed by a musically harmonious universe. Orpheus unites two key elements of Er's story: the Pythagorean concept of a mathematically and musically harmonious universe and the power of music for the spiritual renewal of life (James, 1995, p. 56). Thus the pastoral provides a very powerful set of metaphors for ordering and classifying material and spiritual worlds. This is a complex historical-cultural geography which derives authority from both magical-religious and mathematical-scientific sources.

Music presents particular problems for critical cultural analysis. Music tends to lack the clearly defined content plane formed from words on the page or marks on canvas. The interpretation of musical meanings can therefore be very problematic (Revill, 1998). A cultural geography of music must recognize that musical meaning requires musical organization which itself necessitates social organization. If we believe that social organization is always spatial then socio-spatial structuring is intrinsic to the language and practices by which sounds are incorporated, merged together and differentiated in music (Revill, forthcoming). Sound is spatially ordered as music (as opposed for example to noise) in discourses of the auditory, in musical criticism and appreciation, in noise pollution legislation, in codes of etiquette and behaviour, in advertising, in education, in anything which values sound as good or bad, appropriate or inappropriate, in the right or the wrong place. In this sense the pastoral is a key set of aesthetic criteria by which certain kinds of sound are judged in place or out of place. This is evident in the controversy surrounding the appropriateness of rock, pop and dance music in the countryside, in for example the conflict surrounding the contemporary Glastonbury Festival, or the attempts by government and police to stop rural rave parties (Sakolsky and Wei-Han Ho, 1995; McKay, 1996; Leyshon, *et al.*, 1998, pp. 1–5, 21–5). In all these cases, the pastoral both sanctions appropriate sound and links this to conceptions of moral and social order.

The pastoral has been an enduring ideological resource in Europe from its origins in classical antiquity. Though the British isles may appear to be located on the cultural periphery of Europe the 'heritage' of Greece and Rome has played an important part in the development of its elite cultural geography for much of the past five hundred years. From the mid-sixteenth century and England's belated engagement with the Italian renaissance, the pastoral has been a genre central to English landscape, its physical form and artistic representation.[1] In England, the practices of pastoral landscape in terms of garden design, agricultural improvement, painting, literature, poetry and music have all been informed by aesthetic and moral codes derived from the classical pastoral. Studies by geographers and historians of the designed landscape in England from the seventeenth century have, for example, demonstrated the enduring importance of some form of rural myth for the legitimation of social elites and their economic wealth as well as the establishment and maintenance of national identity (for example, Barrell, 1972; Williams, 1973; Wiener, 1981; Marsh, 1982; Colls and Dodd, 1986; Daniels and Seymour, 1990; Daniels, 1993; Matless, 1998).

The musical pastoral is a powerful cultural form which is able to carry a wide range of ideological, social and moral values. The discursive rhetoric of pastoral landscape in music may be considered under two headings which examine differing aspects of its representational geography. These will be considered in turn before examining how these are combined in the discursive strategies of Elgar's pastoral cantata *Caractacus*.

## A *terrain of moral ambiguity*

Though located in very specific historical, geographical, cultural and political contexts, the geography of the pastoral is one of moral, social and locational ambiguity. Thus the cultural resources of the pastoral are available to a wide range of ideological positions. The ambiguity of the pastoral may be traced back to perhaps the most important texts of the classical pastoral, Virgil's *Eclogues* and *Georgics*. Though Virgil's work itself refers back to earlier versions of the pastoral, it above all others has formed a model to be copied, a set of metaphors and a symbolic language for countless subsequent reinterpretations of pastoral ideas (Lyne, 1983, p. Vxiii; Patterson, 1988, pp. 1–17; Coates, 1998, p. 35). Virgil's *Eclogues* are a series of 10 poems written between 42 and 39 BP. Ostensibly the poems relate stories of shepherds, goatherds, farmers and other rural characters, their singing competitions, amorous adventures and farming lives. Following Theocritus, the rustic characters of Virgil's *Eclogues* inhabit a highly idealized terrain; they speak with an elegant and learned turn of phrase. Yet, Virgil's poetry is not simple escapism, it 'smacks of present Roman reality' (Lyne, 1983, p. Vxvi), *Eclogue I* relates the conversation between two farmers: Meliboeus who has been dispossessed of his land, and Tityrus who has won reprieve from dispossession at the behest of Octavian in Rome. During the period in which Virgil was writing the *Eclogues* Rome was undergoing the military and political upheavals which resulted in the overthrow of the Roman Republic and the establishment of the Roman Empire under Octavian. Virgil was both associated with Octavian's circle and probably himself a victim of the process by which landowners were dispossessed in order to reward and resettle returning soldiers. As *Eclogue I* demonstrates, Virgil's poetry is not simple Augustan propaganda. It soon becomes clear that the dispossessed Meliboeus is associated with the most precious of Roman values, *patria*, love of country, and the poem tells how he has been expelled by an unjust military force (Patterson, 1988, pp. 2–3; Schama, 1995, p. 528).

Patterson (1988) argues that the ethical indeterminacy apparent in these early lines, and compounded in later verses, forms the basis from which the classical pastoral has come to form a potent force in cultural politics. She shows, for example, how Virgil's ambivalent stand on the relative merits of the Republic and the Empire was used in translations of Virgil which for example legitimated alternately Commonwealth and Royalist claims to rule England in the seventeenth century and Republican and Royalist claims to rule France in the eighteenth. The pastoral makes connections with a wide variety of ideological positions central to European thought. *Eclogue IV*, for instance, tells of a forthcoming Golden Age presaged by the birth of a male child which Christian interpreters read as a prophesy of the coming of Christ (Lyne, 1983, p. Vxx). This conjunction is reinforced by the prominent role played by shepherds in

Virgilian poetry and the descriptions of both idyllic garden landscapes and stony wildernesses, all of which are key metaphors in Christianity. Ambiguity regarding the moral worth alternately of agricultural work and leisured play may lend validity to utilitarian theories which privilege physical labour and material wealth. However, such ambiguity may equally validate romantic theories in which artistic and intellectual endeavour is either necessary for a balanced and good life or itself a legitimate form of labour. Thus the ambivalence and ambiguity of Virgil's poetry might condone alternately democracy or despotism, Christian values of work and duty or pagan hedonism.

In England the multiple interpretations of landscape and society available through the pastoral have been traced by Barrell (1972), Bermingham (1986), Patterson (1988), Daniels (1991, 1993), and Williamson (1995), amongst others. Most notably, Raymond Williams has traced the longstanding importance of the idea of a 'golden age' and its fusion of classical and Christian in ideas of England and Englishness. He argues that recourse to an idealized rural past has been a recurring characteristic of English rural literature since the fifteenth century. Given the complex history of the transformation of English rural society from feudalism in the middle ages, the 'golden age' may celebrate alternately the paternalism and social hierarchy of feudalism, the plebeian liberties of the rural commons, or the individual freedoms of the yeoman farmer (Williams, 1973, p. 36).

For the period covering the English musical renaissance of the late nineteenth and early twentieth centuries, the political implications bound within discourses of rural social order have been examined most notably by Martin Wiener. It is easy to disagree with Wiener's overall thesis. The idea that a 'rural myth' diverted the energies of a Victorian entrepreneurial class away from enterprise and into rural leisure has been criticized on historical grounds no less than for its own political agenda at the vanguard of Thatcher's 'new right' in the early 1980s (see, for instance, Rubinstein, 1993). However, Wiener's work still constitutes the most useful survey of the multiple versions of lost English rural life as competing ruralist discourses. Despite his very different ideological stance, Wiener (1981) draws substantially on Raymond Williams. He shows how the idea of a lost golden age formed a metaphorical resource on both the left and right of the political spectrum. For those on the political right the countryside could represent a model of, for example, stable social hierarchy, moral rectitude and physical health in contrast with the social disorder and degeneracy of the town. For the left, the rural represented a landscape of community support, mutual organization, common ownership and wholesome craft-based work in contrast with the dehumanized individualism of urban capitalism. Most recently work by Matless (1990a, 1990b, 1991, 1995, 1997, 1998) has shown how such highly politicized ideas of the rural intersect in this period with a wide range of ideologies, including modernist progressivisms, romantic historicist mysticisms and also with the roots of 'the green movement', in ecology and organic farming. The 'English' musical renaissance inhabited a Britain whose rural representations were framed by these ideological currents and music from this period played an important part in these discursive formations.

# A place of realistic representation

Given its moral, social and political ambiguity, it is perhaps not surprising that authors find it very difficult to classify the pastoral as an artistic form (see, for instance, Empson, 1968). Nevertheless, the pastoral is perhaps one of the most easily readable of all musical genres. Its iconology is one of the most formalized within music. Derived from classical sources and reworked in the renaissance, baroque and even in nineteenth century nationalism there remain still a widely agreed set of musical symbols for depicting the pastoral in music. These include:

(a) *Imitation*. This includes the musical depiction of nature, in particular, bird song, animal calls, sounds of running streams, rain, wind, etc.

(b) *Quotation*. This may involve, for example, directly quoting folk songs and dances within works of art music. Sometimes the pastoral village scene, symbolized by rustic dance, depictions of woodland or garden may form a 'realistic' interlude used to anchor or locate the didactic or ideological trajectory of a work.

(c) *Allegory*. This may relate either to musical instrumentation or to dramatic characterization. The use of instruments associated with the classical pastoral is important, for example, flutes, recorders (which suggest the reed pipes played by Pan) and the harp (as a modern equivalent of the lyre). Narratives may also be based around or feature as supporting cast classical figures associated with music: Orpheus, Pan (musician and god of shepherds), Bacchus, Apollo, Ariel, etc.

(d) *Convention*. Composers and librettists may also adopt the formal characteristics of classical pastoral. These include heavy dependence on stylized 'unreal' characters which carry the allegorical content of pastoral drama. Narratives often follow a dramatic trajectory founded on the pastoral quest for love, whether this is related to people, or place, or both. Settings frequently involve contrasting scenes, usually opposing the decadence of the town to the good life in the country. Pastorals may also draw on the formal strategy of the 'Greek' chorus. Drawing on classical theatre, a chorus of singers may be used in dialogue with the actions on the stage to either comment on the narrative or carry it forward.

Thus the symbolic language associated with the pastoral is able to make very clear connections with the world outside music. This might be in terms of associations with history and literature through *convention* and *allegory* or actual events and people through the medium of *quotation* and *imitation*. Throughout the history of western civilization these connections have been made as legitimations for political and social ideologies. In England, for example, references in Virgil to Rome as a fighting and seafaring nation encouraged comparisons to be drawn with England as the inheritor of classical civilization and a legitimate imperial power. Pastoral stories were transferred from Arcadia in Greece to English woodland landscapes such as the forests of Arden and Sherwood.

Descriptions of actual farming techniques in the *Georgics* encouraged aristocratic landowners to identify with and justify their agricultural 'improvements' in terms of Virgil's idealized rustic characters. The rise of nationalism in nineteenth century Europe provided further occasion for the use of pastoral imagery to make connections between an idealized and fantastic history and actual people, places and experiences for political ends.

Anthony Smith (1991, pp. 84–91; 1997) highlights two main ways of imagining the nation for the purpose of forging national identity. Poetic spaces are specific historic spaces which constitute the historic home of the people. These may include natural features such as lakes, mountains, rivers and valleys which can all be turned into symbols of popular virtues and 'authentic' national experience. In contrast, the cult of golden ages provides both moral examples of the nation's ethnic past and 'vivid' re-creations of the glorious past of the community. Focused on the memory of a golden age and set within a range of idealized though metaphorically translocatable places – garden, orchard, pasture, city and village – it is clear that the pastoral provides a range of representational resources suitable for the musical culture of European nationalism. This is illustrated in a key work of European national music *Ma Vlast* (*My Country*) (1874–9), a set of six symphonic poems by the Czech nationalist composer Bedrich Smetana (1824–1884). This work depicts a series of key locations and natural landscape features important to the Czech nation, for example, Vysehrad, the great rock overlooking the river Vltava which guards the entrance to Prague and the river Moldau, a symbol of national integration. The work further grounds its patriotic message through a variety of overtly pastoral references. In the final part of the work the Czech landscape and Hussite history become fused with pastoral conventions of exile, wilderness and the symbolic role of the shepherd (Jan Huss) as protector of the Czech nation.

Opera was fundamentally important to European nationalist music. The rustic and pastoralized conventions common within opera constituted an ideal vehicle for the musical realism necessary for a national music.[2] This may be symbolized by the use of vernacular language, the integration of popular and folk songs and dances into 'everyday' scenes in taverns, on streets and village greens, fields and woods (Dahlhaus, 1989). Opera was also the primary musical medium for staging specific historical stories, events, myths and legends. Evidence may be found from the beginnings of a Russian national music in Michael Glinka's (1804–1857) *A Life for the Czar* (1836) and *Russlan and Ludmilla* (1842). Later examples include Smetena's Czech patriotic works, *The Bartered Bride* (1866) and *Dalibor* (1868) and Wagner's epic settings of Germanic folk myths *Der Ring des Nibelungen* (completed 1876) (Mellers, 1988, pp. 851–93).

In Britain the attempt to build a national school of composition, begun in the 1880s, looked to an idealized rural life as a subject matter. This found an historical source not only in literature and folklore but in English music of the period 1540–1740, particularly masques, semi-operas and operas of the seventeenth and eighteenth centuries. These themselves formed part of an engagement with political issues of the day and the complex cultural matrix by which the nation was represented to itself. Harris has argued that from its beginnings in the sixteenth century, the English pastoral was more concerned with the depiction of 'real' rural life (however stereotyped and idealized these representations may have been) than in any other European country. It was therefore more concerned with social and political issues than the pastoral elsewhere. Later seventeenth century English semi-operas (a form closely related to the masque) also linked Arcadia and contemporary politics quite overtly: for example, Henry Davenant's operas *The Cruelty of the Spaniards in Peru* (1658) and *The History of Sir Francis Drake* (1659). The former concludes with a fantastical scene in which the Spaniards are expelled by a combined Peruvian–English army. This opera was intended as a kind of lecture-recital, with music, dancing and scenery (Harman and Milner, 1988, pp. 427–8). As such it not only reflects a reinvented Greek drama grafted on to English patriotism but perhaps also presages the English fashion for classically inspired didactic political landscaping in the 1740s.[3]

Ralph Vaughan Williams (1872–1958) was leader of the group of composers most closely associated with the development of a consciously national music (Kennedy, 1964, pp. 23–40). With Vaughan Williams as elder statesman, a movement for a national music developed most articulately after 1910. Many of those involved in the movement were either students at the Royal College of Music under Parry, Stanford, and later Vaughan Williams, or, they were members of the English Folk Song and Dance Society (EFDS) of which Vaughan Williams was a leading member along with the folk song collectors Cecil Sharp and Maud Karpeles.[4] After his initial exploration of the field early in the century Vaughan Williams had written on 'The foundations of a National art' in *The Music Student* in 1914 and published a series of lectures given in the USA on the subject of national music during 1934 (Vaughan Williams, 1996).

Vaughan Williams believed that the key to a truly national English music lay in folk song. In the last of his Pokesdown lectures of 1903, Vaughan Williams justified the supreme importance of folk song to liberally minded musicians, enabling the composer to 'make his art an expression of the whole life of the community' (Vaughan Williams, 1996, pp. 9–10). There are many ambiguities and inconsistencies in Vaughan Williams' approach to folk music. However, folk music was seen to be a true expression of ordinary country life for a variety of reasons. Folk songs are transmitted orally, without a definitive version; such music was viewed as democratic, freely adaptable to the desires of the people. Songs and dances are frequently based on the labour process, sowing and reaping and the social activities of community life such as courtship dances. Most importantly, folk songs and dances were believed to be founded on the rhythmic and melodic traits of native language. Vaughan Williams said that folk music is the most direct expression of personal and intimate emotion because 'it is the natural development of excited speech' (Kennedy, 1964, p. 31). Vaughan Williams wrote a number of works based on British folk song, for example, *In the Fen Country* (1904), *Norfolk Rhapsodies No.1 E min, No.2 D min* (1906). He was later to develop a distinctive melodic language based on an abstracted synthesis from elements of folk melody (Revill, 1991). This is evident in his pastoral fantasia for violin and orchestra *The Lark Ascending* (1914, rev.1920) and his rustic pastoral opera *Hugh the Drover: or life in the Stocks* (1924).

## *Elgar's* Caractacus: *pastoral music and the English musical renaissance*

The conjunction of rustic 'realist' references to the British countryside with allusions to both classical and Christian sources formed a powerful medium for composers imagining Britain during the period 1880–1940. It was argued, most notably by the composer Rutland Boughton, that classical theatre and its renaissance reinterpretations in the masque and semi-opera could form the basis for a distinctly English form of opera. As large scale dramatic choral music was perceived to be central to the development of a mature musical culture, this was viewed as a particularly pressing task for the nation (Kennedy, 1964, pp. 9–10; Trend, 1985, pp. 63–7; Hurd, 1993).[5] Taking their lead from the Darwinian rationalist Sir Hubert Parry, many composers believed that the Anglican Christian oratorio tradition leading from Handel through Mendelsson was stifling British music. A number of composers, including Vaughan Williams and Holst, were either atheists and rationalists or were experimenting with non-Christian forms of spirituality. Classical thought provided one route out of the Puritanism of Victorian respectability for composers with a wide variety of theological opinions. Gustav Holst's song cycles *Choral Hymns from the Rig Veda* (1908–12) exemplify this as does his popular orchestral suite *The Planets* (1918). This closely follows a neo-platonic conception of the character of the heavens based on the mathematical relationship between them (Dickinson, 1995, pp. 117–31). Thus the pastoral may be discerned informing work by composers not normally associated with the pastoral school itself.

Most interesting in this regard must be the cantata *Caractacus* (1898) (a work for orchestra, chorus and soloists) by Edward Elgar. Though such composers as Sulivan, Stanford and most of all Parry are associated with the renaissance and its institutional respectability, it was an outsider, the Catholic, provincial Edward Elgar (1857–1934), who was most influential in marking out the agenda for subsequent British composers up to World War II. In spite of his loss of popular appeal during World War I, because of his German romantic harmonic language and German friends, his music formed an enduring yardstick which younger composers reacted both for and against (Crump, 1986). Elgar was a composer of conservative sympathies; he supported the Tories at several elections, yet he was acknowledged by musicians of sometimes radically left wing views as the key to the Englishness of English music. Vaughan Williams, an upper middle-class liberal-radical, said in 1930,

> When I hear the fifth variation of the *Enigma* series I feel the same sense of familiarity, the same sense of something peculiarly belonging to me as an Englishman which I also felt when I first heard *Bushes and Briars* or *Lazarus*

[two folk songs collected by Vaughan Williams]. (Vaughan Williams, 1996, p. 42; see also 1934, pp. 129–30).

Though Elgar did not directly use, like or even acknowledge the use of folk song in music his compositional programme was a model for the pastoral or folk song school and a national music more generally. Elgar believed that the composer should play a social role for the nation. Partly derived from his interest in English history, he was fascinated by the idea of the composer as a modern form of medieval minstrel. Like composers of the so-called folk song and pastoral schools, Elgar was concerned with the composition of music from the direct experience of landscape and nature (Kennedy, 1964, pp. 34–6; Revill, 1991).

Written towards the beginning of Elgar's rise to international acclaim, *Caractacus* highlights major themes in his mature work, a deep concern with individual character, personal morality and a desire to write music directly from the experience of landscape. Frank Howes sees *Caractacus* as answering the call for a nationalist school of music expressing Englishness (Howes, 1966, p. 170). Andrew Blake sees it as an important forerunner of a number of musical attempts to provide a British communalizing narrative based on a heritage of folk tales and romance rooted in Celtic legend (Blake, 1997, p. 42). As an ostensibly secular dramatic oratorio Elgar's *Caractacus* presages attempts to produce a secular choral drama by at least a decade. Elgar had considered the possibility of turning it into an opera in 1901 (Kennedy, 1993, p. 74). Like Elgar, Boughton was a devotee of Wagner and visited Bayreuth on a number of occasions. *Caractacus* demonstrates a use of leitmotivs, thus the work shows clear affinity with other works of the musical renaissance which draw heavily on European conceptions of nationalism.

Much of Elgar's early work makes overt connections between notions of chivalry, patriotism and attachment to local countryside, for example *The Black Knight* (1892), *King Olaf* (1896) and *The Banner of St George* (1897). However, few of Elgar's compositions have excited as much controversy as *Caractacus*. Following its first performance at the Leeds Festival in 1898, the imperialistic sentiments attached to the work by successive commentators resulted in its subsequent neglect. When in 1970 Sir Adrian Boult accepted an invitation to conduct the work at the Cheltenham Festival, he did so on condition that the libretto was rewritten (Kennedy, 1987, p. 266).

The story is set in the Malvern Hills and tells how the Ancient Britons defended their country to the last against the overwhelming might of the Roman Empire. After the British defeat, Caractacus is taken to Rome where he is taken to meet the Roman Emperor Claudius. The crowd call for Caractacus to be killed but Claudius is persuaded by Caractacus's love of homeland, his selflessness and the pleading of his daughter Eigen and her lover the Druid bard Orbin. Claudius decrees that they should all live henceforth in Rome in peace and safety. The libretto is rounded off with an epilogue which conveniently foretells the passing of Rome's power and the coming glory of Britain. The librettist H.A. Ackworth was a Malvern neighbour of Elgar's and a retired civil servant, who had assisted with the words for the successful *King Olaf*. The story is based on the writing of the Roman historian Tacitus. Whereas Tacitus tells of the British king being betrayed by the queen of the Brigantes with whom he sought refuge after the battle, Ackworth ignores this and introduces the Arch-Druid as the villain whose false prophecy leads to Caractacus's downfall. Ignoring the portents and rejecting advice to the contrary, the Arch-Druid claims 'Mine is the ancient wisdom' and urges the British leader to face the might of the Roman legions. Notwithstanding the consequent noble defiance of the British king, the fact remains that the defeat reflected badly on British military prowess and a triumphalist chorus was called for, in which as Elgar said, 'we should dabble in patriotism' (Northrop Moore, 1984, p. 76).

> For all the world shall learn it
> Though long the task shall be,
> The text of Britain's teaching,
> The message of the free.
> And when at last they find it,
> The nations all shall stand
> And hymn the praise of Britain
> Like brothers, hand in hand.

This is the jingoistic message that commentators from Elgar's friend August Jaeger to the present day have found offensive. Frank Howes (1966, p. 170), historian of the English musical renaissance, described this passage as 'something of an embarrassment'. This seems to confirm the stereotyping of Elgar's music as a celebration of Edwardian pomp. However, closer examination of the work and its use of pastoral devices suggests a rather different cultural geography.

*Caractacus* is not a simple story of imperial triumph. Caractacus is saved through his own personal integrity in the face of defeat and the pleading of his daughter. Nevertheless, Elgar did claim to be unapologetic in his setting of the jingoistic finale. However, evidence suggests he was rather less than convinced by Ackworth's over-enthusiastic patriotism. In a letter to August Jaeger he placated the charge of jingoism made by his friend and publisher acknowledging that the 'worder . . . wallows in it' (Northrop Moore, 1984, p. 239). In Ackworth's interpretation the Druids are the villains who betray Britain from within, yet in spite of this the British nation rises in triumph to inherit the mantle of Empire and civilization from Rome. It is clear that Elgar was far more sympathetic to the Druids and was able to use his music to subvert this rather conventional establishment reading of the story.

It is interesting that Elgar reserves the most triumphal music for the scene in Rome. It is at this point (the beginning of Scene VI) rather than the final statement concerning the triumph of the British Empire that the music most resembles the music later written by Elgar for British state occasions. The central role and patriotism of the Druids complicate a simple picture of conservative (Anglican) national identity. Whilst the British heroes are Druids, not

Christians, it is significant that the Druids as nature worshippers symbolize a deep love of native land. The story resonates also with a distinctly Christian history. In the story of Caractacus and Claudius there are obvious echoes of the story Christ brought before Pilate. Perhaps also the Druids represent for Elgar something of his own unofficial faith and his claim for Catholics to feel part of the nation. Evidence for Elgar's feelings on this matter, for example, might be found in his opposition to home rule in Ireland.

It is perhaps significant that the 'Britain' motif is noticeably used against pessimistic passages in Ackworth's text with two significant exceptions.[6] The first is the setting for the Arch-Druid's false prophecy in Scene II.23, 'Go forth, O King, to conquer'; the second is Elgar's setting of Ackworth's imperialistic epilogue (Scene VI.61), which is the only instance of the 'Britain' theme being used in a triumphalist sense. Elgar's interest in cryptograms and musical humour has been widely acknowledged; these are most notably celebrated in his *Enigma Variations* written the year after *Caractacus*. It is arguable therefore that Elgar is being both cryptic and ironic here. The use of this motif signals not jubilation and conquest but doubt and deceit, the empty pomposity of pageantry. Furthermore, as Anderson points out, the music accompanying Caractacus's entry before the Arch-Druid's prophecy is inverted, as if to mock the promised triumph (Kennedy, 1993, p. 74).

Elgar's own attachment to the landscape connects with a sense of patriotism very far from the jingoistic militarism with which he might be associated. Much of the action is set in the Malvern Hills, a landscape of much personal significance and affection for Elgar. The choice of subject, for example, appears to have originated with Elgar's mother with whom Elgar had a particularly strong relationship (Northrop Moore, 1984, p. 225). Importantly, Elgar reserves two of his most personal touches as motifs for the Britons and the Malverns. The composer told Herbert Thompson, who was to write the programme notes, that the forest sounds were 'written in our own woods'. Similarly, of Scene V, which takes place by the River Severn, Elgar said of his music 'I made this on the banks' (Anderson, 1990, p. 35). The British motif consists of descending steps, with a rise in the middle, repeated in sequence. Descending steps and sequential repetition are most characteristic of Elgar's music language and can be linked to his habit of composing whist walking in the countryside. In order to write the work Elgar was recalled by a friend having 'walked all over the ground, he tramped over the hills and went along the Druid path from end to end'. The motive used to describe the scene where Caractacus and his group are entering the British Camp is derived from a tune written at the age of 10. The melody which sets the mood for Scene III, 'The forest near the Severn' derives from a melody inspired by the woodlands around the cottage at Birchwood which Elgar rented whilst writing the music. He wrote to August Jaeger (August 1898): 'I made old Caractacus stop as if broken down and choke and say "woodlands" again because I'm so madly devoted to my woods' (Kennedy, 1993, p. 65).

*Caractacus* fuses British history and the portrayal of a specific British landscape with a range of classical and Christian references and symbolism. As such it is clearly in line both with the kind of nationalist pastoral developed in other

European countries during the nineteenth century and the characteristics of the English pastoral. This is in terms of the English pastoral's particularly rustic and overtly political qualities. These may be traced to the beginnings of English pastoral in Spencer's *Shepherd's Calender* and to the thinly veiled political satire represented by John Gay's *Beggar's Opera* (1728). Unlike Vaughan Williams, Elgar was not part of a school of nationalist composition, its aims and aesthetics consciously articulated in writings and lectures. However, Elgar's interest in history and literature would suggest that the pastoral qualities of *Caractacus* were more than coincidental. Elgar quotes Virgil at the head of the score of *The Dream of Gerontius* and on early proofs of the vocal score of *Caractacus* (Anderson, 1990, p. 35). Like Virgil, Elgar was in an ambiguous position as an ambitious outsider. His personal ambition and his anxiety to ingratiate himself with royalty and aristocracy have been dealt with by Merion Hughes in his study of Elgar's insecurity (Hughes, 1989). Elgar wished to dedicate *Caractacus* to Queen Victoria and would not have wished to prejudice his chance of gaining that most prestigious of accolades, royal assent.

## English pastoral: an herbaceous border

As a provincial Catholic from a lower-class background the terrain of Elgar's conservatism could never have been simple. The example of *Caractacus* demonstrates the complex of English, British, European, pagan and Christian, archaic and modern ideas frequently found in musical depictions of English landscape. Yet Elgar is far from unique in this respect. Ralph Vaughan Williams, Elgar's successor as 'father figure' of British music, is both a contrast and a parallel. Vaughan Williams' musical nationalism fuses academic historicism, anthropological folklorism, and a mystical-mathematical sensibility with a paternalistic liberalism (Harrington, 1994; Revill, 1991). These are evident in, for example, his settings of Walt Whitman *Towards the Unknown Region* (1907) and the choral *Sea Symphony* (1909). These works explore liberal aesthetic *terra incognitae* of self-expression and self-discovery. Abstract as they are, they are no less grounded than Elgar's *Caractacus*; they resonate with a British history of polar exploration and seafaring. In this sense they are just as located as the opera *Hugh the Drover* (1924), set in the Cotswold village of Northleach. This opera itself has a clear liberal message, telling of the oppression of the village and the freedom of the road. For British composers attempting to create some form of national music in the period 1880–1940 the pastoral formed an ideal space unified by a set of symbolic structures and conventions. Into this territory were inserted competing and contrasting sets of ideas concerning the nature of the nation. Evidence of this might be found in work by the atheist socialist Holst *The Cloud Messenger* (1909–10), *First Choral Symphony* (1923–4); the Christian socialist Rutland Boughton *The Immortal Hour* (1912–13), *Bethlehem* (1915); in the Celtic romanticism of Arnold Bax *The Garden of Fand* (1913), *Tintagel* (1917); and the avant-garde modernism of Authur Bliss *A Colour Symphony* (1922), *Pastoral*: '*Lie strewn the White Flocks*' (1928), just as it may be found with the 'conservative' Elgar and the 'liberal' Vaughan Williams. In this respect the music of the 'English' musical renaissance demonstrates the manifold political perspectives of the pastoral in the late nineteenth and early twentieth centuries highlighted by Wiener, Matless and others.

The persuasive character of the pastoral rests on a fusion of artifice and naturalism, realism and fantasy. Shepherds talk of real places and actual events in the same breath that they discuss the activities of deities and mythological figures as if all these were part of the quotidian world. Characters alternately personify the human frailties of avarice, lust, deceit and the magical and super-human qualities of heroism, duty, music and poetry. Descriptions of an idealized land of plenty in which goats lie down to be milked and trees bend over to have their fruit picked seem to have equal status with texts advising on the tending of bees or the construction of a plough. The result is a powerful fusion of utopianism and realism that gives to the pastoral as a cultural genre its capability to naturalize ideology. In music, as in landscape design, theories of natural order justify aesthetics and together support a moral order and a theory of society. One may view this operating both in seventeenth century English masque and semi-opera and in the work of Elgar and Vaughan Williams. Utopian as the pastoral obviously is, it can only work if it is constantly interrupted by 'reality'. Like the carnival–lent dichotomy, it is a dialectic of mutual otherness whose eventual outcome is the re-establishment of a status quo. Fundamental to the English masque was the act of unmasking the participants at the end. Increasingly masques were interrupted by the anti-masque as members of the audience were sent on stage to dance. This destroys the illusion of the spectacle at the same time that their very participation formed a supporting legitimation to the drama itself. In nineteenth and twentieth century musical nationalism the combination of realism and fantasy is equally central to music's ideological rhetoric. To play an effective social role nationalist art needs to combine both abstraction and realism as characteristic elements of modernity. It needs to have populist appeal and reflect the 'real' circumstances of a people's history, culture and social condition. At the same time it needs to 'abstract' these from current circumstances and project an idealized path from past to future, a form of justification grounded in an abstract logic. In pastoral music, music's theoretical structures, its form and harmony combine with classical philosophy to provide an abstracted legitimation for social and political ideas. Complementary to this process, pastoral conventions provide the 'realistic' references to landscape and society. The musical pastoral seems to be a 'natural' adjunct to the production of national identity.

This essay has only hinted at the questions concerning the location of the musical renaissance in Britain. It is obvious that use of the term 'English' is itself political. The most obvious explanation for the use of this term is that both the renaissance and the writing of its history were generated within new London-based musical institutions, particularly the Royal College of Music and the ambitious group of new staff and students intent on creating a distinctive place for themselves against the pre-existing musical establishment. It is also true that identification of the musical renaissance with the pastoral was also political. Just as supporters of a national music tried to read complex works as a triumph for bucolic simplicity, so modernist critics endeavoured to dismiss such works as naively rustic. As leader of the nationalist school, Vaughan Williams was frequently a victim in this regard. When *Flos Campi*, an abstract choral war requiem, was first performed the naively rustic associations were emphasized by folk movement enthusiasts. Vaughan Williams claimed to the contrary that 'this work has nothing whatsoever to do with buttercups and daisies'. Similarly Vaughan Williams' *Pastoral Symphony* (No. 2) was rubbished by the composer and critic Constant Lambert as being from the 'cow pat' school. Yet this work was based on Vaughan Williams' experience in the Ambulance Corps during World War I and is set in France. Here the pastoral can itself be seen as subject to a process of spatial ordering whereby particular music is located in the countryside and therefore dismissed as rural and irrelevant.

It is evident that the 'English' musical pastoral is a vehicle for diverse forms of aesthetic and social ordering. It would be easy therefore to interpret the pastoral as some form of heterotopia, a meeting point and a shared space. However, it seems to me that it is always socially exclusionary even when its conventions are used for populist ends. One only has to consider the inherently middle-class assumptions made by composers dedicated to creating an 'authentically' popular form of 'English' opera or music drama for an example of this (Lambert, 1934, pp. 100–15). The moral and political ambivalence of the pastoral suggests perhaps the marginal location of pastoral landscapes. In terms of the classical pastoral this is exemplified by the idea of the garden not so much as a clearly defined refuge but as a contested space between country and city, civilization and wilderness. It is significant therefore that Rob Stradling (1998) has located the country around the River Severn (the setting for *Caractacus*), Shropshire, Hereford, Worcestershire and Gloucestershire rather than Surrey or Sussex as the key landscape for composers of the musical renaissance. It is perhaps within the borderland between upland and lowland, England and Wales, Celtic 'myth' and English 'history' that contesting versions of 'Englishness' find the space to define themselves most fully.

## Conclusion

Cultural genres, like the pastoral, which have a longstanding historical currency can be extremely useful for a cultural geography of music. The pastoral, for example, can help us examine how relationships to land, landscape and society change over time and how specific works of art at particular times both carry and transform cultural meanings carried over from the past and brought in from elsewhere. At the same time this can help us understand how particular art works, cultural practices and artefacts can themselves be read as formed from complex and sometimes contradictory sets of meanings, symbols and readings. The example of *Caractacus* shows how Elgar used a complex fusion of Celtic, classical and Christian imagery combined with nostalgic depictions of the Malvern Hills to produce a piece of modern music that is part of a progressive movement in British music. The juxtaposition of symbols and meanings does not necessarily produce a unified image. One only has to remember the divergent views of Ackworth and Elgar over the role played by the Druids, their relationship to the Romans and to ideas of Britishness and Christianity.

The lack of an obvious content plane as a location for musical meaning in contrast to words on the page in literature, or marks on canvas in painting, is important here. A geography of musical meaning which includes but also goes beyond the study of lyrics and audience reception must move outside the individual work itself. It should trace the multiplicity of ways in which the symbols and practices of music gain meaning outside the work of music in other artistic forms and other spheres whether these are economic, social or political (Revill, 1998). In the case of pastoral music these might include poetry, politics, landscape design, or even (in the case of Elgar) walking and cycling. More generally it shows us that the cultural geography of art works themselves (and not just music) cannot be represented by a flat plane but must be thought of as a multiplicity of traces, circulations and structures which are deeply contoured, layered and intricately interwoven. Because of this, a cultural geography of music should be as historically sensitive as it is geographically aware. This may seem both banal and self-evident, but it is more difficult to achieve than it is to say. A geography of music studied through the cultural lens of the pastoral is just one way of enabling such an approach.

## Notes

1. The first works of the English pastoral were probably *The Shepherd's Calender* (1579) by Edmund Spencer (which carries a very early reference in English to the term 'landscape'), and *Arcadia* (1580) by Sir Philip Sidney. Raymond Williams (1973, p. 22) believes that it is possible to follow a direct line in English literature from Virgil through Spencer *et al.* to the work of Alexander Pope in which the eclogue becomes a disguise of simplicity and a justification for 'naturalized' styles of landscaping.
2. Most importantly it was also the form of 'classical' music with a widespread popular appeal in many European countries. At the very least on the European mainland opera had an important constituency amongst the bourgeoisie who were also central to the nationalist movements of the middle and later years of the nineteenth century (Hobsbawm, 1975, pp. 21–40).
3. The English semi-opera was itself the product of political circumstances. During the period of the English Commonwealth theatrical entertainment was banned by the Puritan authorities as decadent. In contrast, music was actively encouraged as a diversion and entertainment. It was not Davenant's idea to create opera, merely to set his plays to music and thereby overcome the prohibition (Harris, 1980, pp. 110–13).
4. It is the London metropolitan institutional basis of the musical renaissance which as much as anything has resulted in the epithet 'English', however partial and misleading this title may be.
5. It is thought that the lack of a British operatic tradition severely restricted opportunities for the development of a high standard of professional musicianship and

hindered the growth of a body of trained musicians able to play complex music. It reduced the possibilities to cultivate an 'educated' audience for serious new music as well as denying composers the opportunity to compose works within an important musical genre. The result was the charge made against Britain in the nineteenth century that it was 'a land without music' (Ehrlich, 1989; Beedell, 1992; Stradling and Hughes, 1993; Blake, 1997).

6. This section relies heavily on the work of Alan Bartley who has very kindly allowed me to quote from his essay 'Elgar's *Caractacus*: an imperialist tract, or a pastoral cantata', term paper for MA in Music, History and Culture, special subject Music, Nationalism and National Identity, 16 March 1999.

## *References*

Anderson R. (1990) *Elgar in Manuscript* (British Library, London).

Barrell J. (1972) *The Idea of Landscape and the Sense of Place* (Cambridge University Press, Cambridge).

Beedell A.V. (1992) *The Decline of the English Musician 1788–1888* (Clarendon Press, Oxford).

Bermingham A. (1986) *Landscape and Ideology: The English Rustic Tradition, 1740–1860* (University of California Press, Berkeley).

Blake A. (1997) *The Land without Music: Music, Culture and Society in Twentieth-century Britain* (Manchester University Press, Manchester).

Burke J. (1983) *Musical Landscapes* (Webb & Bower, Exeter).

Coates P. (1998) *Nature: Western Attitudes Since Ancient Times* (Polity Press, Cambridge).

Colls R. and Dodd P. (eds) (1986) *Englishness: Politics and Culture 1880–1920* (Croom Helm, London).

Crump J. (1986) 'The Identity of English Music: The Reception of Elgar 1898–1935', in Colls R. and Dodd P. (eds) *Englishness: Politics and Culture 1880–1920* (Croom Helm, London).

Dahlhaus C. (1989) *Nineteenth-Century Music* (University of California Press, Berkeley).

Daniels S.D. (1987) 'Marxism, Culture and the Duplicity of Landscape', in Peet R. and Thrift N. (eds) *New Models in Geography, Volume II* (Unwin Hyman, London), 196–220.

Daniels S.D. (1991) 'Envisioning England', *Journal of Historical Geography* 17: 95–9.

Daniels S.D. (1993) *Fields of Vision: Landscape Imagery and National Identity in England and the United States* (Cambridge University Press, Cambridge).

Daniels S.D. and Seymour S. (1990) 'Landscape Design and the Idea of Improvement 1730–1914', in Dodgeshon R.A. and Butlin R.A. (eds) *An Historical Geography of England and Wales* (Academic Press, London).

Dickinson A.E.F. (1995) *Holst's Music, a guide* (Thames Publishing, London).

Ehrlich C. (1985) *The Music Profession in Britain since the Eighteenth Century* (Oxford University Press, Oxford).

Ehrlich C. (1989) *Harmonious Alliances – A History of the Performing Rights Society* (Oxford University Press, Oxford).

Empson W. (1968) *Some Versions of Pastoral* (Chatto & Windus, London).

Glacken C. (1967) *Traces on the Rhodian Shore: Nature and Culture in Western Thought from Ancient Times to the End of the Eighteenth Century* (University of California Press, Berkeley).

Harman A. and Milner A. (1988) *Late Renaissance and Baroque Music* (Barrie & Jenkins, London).

Harrington N. (1994) *English Art and Modernism 1900–1939* (Yale University Press, Yale).

Harris E.T. (1980) *Handel and the Pastoral Tradition* (Oxford University Press, Oxford).

Hobsbawm E.J. (1975) *The Age of Capital 1848–75* (Weidenfeld & Nicolson, London).

Holman P. (1992) 'Music for the Stage I: Before the Civil War', in Spink I. (ed.) *The Blackwell History of Music in Britain: The Seventeenth Century* (Blackwell, Oxford).

Howes F. (1966) *The English Musical Renaissance* (Secker & Warburg, London).

Hughes M. (1989) 'The Duc d'Elgar: Making a Composer Gentleman', in Norris C. (ed.) *Music and the Politics of Culture* (Lawrence & Wishart, London).

Hurd M. (1993) *Rutland Boughton and the Glastonbury Festivals* (Clarendon Press, Oxford).

James J. (1995) *The Music of the Spheres: Music, Science and the Natural Order of the Universe* (Abacus, London).

Kennedy M. (1964) *The Works of Ralph Vaughan Williams* (Oxford University Press, Oxford).

Kennedy M. (1987) *Adrian Boult* (Macmillan, London).

Kennedy M. (1993) *Portrait of Elgar* (Clarendon Paperbacks, Oxford).

Lambert C. (1934) *Music Ho! A Study of Music in Decline* (Penguin edn 1948, London).

Laurie M. (1992) 'Music for the Stage II: From 1650', in Spink I. (ed.) *The Blackwell History of Music in Britain: The Seventeenth Century* (Blackwell, Oxford).

Leyshon A., Matless D. and Revill G. (1998) *The Place of Music* (Guilford/Longman, London).

Lyne A.M. (1983) 'Introduction', in Virgil *The Eclogues, The Georgics* (translated by C. Day Lewis) (Oxford University Press, Oxford).

McKay G. (1996) *Senseless Acts of Beauty: Cultures of Resistance since the Sixties* (Verso, London).

Marsh J. (1982) *Back to the Land: The Pastoral Impulse in Victorian England from 1880 to 1914* (Quartet Books, London).

Matless D. (1990a) 'Ages of English Design: Preservation, Modernism and Tales of their History, 1926–39', *Journal of Design History* III: 203–12.

Matless D. (1990b) 'Definitions of England', *Built Environment* XVI: 179–91.

Matless D. (1991) 'Nature, the Modern and the Mystic: Tales from Early Twentieth Century Geography', *Transactions of the Institute of British Geographers* N.S.16.: 272–86.

Matless D. (1995) 'The art of right living: landscape and citizenship 1918–39', in Pile S. and Thrift N. *Mapping the Subject* (Routledge, London).

Matless D. (1997) 'Moral Geographies of English Landscape', *Landscape Research* XXII: 141–56.

Matless D. (1999) *Landscape and Englishness* (Reaction Books, London).

Mellers W. (1987) *The Masks of Orpheus: Seven Stages in the Story of European Music* (Manchester University Press, Manchester).

Mellers W. (1988) *Romanticism and the Twentieth Century* (Barrie & Jenkins, London).

Northrop Moore J. (1984) *Edward Elgar: A Creative Life* (Oxford University Press, Oxford).

Patterson A. (1988) *Pastoral and Ideology: Virgil to Valery* (Clarendon Press, Oxford).

Platt R. (1990) 'Theatre Music I', in Johnstone H.D. and Fiske R. (eds) *The Blackwell History of Music in Britain: The Eighteenth Century* (Blackwell, Oxford).

Price C. 'Music, Style and Society', in Price C. (ed.) *Music and Society: The Early Baroque Era, from the late 16th century to the 1660s* (Macmillan, London).

Revill G. (1991) 'The Lark Ascending: Monument to a Radical Pastoral', *Landscape Research* 16 (2): 25–30.

Revill G. (1998) 'Samuel Coleridge-Taylor's Geography of Disappointment: Hybridity, Identity and Networks of Musical Meaning', in Leyshon A., Matless D. and Revill G. *The Place of Music* (Guilford/Longman, London).

Revill G. (forthcoming) 'Soundscapes? History, Culture and the Spaces of Musical Meaning', *Environment and Planning D: Society and Space.*

Rooley A. 'Introduction' *Music from "Cupid and Death"* (Deutsche Harmonia Mundi 5472-77428-2).

Rubinstein W.D. (1993) *Capitalism, Culture and Deline in Britain 1750–1990* (Routledge, London).

Sakolsky R. and Wei-Han Ho F. (eds) (1995) *Sounding Off! Music as Subversion/ Resistance/Revolution* (Autonomedia, New York).

Schama S. (1995) Landscape and Memory (Simon and Schuster, New York).

Smith A.D. (1991) *National Identity* (Penguin, London).

Smith A.D. (1997) 'The "Golden Age" and National Renewal', in Hosking G. and Schopflin, G. *Myths and Nationhood* (Hurst & Company, London).

Stradling R. (1998) 'England's Glory: Sensibilities of Place in English Music 1900–1950', in Leyshon A., Matless D. and Revill G. *The Place of Music* (Guilford/ Longman, London).

Stradling R. and Hughes M. (1993) *The English Musical Renaissance 1860–1940* (Routledge, London).

Thomas K. (1984) *Man and the Natural World: Changing Attitudes in England 1500– 1800* (Penguin, London).

Trend M. (1985) *The Music Makers: Heirs and Rebels of the English Musical Renaissance, Edward Elgar to Benjamin Britten* (Weidenfeld & Nicolson, London).

Vaughan Williams R. (1934) *National Music and other Essays* (Oxford University Press, Oxford).

Vaughan Williams R. (1996) *National Music and other Essays* (Clarendon, Oxford).

Wiener M.J. (1981) *English Culture and the Decline of the Industrial Spiret: 1850–1980* (Cambridge University Press, Cambridge).

Williams R. (1973) *The Country and the City* (Chatto & Windus, London).

Williamson T. (1995) *Polite Landscapes: Gardens and Society in Eighteenth-Century England* (Allan Sutton, Stroud, Glos.).

# PART III

# Culture and political economy

London. Homeless man sells 'The Big Issue' next to a cash machine: he keeps half the sale price of the magazine. 1995. © Chris Steele-Perkins/Magnum Photos

# Introduction

*Ian Cook*

Since the seventeenth century, when philosophical discussions of what we understand today as 'political economy' began to emerge, there have been massive changes in the way that the global economy works – i.e. in what there is to study – and in the ways in which academics, politicians and other commentators have framed their understandings of this – i.e. how they study, make sense of, and represent (bits of) 'it'. In its 'traditional' form, political economy involved 'the study of . . . economic surplus, however defined, as it is produced, distributed and accumulated within a class divided society' (Barnes, 1995, p. 425). These issues were fundamental to (framings of the) relatively stable political economic processes that characterized the period between the 1870s and the fall of Bretton Woods in the late 1960s (Thrift and Olds, 1996; Thrift, 1989).

But late twentieth century political economy, in practice and in theory, has become a more fragmentary, unstable and heterogeneous beast. For economic and cultural geographers, the bringing together of 'culture' and 'political economy' has been one important way to begin to deal with these changes. But this has been no easy task, particularly because they have so often treated the relationships between 'the cultural' and 'the economic' as either/or relationships (Crang, 1997). Discussions of 'the cultural' and 'the economic' as more or less inseparable have, however, been central to the recent emergence of a 'new economic geography'. Its textbooks and readers are perhaps the first places to go for detailed reviews and representations of this emerging and heterogeneous literature (e.g. Bryson *et al.*, 1999; Lee and Wills, 1997). But, to set the scene for the five chapters which follow, thumbnail sketches of these recent changes and the heterogeneous ways in which new economic geographers are studying them may be useful. These will allow us to put the chapters in a wider context.

In the 'new economic geography' literature, perhaps the major 'real world' transformation said to have provoked studies combining the 'cultural' and 'the economic' has been the decline of 'Fordism' and the rise of 'post-Fordism' over the past twenty or so years. The Fordist regime of mass production and mass consumption is said to have given way to a more flexible and diverse regime of 'post-Fordist' production and consumption. Here, fewer people are involved in designing, making and selling tangible things and more are involved in designing, making and selling less tangible impressions, representations and connections; the norm of steady employment for a predominantly male workforce has given way to one of unsteady employment for a much more feminized workforce; mass production has been exported to poorer countries, producing a new international division of labour; the inequalities in all of these places are increasing, but the 'underclasses' thereby (re)produced are more diverse than ever; and, more generally, 'traditional' distinctions between what counts as 'economic' or 'cultural' make less and less sense, particularly with the rise of the so-called 'culture industries' (see Allen and Massey, 1988; Amin, 1994; Corbridge *et al.*, 1994; Dicken, 1992; Leyshon and Thrift, 1997; McDowell, 1997). The extent, significance and exact nature of these changes – e.g. how 'new' these post-Fordist trends are, and the extent to which post-Fordism can be said to have replaced Fordism as the dominant mode of production – have been the subject of much debate. Indeed, the meanings of even the most fundamental terms in these debates – i.e. 'culture' and 'economy' – along with how they fit together, and whether they can be prized apart in the first place, have also become subjects in these debates, as the chapters which follow will show.

These debates have been provoked not only by changes in the economy. Rather, they have also been shaped by the ways that the 'cultural turn' (as discussed in the first part of the book) has affected theories and research practices across social and cultural studies. So what effect has this 'turn' had on the theories and practices of economic geography? In a nutshell, cultural and economic researchers from a variety of disciplines sharing 'traditional' political economic concerns have begun to explore them:

- using more 'interpretive' methodologies (e.g. corporate interviewing and discourse analysis);
- in previously neglected and/or 'new' areas of 'economic' activity (e.g. consumption and management consultancy);
- to highlight more 'qualitative' aspects of this activity (e.g. meanings, identities, knowledges, moralities and relationships of trust);
- placing findings from these studies within wider social, cultural, political, economic, etc. relations (e.g. in local/global and actor network relationships); and
- situating this within a heterogeneous and hybrid body of theoretical work (e.g. in poststructuralism and postmodernism).

In the process, then, the 'traditional' approaches and concerns of political economy have not necessarily been *replaced* by 'new' ones. As the following chapters illustrate, they are usually still in there somewhere: usually centre stage, usually bounced off, usually reworked, and usually returned to in a different form.

The first chapter is by Andrew Sayer, a sociologist whose work on social theory and methodology has had an important and sustained influence on the work of human geographers. One of his major concerns has been to assess critically what the 'cultural turn' has meant for political economic analyses. Here, he argues, this 'turn' has largely been from what he calls a 'vulgar materialism' to a 'vulgar culturalism'. Contemporary cultural studies, he believes, takes an approach to the economy in which issues of 'style and taste' are prioritized, in which nothing concrete can be said about anything, and in which everything is relative. So, he argues, this 'cultural turn' plays right into the hands of a neo-liberal world view because it often involves an 'exaggerated wariness of normative judgements', prioritizes the 'vanity' and 'self-interest' of consumers, and undermines a fundamental strength of more 'traditional' political economic work: i.e. its ability to separate right from wrong, just from unjust, truth from lies. Critical academics, he argues, need to be able to take principled moral political standpoints in a world which needs them perhaps more than ever.

The second chapter is by Linda McDowell, a geographer who has been at the forefront of feminist critiques of economic geography. She picks up Sayer's concern with principled moral political standpoints but argues for a political economy which not only focuses on 'traditional' class relations but, more widely, on 'the production of difference' by gender, ethnicity, sexuality and so on. Thus, she offers a constructive critique of 'traditional' political economic theory by drawing on the literature on the 'politics of difference' which has been central to the 'cultural turn' in geography and elsewhere. After thinking through the implications of these debates, she asks what they might mean for the political/academic practice of cultural/economic geographers who are still driven by 'traditional' political economy's concerns with social and economic justice. How might these revised moral political standpoints work in practice?

The third chapter is by Daniel Miller, an anthropologist whose wide-ranging work on material cultures, political economic theory, and consumption has become increasingly influential in human geography. He argues here that changes in the global economy and the ways in which (some) academics study it have become intimately linked through the discipline and practice of Economics. What this has meant is that (political) economic theory no longer *reflects* the ways that 'real' economies work, but constitutes a 'virtual economy' (e.g. of the perfect 'free market') which is put into practice in such a way as to *shape* the world in its own image. Economics is a powerful set of discourses put into practice through, for instance, the structural adjustment policies of powerful organizations such as the World Bank and International Monetary Fund. And central to this process of 'virtualism' is, he argues, an equally 'virtual consumer' who will reap the benefits of these changes. Having said this, Miller goes on to argue that this process of 'virtualism' is more widespread – in food retailing, in educational auditing and, indeed, in some of the literature relevant to this part of the book. So, when reading intricate discussions of political economy, for instance, we perhaps need to ask ourselves whether we are reading about it in a 'virtual' and/or 'actually existing' form?

This question could easily be asked of our next authors, David Clarke and Marcus Doel. In recent years, they have played significant roles in the development of 'poststructural geographies' (Doel, 1999). In a highly abstract and sometimes surreal chapter, they attempt to work through the 'false distinction' of 'culture' and 'economy' in theorizations of the commodity stretching from nineteenth century Marxism to late twentieth century French poststructuralism. In particular, they concentrate on Marx's distinction between the 'use-' and 'exchange-value' of a commodity, which is a distinction between its singular material/functional value and its comparative cultural/signifying value. However, they argue that, given that commodities can be so difficult to isolate from one another – to see them alone and just in terms of their 'intrinsic function' – this is where 'the critique of political economy is at its shakiest'. So, building on the kinds of questions posed by McDowell, they ask how could any future/utopian society based on distributive justice be put in place when its necessary distinction between use- and exchange-value cannot exist? If this is the case, they ask, what alternatives can be suggested?

Finally, there is the chapter by James Sidaway, a political geographer who is best known for his work on critical geopolitics, geographical knowledges and, more recently, on transnational trading and linguistic communities. In his chapter, he examines the formation and meanings of transnational trading blocs such as the European Union and the North American Free Trade Association. He argues that these are at once 'political', 'economic' and 'cultural' processes from which member states have gained certain forms of legitimacy and are, in a sense, created. And, to echo Miller's arguments, in a sense they are 'created' through a form of 'virtualism' because of the economic, political and (latterly) cultural theories which member and other governments use to frame and implement these processes. Thus, he argues that the virtual and the 'actual' political economies of these trading blocs have become extremely difficult to disentangle and their 'undecidable' combinations are what characterize them.

This introduction, of course, is by no means the be-all and end-all of the cultural turn in economic geography, or of its representation in the five chapters of this section of the book. But, together, these chapters should give a sense of the wide variety of methodological, theoretical and political approaches which are part of the mix in this important area of human geographical enquiry. We hope that these chapters work well together, provide sustained arguments from quite different perspectives, and combine themes from the 'new economic geography' in distinct, overlapping and often contradictory ways. We hope that it won't be too difficult to imagine the debates that these authors would have with each other – there is plenty more between them than indicated here – and the challenges that they would set you in your own research, writing and other political activities.

## References

Allen, J. and Massey, D. (eds) (1988) *The Economy in Question*. Milton Keynes: Open University Press.

Amin, A. (ed.) (1994) *Post-fordism: A Reader*. Cambridge: Polity.

Barnes, T. (1995) 'Political Economy I: "The culture, stupid"' *Progress in Human Geography* 19(3) pp. 423–31.

Bryson, J., Henry, N., Keeble, D. and Martin, R. (eds) (1999) *The Economic Geography Reader*. London: Arnold.

Corbridge, S., Martin, R. and Thrift, N. (eds) (1994) *Money, Power and Space*. Oxford: Blackwell.

Crang, P. (1997) 'Introduction: cultural turns and the (re)constitution of economic geography', pp. 3–15 in Lee, R. and Wills, J. (see below).

Dicken, P. (1992) *Global Shift: The Internationalisation of Economic Activity*. London: Paul Chapman.

Doel, M. (1999) *Poststructural Geographies*. Edinburgh: Edinburgh University Press.

Lee, R. and Wills, J. (eds) (1997) *Geographies of Economies*. London: Arnold.

Leyshon, A. and Thrift, N. (1997) *MoneySpace: Geographies of Monetary Transformation*. London: Routledge.

McDowell, L. (1997) *Capital Culture: Gender at work in the city*. Oxford: Blackwell.

Sayer, A. (1992) *Method in Social Science: A realist approach* (2nd ed.). London: Routledge.

Thrift, N. (1989) 'The geography of international economic disorder' in Johnston, R. and Taylor, P. (eds.) *A World in Crisis?* (2nd ed.) Oxford: Blackwell, pp. 16–78.

Thrift, N. and Olds, K. (1996) 'Refiguring the economic in economic geography' *Progress in Human Geography* 20(3) pp. 311–37.

# Critical and uncritical cultural turns

*Andrew Sayer*

## Introduction

A turn to culture was long overdue in social science. In the period when the new left was at the height of its influence, culture was frequently reduced to ideology, a mere reflection of material, especially economic, circumstances and its content ignored. Despite the opposition of thinkers like Raymond Williams, economic determinism and vulgar materialism were common. Few registered the necessary hermeneutic dimension of social science, since few saw that meaning could be constitutive and not merely reflective of social practice. The shift to a treatment of culture centring on signifying practices (Hall, 1997) and meaning as performative as well as denotative required its meanings to be interpreted. The cultural turn also helped to highlight the ways in which many forms of oppression in society had a cultural rather than an economic character, depending 'in the last instance' on ascribed characteristics of social groups and cultural meanings. This has freed us from treatments of patriarchy, racism and other forms of non-class oppression as secondary to or derivative of capital. There are more specific gains too. Consumption was viewed in one-sidedly negative terms in Marxist and much cultural materialist writing as passive and individualistic but recent research has shown how it is often active and shared. Relatedly there is also a much more open and less elitist approach to popular culture.

However, at times it seems as if there has been a shift from vulgar materialism to a 'vulgar culturalism' which is as dismissive of or reductive about economy as vulgar materialism was about culture. If cultural studies is concerned with signifying practices, then anything that society registers can be seen to have a cultural dimension since it can signify something. Yet, it does not follow from this that there are no other dimensions, so that social life is reducible to texts or text-like objects, whose signifying qualities are the only aspects that matter. Where culture is defined as 'a whole way of life' the scope for culturalist imperialism is even greater, for it allows one to pass off selective accounts focusing on signifying practices as if they were exhaustive. Similar sleights of hand allowing the economic to be marginalized were encouraged by arguments that culture and economy are no longer distinguishable (e.g. Hall, 1988; Lash, 1990; Jameson, 1990). (If culture and economy were really no longer distinguishable we could use the two terms interchangeably, for example, renaming cultural studies courses economic studies, and speaking of the economic turn instead of the cultural turn, without anyone being misled! The fact that even advocates of this position have had to continue using the terms in standard ways shows that the distinction is still viable and needed.)

There is no reason why a focus on cultural dimensions of social phenomena cannot be combined with analysis of other dimensions. It's therefore a pity that the turn was accompanied by either a wholesale marginalization of economic issues and theory or else a 'dumbing down' of economic analysis to the level of token references to globalization or potted versions of the simplistic, and diversionary grand narrative of Fordism and post-Fordism. Ironically this has happened just at the time when neo-liberal capitalism has been in the ascendancy, so that it has had to face remarkably little opposition from the more avant-garde reaches of social theory. Even Stuart Hall, one of the main driving forces behind the cultural turn, and one who once argued that culture and economy had become indistinguishable, has recently acknowledged that the cultural turn has actually involved a turn away from economy (Hall, 1996):

> What has resulted from the abandonment of this deterministic economism has been, not alternative ways of thinking questions about the economic relations and their effects, as the 'conditions of existence' of other practices, inserting them in a 'decentred' or dislocated way into our explanatory paradigms, but instead a massive, gigantic and eloquent *disavowal*. As if, since the economic, in its broadest sense, does *not*, as it was once supposed to do, 'determine' the real movement of history 'in the last instance', it does not exist at all! (p. 258, emphasis in original)

I shall argue that this disavowal has led to a cultural turn which is in crucial respects uncritical of its object, not only because it ignores or marginalizes economic matters and neo-liberal hegemony, but because its treatment of culture is nevertheless highly compatible with a neo-liberal world view.[1] I shall approach this via discussions of the role of cultural values, particularly moral-political and aesthetic values in contemporary society, and of the way in which cultural studies treats these. In doing so, I shall draw upon some modernist critiques of the cultural aspects of capitalism, some of which date back to the eighteenth and nineteenth centuries. Though old and often forgotten, in my view they still offer crucial critical insights into contemporary society. By way of illustration, I shall then turn to the recent work of Pierre Bourdieu and his influential economic analysis of culture and its 'soft forms of domination'.

However, first I want to note some other tendencies which have blunted the critical edge of recent cultural studies, namely 'ethical disidentification' (Connor, 1993) and relativism. The former term refers to a reluctance to acknowledge that behaviour can be influenced by values rather than mere convention or the will-to-power, and by a refusal to indicate any particular normative standpoint regarding social practices. This gives rise to a relativistic 'god's-eye view' which appears simultaneously to be both critical and uncritical – critical in challenging the view that behaviour can be based on values by presenting it as based on power, but uncritical in that if things cannot be otherwise there cannot be said to be a problem. Thus if truth is always a product of 'regimes of truth', implying some kind of authoritarian and instrumental imposition, then while that sounds very critical of received views, it undermines any basis for criticism, since the criticism invites itself to be dismissed as just another regime of truth. This 'crypto-normativity', as Habermas terms it in Foucault's early work, appeals to a kind of adolescent iconoclasm which assumes that the most cynical view of the world must automatically be the best (Habermas, 1987; see also Fraser, 1989; Hoy, 1986; McNay, 1994; Sayer, 1999). Relativism is the dual of dogmatic foundationalism: on discovering that there are no absolute foundations to ethical and other judgements, the relativist assumes that they therefore cannot be the subject of reasoned judgement about what is at least better or worse. As we shall see, exaggerated wariness of normative judgements makes social theory apologetic rather than critical.

Further tendencies in postmodernism which blunt its critical edge are its exaggerated suspicion of distinctions between appearance and substance, words and deeds, the apparent and the actual, and of treating the first term in each pair as less important than the second for fear of epistemological dogmatism or illegitimate 'normativity'. Now appearances need be no less real than substance, but it does not follow that they are necessarily equally important. Speaking or writing are themselves a kind of deed, but just as the fact that all sheep are animals doesn't mean all animals are sheep, this doesn't mean that there is no difference between words and deeds in the everyday sense, or that words and deeds cannot belie one another. Nor can anyone avoid a distinction between the actual and the apparent, for even to oppose it involves saying that while there may apparently be such a distinction, actually there isn't (a performative contradiction). To explain social life we have to distinguish what is only apparently the case from what is the case, and fallible though such claims must always be, in making them we are obliged to be critical. We cannot avoid making judgements of value, be they about epistemic, practical, aesthetic or moral-political matters, in everyday life or in social science. In order to do social research we have to decide whether and how far to agree with the accounts of what is going on offered by those whom we study. If forms of deceit or illusion figure significantly in social practices, then not to point this out is to fail to explain what is going on. We therefore have to be critical in order to explain.

## Culture and values

Cultures include values among their signifying practices. These may involve judgements and sentiments regarding utility, aesthetics and moral-political matters

(Sayer, 1999a). Contemporary cultural studies' preoccupation with aesthetic values is evident in its focus on style and taste, indeed in the definition of its object of study as 'the stylization of life' (Featherstone, 1994). There is less interest in moral-political values. These concern how society should be organized, how others should be treated, responsibilities to others, relationships with other species and the environment, and so on. I wish to argue that one of the effects of the continued development of capitalism is that in some spheres of life – though not all – moral-political values have become increasingly aestheticized and behaviour has become more self-interested and instrumental, although the effects of this are not entirely negative. The prioritization of aesthetic values over moral-political values in recent cultural studies uncritically reflects these tendencies.

It has recently been argued that politics is not only about economic distribution but about recognition (Taylor, 1992; Fraser, 1995). People may be oppressed not only through lacking or being denied resources but through a denial of recognition; individuals or groups may be excluded from 'participating on a par with others in social interaction' (Fraser, 1999). While this highlights an aspect of what might be termed cultural politics which is of growing importance, such a politics is by no means exhausted by issues of identity and recognition. Both the politics of distribution and the politics of recognition involve moral-political values and the latter are not only about recognition. Thus, as Szerszynski argues by reference to the case of anti-road protests, while these often involve minority subcultures and make considerable and innovative use of theatre and spectacle, they are not primarily about recognition, but about changing the cultural values of the entire society (Szerszynski, 1999). They are trying to make more environmentally sensitive values universal. Indeed, this immanently universalizing property is characteristic of most moral-political values (Benhabib, 1992).

The scope and nature of moral-political deliberation is strongly influenced by the changing organization of society. As capitalism has developed, economic activity becomes dominated by systems which, despite being always ultimately dependent on the behaviour of individuals and organizations, develop a logic, autonomy and momentum of their own to a significant extent, through the interplay of unintended consequences and anonymous interdependencies. As Habermas puts it, system comes to colonize and dominate lifeworld (Habermas, 1990), and as Marx put it, economic processes increasingly work 'behind our backs'. What might otherwise be moral-political issues regarding our economic rights and responsibilities to others become matters of fitting in with the system. Thus an unemployed person might, on moral-political grounds, want a job building houses for the homeless, but she will only get one if there happen to be profitable opportunities for doing so. To get a job she has to find a niche in the prevailing economic systems and conform to their rules. The expansion of capitalist forms of organization of economic activities depoliticizes them, making them the product of so-called 'market forces', subjecting them to what Marx called 'the icy waters of egotistical calculation'. In this way, as Habermas puts it, questions of validity are turned into questions of behaviour. Instead of asking what we ought to do, we ask how the system works and how we can fit into it.

Correspondingly, political economy has changed in the last two centuries from a subject which was continuous with moral philosophy and which engaged centrally with questions of validity, to one which attempts to expel normative questions and instead focuses on 'engineering' questions, as Sen calls them, about the operation of economic systems (Sen, 1987). This de-valuation or de-moralization of political economy and other social sciences has been accompanied by a de-rationalization of values, so that they are no longer seen as a matter for rational deliberation. As science pursues value-freedom, so values come to be seen as 'science-free', to use Bhaskar's term (Bhaskar, 1986). As people lost control over their economic lives, the competitive forces of global economy tended to reduce the purchase of normative standpoints in political economy, correspondingly making philosophical discourse on ethics appear irrelevant (Bauman, 1995, p. 211). Values regarding moral-political questions are supposed to be kept private. To raise them, even in avowedly 'critical' social science, is to invite incredulity that one has forgotten science and theory and descended into 'moralizing'. Hence the massive disproportion between the amount and sophistication of our explanatory efforts and our rare and usually private and embarrassed fumblings regarding normative questions about the organization of social life (Sayer and Storper, 1997). The divorce of normative and positive thinking, and the imbalance between them, are themselves symptomatic of the emergence and separation out of systems from lifeworld, and the domination of the latter by the former.

A critical analysis of contemporary society would have, among other things, to turn some of these questions of the behaviour of systems back into questions of validity. Thus, in relation to the current instability and crises of global financial markets, many are asking 'engineering' questions about how the system can be repaired. But one can also ask questions regarding the validity of allowing the fate of economies to be determined in this way rather than through a democratically regulated process. The pressure of markets themselves, backed by the fatalistic rhetoric of the World Bank and allied interests, compels countries to submit to the system, and effectively to disregard moral-political questions about economic organization.

Before making further critical points, I should perhaps pause to make several caveats. First, I do not want to imply that all of the problems discussed here are unique to capitalism (for example, some non-capitalist practices and forms of organization can also encourage callousness); I don't wish to revive the bad tradition of blaming each and every social ill on capitalism alone. Second, it should be noted that, from a normative point of view, we may regard some existing moral norms, for example, those associated with gender, as actually 'immoral'. Correspondingly, it should not be assumed that instrumentalization and even 'de-moraliation' are always bad – it depends what is being instrumentalized. It also depends on what the consequences are; instrumentalization may bring about goods which outweigh any negative consequences. This point is often associated with Adam Smith, though he was equally aware that this isn't always the case, indeed he was most concerned with the moral implications of the rise of commercial society (Smith, 1759). Third, and following from this, I want to acknowledge that the development of capitalism has brought huge benefits for many, and not only in material terms. At the same time as capitalism de-values some practices, its continued erosion of traditional relationships frees them up to be determined by actors through deliberation and choice rather than by convention, thereby allowing the possibility of a re-moralization in some cases.[2] The clearest example of this is in the area of gender relations. Here, the cultural turn has been much more effective in contributing to a critical analysis of contemporary society. The feminist critique of the inequities of patriarchal domination is hardly ever referred to as being an ethical matter, perhaps because, strangely, morality or morality-talk is seen as inevitably conservative (Sevenhuijsen, 1998). But since this literature is about how people treat and ought to treat one another it does of course have a moral dimension. This re-valuation or re-moralization is encouraged by similar processes to those which produce a de-valuation or de-moralization of social relations.

Having made these points I now want to focus on the more problematic cultural aspects of the rise of capitalist societies.

To find a place within contemporary economic systems requires instrumental behaviour: as sellers, whether of labour power or other goods and services, we have to appeal to the self-interest of buyers. Equally, as buyers, we expect sellers to appeal to our self-interest. This of course was Adam Smith's famous point about 'commercial society', but it has major implications for cultural values, many of which Smith himself examined and indeed criticized. The motivations are of self-interest, not moral-political concerns regarding our responsibilities. Thus, once people have to get private health insurance, the dominant consideration becomes not what would be a good and fair way of providing health services based on the assumption of equal moral worth of all, but what suits our individual self-interest, or our 'lifestyle' – and of course, our pocket. If we can get a lower premium than someone who has the misfortune to have a genetic susceptibility to a certain illness, then market logic will encourage us to do so rather than cross-subsidize them on moral grounds. Only exceptionally, and if it fits with sellers' self-interest (profitability) will sellers appeal to moral-political values, as in the case of The Body Shop. Otherwise it is the buyers' self-interest and vanity which is appealed to.

Notions of responsibility for others are attenuated as responsibility for oneself in the marketplace takes priority. Not surprisingly, an extreme pro-market ideology such as that of neo-liberalism encourages the displacement of compassion by self-interest. Markets also encourage dissembling, flattery and indeed lying – sometimes little lies, such as the carefully crafted exaggerations and omissions of the job application, or the hints of advertisers at newfound pleasure and sociability associated with the consumption of mundane products (Keat, 1999), but sometimes serious lies, such as those about the safety of products. These are to be expected in the culture of capitalism.[3]

Capitalism, as an amoral system, finds aesthetic values and vanity much easier to deal with than moral-political values. Even if we note that most shopping is done for others and hence allow that moral considerations come into shopping, advertisers will only appeal to them for non-moral reasons of profitability. Whereas aesthetic values can fit neatly with the appeal to self-interest, moral-political values are decidedly awkward, having a tendency to put judgements of intrinsic value before price.

As capitalism develops, the logic of exchange-value also gains ground, overriding concerns with use-value or 'existence value'. When football clubs are set up as businesses and quoted on the stock market, they become permanently up for sale and subject to the offers and wishes of those who are only interested in them as a source of money. The owners or shareholders need only take any notice of the interests of supporters – who value the team and the game itself rather than the financial reward – insofar as their actions might affect its market value. Moreover, we are continually encouraged to value things which are not currently marketed, as if they were commodified. Environmental economists who use 'contingent valuation' ask people how much they would be willing to pay to retain some environmental 'good' or how much they would accept in compensation for its loss (Foster, 1997). Thus, even environment comes to be valued in exchange-value terms, entering the cost calculus, to be weighed against the market demand and supply for millions of other commodities, as if it were commensurable with them. As Michael Walzer has argued, one of the major malaises of contemporary society is the encroachment of market criteria into practices and situations where they are inappropriate (Walzer, 1983). In parallel, market metaphors come to pervade more and more of our language, and are applied to situations where they are inappropriate, such as when we talk of selling someone an idea rather than reasoning with them. This is particularly evident in mainstream politics, where discourse increasingly follows the model of selling, so that political practice comes to emulate public choice theory. Instrumental attitudes are further invited by calling behaviours which properly are moral in character, such as how we treat others in conversation, as 'skills', as in 'communication skills'. Of course, dissembling and lying can be encouraged by other (non-capitalist) social forms too, but a critical cultural studies should surely be alert to whatever encourages it instead of paralyzing critical impulses by refusing any judgements as 'foundationalist', 'elitist' or 'paternalist'.

There is a tendency noted by theorists such as Rousseau, Smith and Marx, for identity in commercial society to become a matter of appearance which is divorced from the qualities a person actually has, a complicity shared by postmodernism (O'Neill, 1998). We are increasingly encouraged to judge people in aesthetic rather than moral terms – as 'cool' or as 'anoraks', etc., as if style were a measure of moral worth. As cultural theory notes, the most sophisticated advertising highlights the sign-value of commodities and how this may contribute to the construction of identities – as if identities were merely matters of appearance. If they are more than this and are related to what we are and do, then it has to be said that the sign-value aids pretension and vanity.[4] A cultural studies which reduces action to discourse and is afraid of such 'normativity' endorses these superficial identities by default. The manipulation of sign-values may be about recognition, particularly recognition of difference, and this appears to be what many students of culture find interesting, but it remains an open but crucial question whether the prestige or recognition is deserved (see Taylor, 1992).

In a widely cited passage, Adam Smith discussed the ways in which moral sentiments – or recognition – can get distorted by inequalities:

> This disposition to admire, and almost to worship, the rich and the powerful, and to despise, or, at least, to neglect, persons of poor and mean condition, ... is ... the great and most universal cause of the corruption of our moral sentiments. (Smith, 1759)[5]

We now know that there are other important sources of distortions of moral sentiments and judgements besides those of inequality of wealth. They concern gender, race, age, sexuality, cultural difference, style, beauty and ugliness, all of which are associated with double standards and undeserved kinds of recognition; what is acceptable in a man is unacceptable in a woman, what the beautiful can get away with the plain cannot, and so on.[6]

Moral evaluations of individuals and social groups and the recognition they are due discriminate between what they have through luck and what they have achieved through their own efforts, taking into account the circumstances in which they live. But market valuations of people make no such distinction. Those who, through no effort of their own, already have an advantage of good looks, get an additional advantage through a high market and social valuation. Married heterosexual couples with children enjoy 'the profit of normality', as Bourdieu puts it, though the profit may come in non-monetary form. Equally, those who, through no fault of their own, already have a disadvantage such as a disability, suffer the added disadvantage of a low market valuation, increased insurance premiums, etc. It is bad enough to get a penalty from nature, as it were, without having a further penalty slapped on top of it from society. Market evaluations reflect only scarcity and the buyers' self-interest. Thus the libertarian philosopher Robert Nozick justifies the high incomes of the talented and the beautiful purely on the grounds that that is what others freely choose to pay to enjoy them, regardless of whether they deserve such high incomes (Nozick, 1974). Presumably, by the same logic, the low incomes of the disadvantaged can be justified in that others freely choose not to enjoy or employ them. Interestingly, and in complete contrast to prevailing views today, John Stuart Mill considered that talents are a reward in themselves, not requiring supplementation from others in the form of payments or privileges.

Markets ignore these moral considerations not only because they encourage self-interested behaviour but because they are 'reason-blind' (Keat, 1997); they don't require buyers and sellers even to know each others' reasons let alone justify themselves to each other. (In many cases of course, the buyers and sellers, or rather producers and consumers, remain anonymous.) Even where firms conduct market research to discover buyers' reasons they do so only in order to learn how to increase sales, not because they are interested in the reasons themselves and want to evaluate them, as one might evaluate someone's political beliefs. Similarly, as mainstream politics becomes more like a market, it treats voters increasingly like consumers whose votes might be attracted rather than as participants in political argument. Likewise capitalism is only interested in the difference between the appearance of worth and real worth if it affects profitability. For example, whether someone is famous merely for being famous or for having done something exceptional does not matter. A critical cultural studies would challenge this elision, but those students of culture who have learnt to reject distinctions between appearances and the real, words and deeds (despite making such distinctions regularly in everyday life), cannot challenge it for they have undermined any critical standpoint. As O'Neill (1995) points out, a man who claimed to be a feminist would understandably be regarded with suspicion, and his deeds would be scrutinized to see if they contradicted his words. But if we misguidedly refuse distinctions between appearances and substance, the apparent and the real, words and deeds, as involving some kind of epistemological authoritarianism, then no criticism can be made of a man who treated women badly while claiming to be a feminist. Similarly, we could not distinguish between what President Reagan said the US did in relation to Nicaragua and what they actually did. The fact that in their everyday lives, sceptics are as alert as anyone else to whether words and deeds belie one another shows they don't believe their scepticism.

Such agnosticism or relativism derives from arcane – and in my view misguided – postmodernist debates about epistemology, but as already noted, they are entirely compatible with the treatment of values as mere subjective preferences in capitalism and in its economic theory, neo-classical economics. Capitalism encourages relativism. Nothing is intrinsically more important or valuable than anything else. Whatever sells/appeals to the subject is all that matters. And since markets are reason-blind, it's not in the interest of the seller to ask whether you deserve what you have the means to buy, or whether you deserve the envy and prestige it may bring you. Capitalism has nothing to fear from those who refuse to distinguish appearance from substance on epistemological grounds (see Eagleton, 1995).

In criticizing the aestheticization of moral-political issues I may have seemed to have been rather negative about the aesthetic side of culture. However, I appreciate the latter as much as anyone, and I accept that from an academic point of view, styles and tastes, especially exotic ones (and, in an ageist society, those of youth), have their fascination. But to put it bluntly, aesthetic values are less important than moral-political values. What our tastes are in clothes or music, is far less important than how we treat one another. Whether we like or loathe rap or bodypiercing is less important than whether we oppress others or are oppressed by them, or whether we (inadvertently) reproduce systems which treat people unfairly. If you differ with others over style and taste, it matters little, indeed such differences may be seen as something to celebrate. If you disagree over how you should treat one another, over your rights and responsibilities with respect to one another, then that is a more serious problem, one that requires some kind of resolution, if domination, inequity, conflict or worse are not to result. The very elevation of aesthetic matters of style over moral-political behaviour is symptomatic of a society in which instrumental-ization and aestheticization have become dominant.

Certainly, groups may attempt to make political statements through their tastes and styles, but there is nothing inherently progressive about the politics of style; after all, the management of style and spectacle can be seen in the parades of the Orange Order and Apprentice Boys in Northern Ireland, and of course it was an important ingredient of Nazi political success. Whether a particular politics of style is progressive depends on the politics, not the style – except insofar as the style itself prefigures progressive values. It is the moral-political values, not the aesthetic values which matter more.

## Bourdieu's capital and the elision of exchange-value and use-value

In order to pursue this critique in more depth, I now want to focus on Pierre Bourdieu's work on taste and social distinction and his concept of capital (Bourdieu, 1986). This is particularly interesting here, because he offers an economic view of culture – a cultural turn with an economic twist. Thus he applies concepts of exchange, circulation, price, capital, profit and the like to areas of life beyond the domain of conventional economics – in particular to the circulation and valuation of symbolic phenomena.[7] Bourdieu repeatedly argues that actions which appear to be disinterested are actually (subconsciously) instrumental[8] and differences of taste and style can mask differences and struggles over status or distinction. It is tempting to read his work as a devastating critique of the struggles of the social field in capitalist society, the 'soft forms of domination' therein, and of the structuring of cultural values by inequalities, but I shall argue that his economic analysis of culture is too reliant on capitalist economic ways of seeing to succeed in this respect. In particular, it elides a distinction which has been central to any critical view of cultural values, that between use-value and exchange-value, between valuations of goods – and by extension, social relations – for their own qualities and the satisfaction they bring, and valuations in relation to what they will fetch when sold in the market, in terms of rewards, monetary or otherwise.

Bourdieu widens the concept of capital beyond that of conventional or commercial capital, to include other, non-monetary forms – particularly symbolic, social and cultural capital. Symbolic capital brings rewards through association with respected and revered institutions. Thus Oxbridge trades on its symbolic capital. Social capital consists in the advantages that flow from having networks of contacts and others obligated to us on which we can trade. Cultural capital consists in the advantages that flow from having or being associated with certain cultural goods, particularly those that imply or resonate with a leisured lifestyle, removed from the pressures of economic necessity. Other kinds of capital can be identified too, such as linguistic and educational capital. People differ in their holdings of all these kinds of capital and their social influence and power varies accordingly. Largely unwittingly, people make investments in these forms of capital, which may bring them profits, pecuniary or non-pecuniary. The social field is structured by constant struggle over the valuation of these forms of capital, though the struggles are only barely recognized as such by actors. Bourdieu's work is remarkable in revealing how far apparently disinterested judgements involve hidden forms of social distinction.

However, Bourdieu's use of the concept of capital also contains a serious ambiguity, one which compromises the critical thrust of his work. It fails to distinguish between goods – such as education, culture, social relations – from the point of view of their use-value as it were, and from the point of view of their exchange-value, that is, in terms of what advantages (or disadvantages) they bring in the struggles of the social field. In relation to all Bourdieu's forms of capital – cultural, educational, linguistic, social and symbolic – this distinction is vital from both an explanatory and a critical point of view. Thus I may 'value' some people as friends, appreciating their sense of humour, intelligence, sensitivity, loyalty or whatever. Here I am valuing them individually, for what they are themselves. But if I value them instrumentally as social contacts, able to 'open doors' for me and bring me monetary and non-monetary rewards, then they become social capital for me.[9] Getting an education, enjoying music, making friends may contingently give one educational, cultural and social capital, but to treat the former as the same as the latter is a disastrous mistake. Moreover, it is a mistake which capitalism depends on and invites us to make, in that its primary interest is always in exchange-value, use-value being instrumentalized as a mere means to the end of gaining exchange-value. Sellers are more interested in the buyers' money than the buyer. As we have already noted, advertising repeatedly appeals not merely to the use-value of goods but to the advantages or capital – including non-monetary kinds – they bring us *vis-à-vis* others. This is central to the culture of capitalism.

To develop a critical analysis of this, the use-value/exchange-value distinction as developed by Aristotle, and later Marx, is crucial (Meikle, 1995). Marx insisted on distinguishing capital from mere machines, materials or buildings. The latter have use-value, but only become capital when they are acquired in order to command the labour or tribute of others and to earn exchange-value. In equivalent fashion we might insist on a difference between 'investments' – say in education – made for their own sake (for example, learning French) and investments made in order to enhance the possessor's social standing (educational capital). Of course, the use-value of education includes an instrumental dimension – enabling one to understand French speakers, or whatever – as well as a possible intrinsic interest, but this is different from instrumentalization in order to gain social advantage over others. This difference is equivalent to (and carries the same critical implications as) Aristotle's parallel distinction between production for own use (economy) and money making (chrematistics) (Meikle, 1995).

Another important aspect of the distinction between capital and the goods or activities to which it relates is that capital is a positional good, that is one whose value is depleted the greater the number of people who come to have it, whereas the same is not necessarily true of the activities to which capital relates.

Thus, educational qualifications – as educational capital – are devalued as more people come to get them, though the value of their education is not necessarily devalued too (though it is often contingently devalued as a result of reduced staff-student ratios). A particular geography lesson has a certain quality no matter how many other geography lessons of the same kind are given. Another important difference is that while the use-values of different kinds of activity – such as art history and engineering – are incommensurable, the exchange-values of the corresponding forms of capital – of art historians and engineers – are commensurated in markets or social distinction.

Although Bourdieu is primarily interested in the unintended production of effects of distinction, the relations between motives and effects, whether intended or unintended, are important from both explanatory and critical points of view. Consider the possible relations between (A) activities and their use-values, (B) their exchange-value (if any) as capital, and (C) their effects. It is contingent whether A brings B. A can be pursued without regard for B. B can be sought after and indeed be the motive for A, or it can be an unintended consequence of A, or it may even be achieved independently of A through bluff or accident. Thus, one student might gain a first class degree because they may be deeply interested in their subject, though not in any advantages it may bring them, and another might get a first through being motivated by the social advantages they hope it will bring. Of course, the first student may gain these advantages inadvertently too, and indeed may even gain extra status in certain circles for being uninterested in the exchange-value of their capital. Further, no matter how B arises, whether it is deserved or undeserved, sought after or not, it can be used for good or ill effects (C). For example, Diana Princess of Wales had three undeserved sources of recognition or capital – beauty, wealth and royal connections – but she used these as a lever to do good works, where others with similar capital have failed to do so.

That the relation between the use-value of an activity and the exchange-value of the corresponding form of capital is contingent is of vital importance. For those who do want to raise the value of their capital, there are several ways of doing so. In the case of educational capital, they might get more education, but they could also merely acquire its outward signs – a degree gown, lots of books, 'educated' ways of speaking, and so on. They might be 'entitled' to some of these, but the relation between the (use-)value of education itself and the exchange-value of educational capital and associated forms of symbolic and cultural capital is contingent. There is, for example, no necessary connection between the splendour and superfluity of the Oxbridge college and the quality of the education that goes on within it, though the inmates might like to believe that they deserve their privileges, that the high exchange-value of their educational capital is a reflection of the quality of their education and their own ability. The assumed or claimed qualities used to defend the value of the educational capital may even be a sham, as in the case of the Oxbridge MA. (Students who have an Oxbridge BA can get an MA just by waiting a certain number of years and paying a fee! – that some people make excuses for this bogus degree is a reflection of the cultural privilege enjoyed by Oxbridge.)

Thus, a critical analysis of educational capital cannot evade judgements of the use-value or intrinsic quality of the education with which it is associated. In other words it cannot evade distinguishing between deserved and undeserved recognition. Nevertheless, any attempt to make such a distinction is likely to invite suspicion that one is trying to establish an authoritative, indeed authoritarian, basis for judgement, an absolute set of values. I fully accept that judgements of (use-)value are contestable, but this does not mean either that all claims to recognition are of equal merit, or that there must always be some ulterior motive behind the judgements and contestations such that critical distinctions can never be rationally justified.

The contestation of use-value also differs in kind from that of exchange-value: whereas the prime consideration in the latter contest is instrumental – whatever will fetch the best 'price' – the contest regarding the quality of the goods is by reference to the internal goods of the relevant practice. If we were evaluating a course taught by Pierre Bourdieu, our judgements might be products of our position in the social field relative to Bourdieu's – we might be influenced by his gender, race, age, class, bearing and appearance, by the surroundings. All of these things might influence his educational capital and that of the students who have been taught by him. Of course, one would hope that our judgements were based instead on the quality of the education itself, his insights, the rigour of his arguments, his success in communicating his ideas, etc. These qualities might also influence the exchange-value, that is the status, prestige and other rewards of the course as educational capital, but if it's a good course, then it is so regardless of whether it brings him or his students any such exchange-value rewards. Conversely, no matter what the exchange-value of an Oxbridge MA as educational capital is, its recognition is undeserved for it lacks any corresponding process of education. Its market value depends on the success of the illusion that Oxbridge MA students have done something more than a BA, and on the associated exchange-value (prestige) of the symbolic and cultural capital of Oxbridge, which serve as collateral, so to speak. If we failed to note the illusory character of the recognition we would misdescribe the situation. In other words, in failing to be critical we would fail to explain.

Bourdieu retains a modernist distinction between substance and appearance inasmuch as he looks beyond the appearances of the production of social distinction, and shows that things are not as they appear, in particular that supposedly disinterested judgements of taste hide strategies of distinction or pursuit of status. Yet he fails to distinguish between use-value and exchange-value and reduces the former to the latter, emphasizing investment, calculation and profit. Since he is extremely reluctant to acknowledge the possibility of disinterested judgement, so that all disputes over taste are merely disguised struggles for advantage *vis-à-vis* others, then it is hard to see what there is to criticize if things cannot be otherwise. Strangely, although Bourdieu presents 'a social critique of the judgement of taste' he does not explicitly criticize any particular status distinctions as unjustified, for his resistance to the possibility of disinterested judgement disallows this. The critical implications of his work are therefore always uncertain (Sayer, 1999b). This is more like the postmodernist dismissal of contests over intrinsic or use-value as no more than disguised power-play: like the concept of regimes of truth, it loses its critical power by self-destructing.

The subjectivist theories of value of postmodern relativism and capitalism's economic theory, neo-classical economics, could have been made for each other: the only difference is that, as Eagleton notes, postmodernism's focus on discourse produces a 'subjectivism without a subject'.[10] Both treat values as subjective and a-rational. Claims about the value of the environment or particular practices are treated merely as expressions of individual preferences and measurable in exchange-value terms, or as mere discursive constructions: knowing the price of everything but the value of nothing is a distinctively capitalist cultural tendency. Similarly, the treatment of value regarding recognition or respect due to others as purely subjective or discursively constructed and as having nothing to do with the qualities of the thing valued or what those who are valued or recognized have done fits with both neo-liberalism and postmodernist relativism. Postmodernist suspicion of normative claims as authoritarian mirrors neo-liberal hostility to 'absolute values' and paternalism. Furthermore, a third tendency – the practice in some sociology of 'bracketing out' questions of the validity of actors' judgements – produces similar effects. Not only capitalism but Bourdieu's critique of the social judgement of taste treats values as the outcome of market processes or the struggles of the social field.

## Conclusion

The cultural turn could lead to a reinvigorated critique of contemporary society, but in practice it often seems instead to be complicit with many of the most questionable aspects of it, particularly those which relate to its economic organization and its ever more commodified form. In particular, insofar as cultural studies focuses on the 'stylization of life', it reproduces the aestheticization of aspects of life that might otherwise be considered as moral-political issues. In ignoring economic processes or 'dumbing down' their analysis, or by relying on discourses of globalization or Fordism and post-Fordism which do not offer a critique of capitalism, it offers little to worry either capitalist economy or contemporary culture. Although, peculiarly, modernist theory is sometimes portrayed by postmodernists as being uncritical of modernism, the critique of capitalist culture developed by modernist theory remains far more penetrating and relevant. The common postmodern suspicion of 'normativity' discourages criticism of the aestheticization of moral-political values, 'de-moralization' and depoliticization in contemporary society, and disqualifies distinctions between use-value and exchange-value, substance and appearance which are at least a necessary component of any kind of critical stance.

## Notes

1. On postmodern–neo-liberal affinities, see Sayer, 1995, Chap. 9, and O'Neill, 1998.

2.  This is double-edged, for detraditionalization can open a space for selfish individualism and barbarism as easily as it can allow relationships to be based on moral-political deliberation.

3.  This is not to suggest that the situation is necessarily better in a socialist system, for unless better ways are designed of making producers accountable than existed under state socialism, it may afford even greater possibilities for the abuse of producer power.

4.  Implicit in this, of course, is a criticism of the reduction of identity to appearances and discursive construction rather than something based on deeds. In this context, it is worth remembering what an identity parade is for – identifying who did it.

5.  Those familiar with Smith will know that the omitted parts of this quotation are important for understanding Smith. However, delving into this would require a substantial digression which does not affect my argument.

6.  Deciding what would be fair and justified judgements in the context of these forms of difference is often *not* a matter of disregarding difference and attempting to impose a single standard; thus the differences between the young and the elderly may not all be false ascriptions but may require judgements which take them into account. In other words, the issue of deserved and undeserved recognition takes us into the debate over equality and difference, explored particularly in feminism (e.g. Phillips, 1994).

7.  Bourdieu indicates that he does not regard this formulation as merely metaphorical – for him, culture *is* economic (1993, p. 36).

8.  I pass over the problem of how strategies can be unconscious, according to Bourdieu (Alexander, 1995).

9.  In this case, it is significant that the intentional instrumentalization of the good (friendship) also undermines it.

10. This is sometimes lampooned in philosophy as the 'boo-hooray theory of value'.

# References

Alexander, J. (1995) *Fin-de-Siècle Social Theory*, London, Verso.

Bauman, Z. (1995) *Life in Fragments*, Oxford, Blackwell.

Benhabib, S. (1992) *Situating the Self*, Cambridge, Polity Press.

Bhaskar, R. (1986) *Scientific Realism and Human Emancipation*, London, Verso.

Bourdieu, P. (1986) *Distinction: A Social Critique of the Judgement of Taste*, London, Routledge.

Bourdieu, P. (1993) *Sociology in Question*, London, Sage.

Connor, S. (1993) 'The necessity of value', in Squires, J. (ed.) *Principled Positions*, London, Lawrence and Wishart.

Eagleton, T. (1996) *Illusions of Postmodernism*, Oxford, Blackwell.

Featherstone, M. (1994) 'City culture and politics', in Amin, A. (ed.) *PostFordism: A Reader*, Oxford, Blackwell.

Foster, J. (ed.) (1997) *Valuing Nature?*, London, Routledge.

Fraser, N. (1989) *Unruly Practices: Power, Discourse and Gender in Contemporary Social Theory*, Cambridge, Polity Press.

Fraser, N. (1995) 'From redistribution to recognition? Dilemmas of a "post-socialist age",' *New Left Review*, 212, pp. 68–93.

Fraser, N. (1999) 'Social justice in the age of identity politics: redistribution, recognition and participation', in Ray, L. and Sayer, A. (eds) (1999) *Culture and Economy after the Cultural Turn*, London, Sage.

Habermas, J. (1987) *The Philosophical Discourse of Modernity*, Cambridge, Polity Press.

Habermas, J. (1990) *Moral Consciousness and Communication*, Cambridge, Polity Press.

Hall, S. (1988) 'Brave new world', *Marxism Today*, 24–9, October.

Hall, S. (1996) 'When was the "post-colonial"?: thinking at the limit', in Chambers, I. and Curti, L. (eds) *The Post-Colonial Question*, London, Routledge.

Hall, S. (1997) 'Culture and power', *Radical Philosophy*, 86, pp. 24–41.

Hoy, D.C. (1986) *Foucault: A Critical Reader*, Oxford, Blackwell.

Jameson, F. (1990) *The Cultural Logic of Late Capitalism*, London, Verso.

Keat, R. (1997) 'Values and preferences in neo-classical environmental economics', in J. Foster (ed.) *Valuing Nature?*, London, Routledge.

Keat, R. (1999) 'Market boundaries and the commodification of culture', in Ray, L. and Sayer, A. (eds) *Culture and Economy after the Cultural Turn*, London, Sage, pp. 92–111.

Lash, S. (1990) *Postmodernist Sociology*, London, Routledge.

McNay, L. (1994) *Foucault: A Critical Introduction*, New York, Continuum.

Meikle, S. (1995) *Aristotle's Economic Thought*, Oxford, Clarendon Press.

Nozick, R. (1974) *Anarchy, State and Utopia*, Oxford, Blackwell.

O'Neill, J. (1995) ' "I gotta use words when I talk to you": a response to Death and Furniture', *History of the Human Sciences*, 8 (4), pp. 99–106.

O'Neill, J. (1998) *The Market: Ethics, Knowledge and Politics*, London, Routledge.

Phillips, A. (1991) *Engendering Democracy*, Cambridge, Polity Press.

Ray, L. and Sayer, A. (eds)(1999) *Culture and Economy after the Cultural Turn*, London, Sage.

Sayer, A. (1995) *Radical Political Economy: A Critique*, Oxford, Blackwell.

Sayer, A. (1999a) 'Valuing culture and economy', in Ray, L. and Sayer, A. (eds) *Culture and Economy after the Cultural Turn*, London, Sage, pp. 53–75.

Sayer, A. (1999b) 'Bourdieu, Smith and disinterested judgement', *The Sociological Review*, 47, pp. 403–31.

Sayer, A. and Storper, M. (1997) 'Ethics unbound: for a normative turn in social theory', *Environment and Planning D: Society and Space*.

Sen, A. (1987) *Ethics and Economics*, Oxford, Blackwell.

Sevenhuijsen, S. (1998) *Citizenship and the Ethics of Care*, London, Routledge.

Smith, A. (1984) [1759] *The Theory of Moral Sentiments*, Liberty Fund, Indianapolis.

Szerszynski, B. (1999) 'Performing politics: the dramatics of environmental protest', in Ray, L. and Sayer, A. (eds) *Culture and Economy after the Cultural Turn*, London, Sage, pp. 211–28.

Taylor, C. (1994) 'The politics of recognition', in Gutmann, A. (ed.) *Multiculturalism and the Politics of Recognition*, Princetown NJ, Princetown University Press.

Walzer, M. (1983) *Spheres of Justice*, New York, Basic Books.

# Economy, culture, difference and justice

## Linda McDowell

> Socialist notions of equality need to specify racial and sexual equality and not reduce these concerns to economic class concerns.
>
> (Eisenstein 1994)

In this chapter I want to look at the connections between what are broadly defined as cultural issues and those that are seen as economic, and argue that a socialist conception of justice must take account of differences other than those of class. While economic issues are reasonably straightforward to define – related to the production and exchange of goods and resources, to the distribution of the surplus, to the deployment of labour and to class questions – the distinction of cultural issues, as earlier chapters have indicated, is perhaps more complex. In the context of production or waged labour, however – the focus of this chapter – I want to include a range of issues that might broadly be subsumed under the heading of the production of difference in the economic sphere. In this I include gendered and racialized differences, the sets of issues that habitually are discussed under the heading of the culture of organizations, as well as questions about justice or equity in the distribution of the surplus between social groups.

Despite beginning with this binary distinction – one which is also reflected in my title – I do not want to polarize or counterpose 'the economic' and 'the cultural' here, nor to discuss whether or not they are separate social formations, whether Marxism did or did not recognize culture or cultural differences (which, in the long debates in the 1960s and 1970s about working-class culture, traditions of solidarity, restricted linguistic codes and their impact on economic achievement and so on and so forth, it certainly did). Instead I want to argue that it must be taken for granted that all economic actions are social actions and as such are affected by, operate within, impact on and so forth the norms, standards, customs, conventions and meanings of the society or societies within which economic decisions and practices take place. I argue that this is a more appropriate formulation of the relationship between economy and culture than their separation, and implicit hierarchization. Phrasing the relationship in terms such as 'economy trumps culture' as Sayer (1994) has done seems to me to be an unhelpful way to take this debate forward. This criticism, or rather argument about interconnections, also applies the other way, of course, and it is equally important to challenge discourse analyses that separate the symbolic order from the political economy, focusing instead solely on 'the cultural'.

In this chapter, first I speculate about how and why geographers self-consciously have established a debate about the implications of the so-called cultural turn in our discipline in the last decade or so, briefly identifying the forms it has taken and the questions that are now being addressed by those who broadly identify themselves as economic geographers. I then want to raise some questions about what this cultural turn – with its emphasis on a politics of difference rather than class politics – means for conceptions of social justice. Have we abandoned class analysis, as some have claimed, in the haste to recognize and embrace the significance of other divisions or differences? Is it possible, even, to add class into a notion of a politics of difference, or to combine class politics and identity politics in some ways? In the shorthand of contemporary discourse, what I am interested in is what it might mean to try to think through the concept of difference, and act upon it in the economic arena.

The explicit stimulus to this discussion lies in my interest in workplace cultures and equal opportunities policies. I have become increasingly preoccupied with thinking about equity and justice, about redistribution and recognition and how ideas about a non-hierarchical recognition of difference based on group identification as, for example, a woman, or a member of a minority group, of particular religious faith and so on, makes a difference to analyses of what we term economic processes at a range of spatial scales – from interactions within individual workplaces, to a more equitable – just if you like – distribution of economic resources and rewards within and between nation states. In the last five years or so, a great deal of interesting work about equity, difference and social justice has been published – not least David Harvey's (1996) book, *Justice, Nature and the Geography of Difference*. His work and that of, among others, Iris Young (1990, 1997, 1998) and Nancy Fraser (1995, 1997a and 1997b) in particular, has helped me grapple with these questions and influenced my approach. But for a long time, of course, and not just in the last few years, feminist analysts have argued that cultural understandings of what work is, of what it means to be a man or a woman, are bound up in economic processes. In other words, 'the economic' encompasses more than class differences (Phillips, 1983; Pringle, 1989; Walby, 1990, 1997).

## 'The cultural turn' and its antecedents

Before pursuing the construction of difference and its relationship to questions of equity and social justice, it may be helpful to define more exactly what is meant by culture in the economic arena, to try to distinguish the different ways in which economic and cultural issues are related and the different disciplinary perspectives within which these debates that are now absorbing geographers originally occurred. It is insufficient merely to refer to 'culture' and leave it at that – culture, as we know (and see Raymond Williams' (1981) definition in *Keywords*), is a notoriously slippery term, conceptually ambiguous and perhaps best thought about in relation to specific usages in order to grasp the complexity of the interrelationships and the ways in which they change over time and space.

In the list below I differentiate some of the main ways in which economic geographers and others have addressed cultural questions in their recent work. The list includes both macro- and micro-economic issues, research focused on institutions and on individuals, questions both in the labour market and the workplace but also in the community, and, at least to date – the main stimulus seems to have come from sociology rather than cultural studies (as Sayer (1997) also notes). Indeed, as I constructed this list I began to feel that the cultural turn in economic geography was really a rather belated recognition of the fact that geographers had at last accepted that they need to understand a bit more about what used to be termed 'constraints' on how competitive processes are assumed to operate.

Some of the relevant research areas that are being established at present include:

1. *National cultures*: comparisons of national differences in regulatory regimes, institutional practices, banking systems, business practices, etc. Examples here include recent work by Will Hutton (1995), Susan Christopherson (1993), and perhaps Meric Gertler (1997), although his work also falls into my second category. This body of work builds on the recent arguments in economic sociology about the embeddedness of economic process, at both national level and that of the firm (Granovetter, 1985; Granovetter and Swedburg, 1993; Lee, 1997; Zukin and DiMaggio, 1990). It has, however, a long tradition in sociology, most notably in Max Weber's work on the relationship between Protestantism and the development of capitalism.

2. *Institutional cultures*: inter-firm differences in institutional practices. The recent geographical literature includes Erica Schoenberger's comparison of the practices of Lockhead and Xerox (1996), Meric Gertler's (1995, 1997) study of engineering firms in Germany and Canada, and my own work on merchant banking in the City of London (McDowell, 1997). However, geographers are able to draw on the long tradition of organizational studies in industrial sociology and social economics, although interestingly that tends to have a much more significant focus on single organizations rather than adopting a comparative approach. I strongly believe that one of the strengths of a geographical approach is the comparative perspective, uncovering the difference that place makes. Early studies of institutional cultures include, in the US, Rosabeth Moss Kanter's (1977) influential *Men and Women of the Organisation* which was an early recognition of the significance of gender differences and, in the UK, Huw Beynon's (1973) magnificent study *Working for Fords* (although the focus of this latter study places it more

accurately into my category 3 below). There are numerous recent studies of institutional cultures undertaken by a range of scholars working in disciplines from anthropology (see for example Wright, 1995) to sociology where Catherine Casey's (1995) work on a US computer firm is a good example.

3. *Shop floor cultures*: here the focus is at a more micro scale than the studies included above, although the distinction is often blurred. Here, however, the focus tends to be on shop floor behaviour, and the day to day interrelations between different categories of workers: owners, managers and workers, for example, or men and women employees. Again there is a long sociological tradition here, and a more recent feminist one where workplaces as various as the office, the print shop (Cockburn, 1983), a car components line (Cavendish, 1983), hospitals and GPs' surgeries (Pringle, 1998), chip factories and electronic assembly lines in export processing zones (Mitter, 1986; Salzinger, 1997) have been analysed, as well as domestic work and childcare as waged labour in other women's homes (England, 1997; Pratt, 1997; Romero, 1992) and various crafts in women's own homes (Oberhauser, 1997). Work on trade union cultures and other institutions such as workers councils might also be included here, as well as in the category above when the focus is the region or locality rather than a specific plant or shop floor (Martin, Sunley and Wills, 1997; Wills, 1998).

Then there are two sets of work based on class relations but whose spatial focus is usually outside the workplace:

4. *Class cultures*: where the focus is on class as a social relation. Here work both of sociologists and within cultural studies is relevant, including studies of how the social relations of production influence non-workplace-based behaviour. Working-class culture, youth culture, and studies of class dealignment have all been key areas of research and geographers have recently turned to these topics in significant numbers. The collection by Tracey Skelton and Gill Valentine (1998), for example, includes interesting analyses of a geographically diverse range of youth cultures.

5. *Local cultures*: this category shades into the one above, as well as into category 3. It is distinguishable, however, by a focus on the analysis of class-based solidarity and political cultures in a geographic locality. Here a tradition from Raymond Williams through to the locality projects established by the Economic and Social Research Council in the UK in the 1980s may be mapped. Studies in this tradition have been criticized for their dominant focus on class relations, and often those between men, as feminist geographers noted (Bowlby *et al.*, 1986). Peter Jackson (1991) – a geographer who has had an important impact on the 'cultural' turn in the discipline – was an early critic of the so-called locality school for its lack of emphasis on culture. More recent and sophisticated approaches to local cultures which straddle 'the economic' and 'the cultural' include Jane Wills' (1995) study – drawing on E.P. Thompson's (1991) work – of locally-based trade union cultures in Great Britain.

Two further categories of geographical analysis may be distinguished which have a shorter pedigree:

6. *Cultural industries*: work on significance of new range of goods and products, including 'virtual' products which are non-material and, in a sense, placeless. Here the work by the sociologists Scott Lash and John Urry (1994) has had an impact on geography, but there is also a longer sociological tradition of examining the cultural industries – media, arts, etc. Celia Lury's (1993) work is an example here and, in geography, that by Kevin Robins (1996). There is also recent geographical work, although with even longer sociological tradition, on money as a symbol and on place marketing (Leyshon and Thrift, 1997; Philo and Kearns, 1994). The huge expansion of work on consumption, advertising, and marketing might also be squeezed into this category although it is also related to the work in category 4 on style and lifestyle.
7. *Cultural analyses of economic representations, metaphors, language*, etc. and the language of economic theory/economic geography in general. Here the work of the Canadian geographer, Trevor Barnes (1992, 1996), stands out and is perhaps the only undiluted example of that turn to the analysis of text and discourse rather than social and material relations that has been identified as a key element of the cultural turn, although the work on cultural industries and virtual products I have included above shades into this. But, again, there is a longer tradition elsewhere and here the work of Gareth Morgan (1986) on the importance of metaphors in understanding organizations is a good example.

## Reactions to 'the cultural turn'

So, despite the relatively recent but growing interest in economic cultures by geographers, it is evident that it is, in the main, drawing on a very long tradition in other disciplines. Perhaps then it is appropriate to ask why this 'cultural turn' has occurred in the 1990s, and why it has provoked such uneasy, not to say angry, reactions among some geographers. The root of the unease I think lies in a paradox: during a decade or more of the unprecedented deepening of economic inequalities in the west, class analysis and class-based struggles largely seem to have disappeared from the theoretical and political agendas. Instead there has been a hugely significant shift to a language and politics of difference: cultural differences of language, race, gender, different values, perspectives, single issue politics and ideas about self-identity have come to dominate in debates about inequality and justice. Despite the growth in what the right wing dismissively label the underclass (but at least they implicitly recognize it as a class compared with the liberal New Labour double speak of the socially excluded), the language of politics has shifted on to the terrain of difference. Perhaps part of the reason for the paradox lies in the fact that old class divisions formed in the social relations of production with the clear association between types of work, financial remuneration, political attitudes and class struggle have broken down, partly because of the class dealignment that occurred throughout the 1960s and 1970s but now because the so-called under-class is an extremely diverse group of people – old and young, disabled, ex-service personnel, certain ethnic minorities, etc. whose boundaries blur into the working poor in the casualized insecure and very badly paid jobs of post-Fordism, late capitalism or whatever we choose to term it. These people often move in and out of work, work for a variety of employers and so are unlike and, of course, more difficult to organize than the old Fordist working class, which was also, and significantly, predominantly male. Instead, deregulation, privatization, casualization and the political commitment to a 'flexible' labour force have produced a divided rather than solidaristic working class. Thus – although the differences shouldn't be polarized too absolutely – we are in a different era whose class divisions seem more like those of an earlier, almost pre-modern, period than the relative stability of the 1950s and 1960s.

These shifts are, of course, related to the transformation of the economy in advanced industrial societies from a base in manufacturing to one in the service industries. These latter industries demand completely different sets of social and cultural attributes from their workers. As is now well established, and to use shorthand, service sector work is feminized, inter-active, and often depends on the manipulation of emotions and feelings. This affective work, or emotional work as Arlie Hochschild (1983) so perceptively dubbed it in the early 1980s, depends on displaying and selling emotions and intimacy: characteristics which are defined in opposition to hegemonic definitions of masculinity – at least until the election of New Labour and the death of Diana which in combination seem to have released many Britons from their emotional straitjacket. Traditional masculinist skills and talents – both the physical strength of manual work and the stiff upper lip, disembodied masculinity of the professions, the civil service and bureaucracies – seem increasingly anachronistic. To draw on a royalist metaphor: Spencer attributes have replaced Windsor ones in the workplace as well as in the monarchy. Interestingly, in an article on the death of Diana, Linda Grant (1997) asked why the left – or at least that puritanical branch that had dominance for a long time – is uncomfortable with public displays of emotion and the idea that the personal is political: I cannot help but feel that this is related to both the distrust of ideas about culture and the desire to subsume all differences back neatly under class differences and so reassert the pre-eminence of class struggles evident in some recent geographical work. Workplace-based organization is 'real politics' compared with struggles around style and shopping which seem frivolous. I want to emphasize, however, as I shall try to demonstrate later, that I am not denying the crucial, indeed growing, signi-ficance of class differences; but I also want to argue that there are many valid forms of resistance to oppression.

So, in all sorts of ways, as we approach the end of the century, new social and cultural formations are evident that seem to have effected the collapse of older distinctions between the economic and the cultural, and between the economic and the political or rather to have transformed the interconnections.

Sayer (1997) and other geographers have referred to these changes as effecting 'greater embedding' as values, lifestyles, emotions, etc. become more evident in organizations, firms and in the workplace, seeming to suggest that the cultural has become more dominant than the economic in recent decades. However, I prefer to argue instead that what we have identified as 'economic' has itself always been a social or cultural emphasis of certain processes as more economic than others and that we are just rethinking these definitions.

The material changes that are part of the re-evaluation of how economic processes are defined and explained are too well known to need labouring. Briefly, they include the dominance of right wing ideologies in many western democracies in the 1980s, leading to the eclipse of a language of class action for ideological reasons, the collapse of the former state socialist regimes, the widespread shift to market-oriented policies and associated notions of individual responsibility, the growing theoretical and political significance of subaltern groups (all those people that always seem to be enclosed in parentheses: women, ethnic minorities and people of colour, gay men and lesbians, postcolonial peoples, to which we now have to add religious minorities), the fracturing of family and household structures and the rise of lifestyle and identity politics based on issues and struggles based predominantly, although not only, in the arena of consumption rather than production. What these transformations have done is to make absolutely clear the ways in which economic and cultural processes are interconnected. It seems to me that it is now indisputable, for example, that individual and group identities are partly formed within, influence as well as are influenced by, relations in the sphere of production; that knowledge, values and attitudes affect reactions to crises and the ways restructuring works out, as Erica Schoenberger (1996) has shown so well. Workers not only have a class position but are embodied people bearing social characteristics that affect their class location.

In the rest of this chapter I want to pursue this argument about mutual constitution and its relationship to political identification in order to try to bring together contemporary theories of social justice. Is it possible to resolve what in feminism is known as the equality/difference dilemma? Does class politics depend on the denial or submergence of all other differences, or are there any areas of complementarity, in a theoretical as well as practical sense? I am not convinced by the arguments that we must in the end subsume all differences within class divisions, but nor must we neglect the language and politics of class and turn to a new, entirely cultural form of a politics of identity. Can we perhaps, as Diana Coole (1996) has pondered, theorize class as one of the differences that makes a difference?

## Class politics and the cultural left, rather than class versus culture?

In thinking through arguments about the significance of different differences, I draw on feminist political and cultural theorists, while not denying the existence of parallel debates in other literatures. Richard Rorty (1998), for example, has recently challenged the cultural left to readdress economic inequality. It is, however, within feminism that the issue of the relative significance of class and cultural differences has had perhaps its most extended airing. Here the work of Nancy Fraser is particularly provocative. Her current position is outlined in a collection of essays *Justice Interruptus* (1997b) which, with its subtitle *critical reflections on the postsocialist condition*, is a clear indication of her scepticism about postmodern arguments and her desire to continue to reflect on the significance of socialism. Fraser has debated her position with Iris Marion Young, a more committed adherent to the cultural left position (often summarized in the phrase the politics of difference) in an exchange over several issues of *New Left Review*. The outlines of their debate have been helpfully drawn together by Ann Phillips in the same journal and, more recently, it has drawn in the cultural critic Judith Butler (1998), indicating the significance of this exchange across a range of discursive positions.

In their exchange Fraser and Young emphasized their differences: Young lined up as the prophet of a multiple, flexible, context-dependent group-based version of justice whereas Fraser determinedly held on to a distributional, material class-based concept, thus seeming to encapsulate the cultural versus the economic position that I am at pains to collapse here. In fact, however, careful reading of their work reveals the parallels rather than the differences between them. When reading their work together what is striking is the similarities in their assertions. Both of them seem to have a position very close to the one which I have outlined above – that economy and culture, class politics and identity politics, material and cultural inequalities, matter, and not only matter but are mutually constituted, inseparable and politically significant. Young has never denied that economic inequality was important nor that redistribution mattered. In her deservedly well known book *Justice and the Politics of Difference* (1990) where she provides a detailed argument for the replacement of an individual distributional concept of justice by a multiple, group-based concept, she distinguishes five axes of oppression. Two of these five axes are economic exploitation and economic marginalization, and it is clear that Young regards cultural and economic injustice as interconnected. She is, however, habitually identified with that set of theorists who have challenged the primacy accorded to political economy, and demanded recognition of inequalities associated with ethnicity, gender, embodiment, skin colour and so on. Thus, these theorists, who together are frequently dubbed the 'cultural left', have questioned 'the underlying hierarchy of causation that had distinguished an economic base from a political and cultural superstructure, or defined "real" interests through location in economic relations' (Phillips, 1997, p. 145).

Fraser's aim is to address ways of resolving the unhelpful rhetorical slanging match between the 'economy, stupid' (Sayer, 1994) and 'culture, stupid' (Barnes, 1995) positions. These evade what Fraser defines as

> the crucial 'postsocialist' tasks: first interrogating the distinction between culture and economy; second, understanding how both work together to

produce injustices, and third, figuring out how, as a prerequisite for remedying injustices, claims for recognition can be integrated with claims for redistribution in a comprehensive political project. (Fraser, 1997, p. 3)

This is an agenda that I wholeheartedly endorse and, I would venture to add, it is one which feminists have been developing for more than a little while. Young too has argued from the same position. Thus responding to David Harvey's argument that the lack of response to tragedies such as the fire in a chicken processing plant in North Carolina was because feminists and others were preoccupied with a politics of difference, Young states very clearly and powerfully that

it is both theoretically and politically counterproductive, it seems to me, to construct an account of our recent history as a fall away from the more correct class-based universalist politics to a fragmented and relativist politics of difference, and to suggest, however qualifiedly, that progressives ought to abandon the latter and return to an earlier more correct road. We should not interpret our current theoretical and political situation as a choice between universal and particular, class unity and the recognition of social difference, but rather as a challenge to move beyond these oppositions. (Young, 1998, 37)

Exactly, and the call to move beyond binary oppositions is surely not an unfamiliar one?

But, it is, of course, a movement that demands some clear thinking about critical theory, as well as about possible political actions. Fraser, for one, is scathing about current limited and relativist theoretical positions and calls for coalition politics. She wants nothing less than 'comprehensive, integrative, normative, programmatic thinking' (p. 4) and, like Young, argues that 'critical theorists must rebut the claim that we must make an either/or choice between the politics of redistribution and the politics of recognition. We should aim instead to identify the emancipatory dimensions of both problematics and to integrate them into a single, comprehensive framework' (p. 4).

While I agree it is important to develop a comprehensive, indeed utopian framework like this, I also want to emphasize that I do not think it negates the need to think through and act on provisional alternatives to the present social order. The most disappointing aspect of the old left is its continuing belief in revolutionary action as the basis of socialist politics (Harvey, 1996, p. 402). At the 1998 annual conference of American Geographers held in Boston, the sessions which purported to consider the economic/cultural basis of left politics and critical theory were remarkable only for their continued nostalgia for an earlier golden era of class struggles, which apparently reached their apotheosis in the 1960s – exactly the period in which the 'cultural left' discerns the origins of a politics of difference, in the nascent women's movement and in the anti-war movement, for example.

# A *new comprehensive political framework?*

What might be the basis of an intellectually satisfying and politically realistic consideration of the significance of social and cultural difference? To begin to answer this question, I want to turn to a more detailed consideration of Nancy Fraser's arguments to assess the prospects for a coalition politics. I shall briefly indicate the main lines of Fraser's assessment of the redistributive and recognition paradigms of social justice, of their contradictions and complementarities but the interested reader must turn to her essays for the more complete version. As Fraser points out, using as illustration the problem that is such a dilemma for feminists, people who are subject to economic and to cultural injustices need both to claim and to deny their specificity in their claims for fair treatment. Thus in the labour market, women – who are in the main discriminated against on the grounds of their gender – want both to abolish the gender division of labour and to ensure recognition of their specific demands and their particular circumstances. Here the demands for equal treatment depend on asserting the similarities of men and women as workers (but note that this does not imply treating women as if they were men) whereas the introduction of gender-sensitive policies necessitates the recognition of significant differences (maternity being the most significant).

Fraser suggests that whereas some group injustices are primarily cultural and so demand recognition or revaluing as a remedy for the injustices they are subject to (discrimination on the basis of sexuality is her example), most injustices are what she terms bivalent, that is encompassing both political-economic and cultural-valuational dimensions – gender and 'race' are the obvious examples here – and so the paradox between recognition and denial of difference is uppermost. Fraser also distinguishes between two types of remedy for injustice: remedies either involve what she terms affirmation or transformation. Affirmation involves 'remedies aimed at correcting inequitable outcomes of social arrangements without disturbing the underlying framework that generates them' (p. 23), whereas transformation, as the term implies, involves 'restructuring the underlying generative framework' (p. 23). Thus taking the cultural differences as the example, affirmation lies behind contemporary multiculturalism whereas transformation would mean altering everyone's sense of self. In the economic sphere, affirmation is associated with the policies of the liberal welfare state, increasing the consumption share of economically disadvantaged groups, leaving the system of production unchanged, whereas transformation was associated, ideally at least, with socialism. Thus we arrive at the fourfold matrix in Table 1.

Notice the provocative parallel between deconstruction and socialism in Table 1. Thus, in Fraser's words, 'deconstruction is the cultural analogue of socialism' (p. 28), which might make us think again about the deep distrust of many socialists for deconstructive projects. Reading Fraser's work it is quite clear on which side of the diagram her sympathies lie. As she correctly points out, the politics of affirmation bring with it the dangers that the groups claiming recognition begin to appear as 'privileged, recipients of special treatment and undeserved largesse' (p. 28) and here the current backing off from positive

Table 1

| | Remedies | |
| --- | --- | --- |
| | Affirmation | Transformation |
| **Forms of justice** | | |
| **Redistribution** | *the liberal welfare state* surface reallocations of existing goods to existing groups; supports group differentiation; can generate misrecognition | *socialism* deep restructuring of relations of production; blurs group differentiation; can help remedy some forms of misrecognition |
| **Recognition** | *mainstream multiculturalism* surface reallocations of respect to existing identities of existing groups; supports group differentiations | *deconstruction* deep restructuring of relations of recognition; destabilizes group differentiation |

action legislation in many US states is a clear example. As we have seen in the US, the cultural politics of recognizing difference has inflamed the sensibilities of groups who now feel their rights are being trampled on: white men being the prime example. And further the cultural politics of difference, or multiculturalism as it is habitually termed in the UK, seems at odds with the official commitment to liberal individualism – the equal moral worth and rights of all persons – that still underpins the institutions and practices of the welfare state.

If Fraser's logic is impeccable and her conclusions seem inescapable, why do I hesitate to endorse them? I find her conceptual thinking exhilarating but I think that her sense of current political possibilities is unreal, although in fairness to her I do think that she would probably accept this criticism. Indeed she points out, taking the example of gender injustice, that 'both deconstructive feminist cultural politics and socialist feminist economic politics are far removed from the immediate interests and identities of most women, as these are currently culturally constructed' – a statement that left me breathless and demanding to know what are the implications for organizing? Unfortunately Fraser does not (or cannot?) answer this question.

So what is to be done, to recall an earlier, but still relevant, question? I have no doubt that affirmative action and redistributive policies completely fail to grapple with the ways in which economies are, for example, deeply gendered and racialized, and that they fail to engage with the structures that generate these inequalities. Taking gender again as my illustration, in 1997, as if we needed more evidence, yet another survey showed that the gender division of unpaid labour remains as entrenched as ever in western economies. As the authors of the study concluded, using work in its widest sense to include waged and unwaged activities: 'we discover – or rather rediscover, since this is found wherever we look in Europe and North America – that the resemblance between husbands' and wives' work lives is only superficial' (Gershuny, 1997).

## Towards a transformative politics

A transformative politics of deconstruction would be quite different from the current demands for the recognition and celebration of group difference, and from an older version of class politics. In a transformative politics, binary distinctions between, say, men and women, whites and people of colour, Christians and Muslims, queers and straights, would have to dissolve into 'networks of multiple intersecting differences that are demassified and shifting' (p. 31). These words are Nancy Fraser's but here too there are distinct parallels between her work and that of Iris Marion Young. Those who have criticized Young for reifying current distinctions on the basis of, say, gender, skin colour, etc., I believe, misread her. She too argues for flexible and multiple membership of many intersecting networks. But, this demands, according to Fraser, that 'all people be weaned from their attachment to current cultural constructions of their interests and identities' (p. 31) and whatever type of cultural, psychological and political economy view of the world individuals and groups hold on to, it is clear that this is an almost impossible task. I find it hard to imagine the bases of political struggles that would achieve this weaning and destabilization as 'disadvantaged' groups would have to give up their special claims and sign up for socialism, whereas socialists must accept that intersecting social divisions combine to create divisions that cross-cut those of class. It is, however, a positive and progressive version of the future in which neither similarities nor differences are emphasized at the expense of the other. Perhaps if we rephrase Fraser's plea for weaning (with its rather unpleasant associations of giving up childish pleasures) and replace this with what the British cultural critic Stuart Hall (1990) has termed 'translation' in which the inevitable changes in class, ethnic and gender relations that are consequent on the material changes I outlined are positively welcomed rather than resisted, perhaps Fraser's call for a transformative politics is not so utopian after all. But accepting difference and welcoming translation must go hand in hand with a revitalized class politics that recognizes and resists the growing and increasingly visible injuries of class in economies and societies where unemployment, income inequality, insecurity, lack of worker rights and minimum incomes which meet not even the basic needs for a decent standard of living are part of the new economic landscape at the end of this millennium.

## References

Barnes T. 1992 'Reading the texts of theoretical economic geography: the role of physical and biological metaphors'. In T.J. Barnes and J.S. Duncan (eds) *Writing Worlds: discourse, text and metaphor in the representation of landscape.* London: Routledge, 118–35.

Barnes T. 1995 'Political economy 1: "the culture, stupid".' *Progress in Human Geography* 19, 423–31.

Barnes T. 1996 *Logics of Dislocation: models, metaphors and meanings of economic space*. New York: Guilford.

Beynon, H. 1973 *Working for Fords*. Harmondsworth: Penguin.

Bowlby, S., Foord, J. and McDowell, L. 1986 'The place of gender in locality studies'. *Area* 18, 327–31.

Butler, J. 1998 'The merely cultural'. *New Left Review* 227.

Casey, C. 1995 *Work, Self and Society*. London: Routledge.

Cavendish, R. 1983 *Women on the Line*. Basingstoke: Macmillan.

Christopherson, S. 1993 'Market rules and territorial outcomes: the case of the United States'. *International Journal of Urban and Regional Research* 17, 274–88.

Cockburn, C. 1983 *Brothers: male dominance and technological change*. London: Pluto Press.

Coole, D. 1996 'Is class a difference that makes the difference?' *Radical Philosophy* 77, 17–25.

Eisenstein, Z. 1994 *The Colour of Gender: reimaging democracy*. Berkeley and Los Angeles: University of California Press.

England, K. 1997 *Who will Mind the Baby?* London: Routledge.

Fraser, N. 1995 'Recognition or redistribution? A critical reading of Iris Young's *Justice and the Politics of difference*'. *Journal of Political Philosophy* 3.

Fraser, N. 1997a 'A rejoinder to Iris Young'. *New Left Review* 223, 126–9.

Fraser, N. 1997b *Justice Interruptus: critical reflections on the postsocialist condition*. London: Routledge.

Gertler, M. 1995 ' "Being there": proximity, organisation and culture in the development and adoption of advanced manufacturing technologies'. *Economic Geography* 71, 1–26.

Gertler, M. 1997 'The invention of region culture', in Lee, R. and Wills, J. (eds) *Geographies of Economies*. London: Arnold.

Gershuny, J. 1997 *British Household Panel Survey*. University of Essex.

Granovetter, M. 1985 'Economic Action and Social Structure: the problem of embeddedness' *American Journal of Sociology* 91, 481–510.

Granovetter, M. and Swedburg, R. (eds) (1993) *The Sociology of Economic Life*. Boulder, Co: Westview Press.

Grant, L. 1997 'Emotion and politics'. *The Guardian*, 9 September, p. 19.

Hall, S. 1990. 'Cultural identity and diaspora'. In J. Rutherford (ed.) *Identity: community, culture, difference*. London: Lawrence and Wishart (reprinted in McDowell, L. (ed.) 1997 *Undoing Place?* London: Arnold).

Harvey, D. 1996 *Justice, Nature and the Geography of Difference*. Blackwell: Oxford.

Hochschild, A. 1983 *The Managed Heart*. Berkeley and Los Angeles: University of California Press.

Hutton, W. 1995 *The State We're In*. London: Jonathan Cape.

Jackson, P. 1991 'Mapping meanings: a cultural critique of locality studies'. *Environment and Planning A* 23, 215–28.

Kanter, R.M. 1977 *Men and Women of the Organisation*. New York: Basic Books.

Lash, S. and Urry, J. 1994 *Economies of Signs and Space*. London: Sage.

Lee, R. 1997 'Economic Geographies: representations and interpretations'. In R. Lee and J. Wills (eds) *Geographies of Economies*. London: Arnold.

Leyshon, A. and Thrift, N. 1997 *MoneySpace*. London: Routledge.

Lury, C. 1993 *Cultural Rights*. London: Routledge.

McDowell, L. 1997 *Capital Culture: gender at work in the City*. Oxford: Blackwell.

Martin, R., Sunley, P. and Wills, J. 1997 *Union Retreat and the Regions*. London: Jessica Kingsley.

Mitter, S. 1986 *Common Fate, Common Bond: women in the global economy*. London: Pluto Press.

Morgan, G. 1986 *Images of Organisations*. London: Sage.

Oberhauser, A. 1997 'The home as "field": households and homework in rural Appalachia'. In J.P. Jones III, H. Nast and S. Roberts (eds) *Thresholds in Feminist Geography*. New York: Rowman and Littlefield.

Phillips, A. 1983 *Hidden Hands: women and economic policies*. London: Pluto.

Phillips, A. 1997 'From inequality to difference: a severe case of displacement'. *New Left Review* 224, 143–53.

Philo, C. and Kearns, G. (eds) 1994 *Selling Places*. Oxford: Pergamon Press.

Pratt, G. 1997 'Stereotypes and ambivalence: the construction of domestic workers in Vancouver British Columbia'. *Gender, Place and Culture* 4, 159–78.

Pringle, R. 1989 *Secretaries Talk*. London: Verso.

Pringle, R. 1998 *Sex and Medicine: gender, power and authority in the medical profession*. Cambridge: Cambridge University Press.

Robins, K. 1996 *Into the Image*. London: Routledge.

Romero, M. 1992 *Maid in the USA*. New York: Routledge.

Rorty, R. 1998 *Achieving our Country: leftist thought in 20th-century America*. Cambridge MA: Harvard University Press.

Salzinger, L. 1997 'A maid by any other name: the transformation of "dirty work" by Central American immigrants'. In L. Lamphere, H. Ragone and P. Zavella (eds) *Situated Lives: gender and culture in everyday life*. London: Routledge.

Sayer, A. 1994 'Cultural studies and "the economy, stupid" '. *Environment and Planning D: Society and Space* 12, 635–7.

Sayer, A. 1997 'The dialectic of culture and economy'. In Lee, R. and Wills, J. (eds) *Geographies of Economies*. London: Arnold.

Schoenberger, E. 1996 *The Cultural Crisis of the Firm*. Oxford: Blackwell.

Skelton, T. and Valentine, G. (eds) 1998 *Cool Places*. London: Routledge.

Thompson, E.P. 1991 *Customs in Common*. London: Merlin Press.

Walby, S. 1990 *Theorising Patriarchy*. Oxford: Blackwell.

Walby, S. 1997 *Gender Transformations*. London: Routledge.

Williams, R. 1981 *Keywords: a vocabulary of culture and society*. Revised edition. Glasgow: Fontana.

Wills, J. 1995 *Geographies of Trade Union Tradition*, unpublished PhD thesis. Milton Keynes: Open University.

Wills, J. 1998 'Taking on the CosmoCorps? Experiments in transnational labor organization'. *Economic Geography* 74, 111–30.

Wright, S. 1995 *The Anthropology of Organisations*. London: Routledge.

Young, I.M. 1990 *Justice and the Politics of Difference*. Princeton, NJ: Princeton University Press.

Young, I.M. 1997 'Unruly categories: a critique of Nancy Fraser's dual systems theory'. *New Left Review* 222, 147–60.

Young, I.M. 1998 'Harvey's complaint with race and gender struggles: a critical response'. *Antipode* 30, 36–42.

Zukin, S. and DiMaggio, P. (eds) 1990 *Structures of Capital: the social organisation of the economy*. Cambridge: Cambridge University Press.

# Virtualism – the culture of political economy[1]

## Daniel Miller

Many undergraduate students in Britain taking subjects such as sociology, geography and anthropology tend to come across the term Political Economy when they are first introduced to Marxism as part of the degree course. In most cases they will then broaden this knowledge and learn also about figures such as Ricardo and Adam Smith. They may, however, at some point become curious about why it is that in learning about the political economy they seem only to touch upon figures who wrote in the nineteenth century or even earlier. They may well never be told by their lecturers that there exists another sense of the term political economy which is that used by economists and which is almost unrecognizable from anything that they had been taught. Political economy when taught to economists is almost certain to include some mention of Adam Smith but by comparison to other disciplines tends to have little time for the nineteenth century, but instead is largely concerned with the development of an abstract science of relationships and modelling that is characteristic of economics in the twentieth century (Carrier, 1998).

On face value, being at least a century out of date does not seem to be a particularly good starting point for the likes of geography and anthropology. Indeed it has to be said we are constantly losing ground as the economists' version of political economy gains power and influence. In the British secondary school system the last twenty years has seen a rise in the teaching of economics for examination. I suspect this is often at the expense of subjects such as geography and history. Why on earth should we continue to deal in century-old debates when these seem merely to be digging our own graves while the relentless rise of economics sweeps aside all alternatives? The difference between the nineteenth century political economy and that of today is that the former was grounded in the observation and comparison of economies, and in drawing political conclusions. Economies are forms of social action, including production, distribution and consumption, bound up in social relations such as differences in wealth and power, and their politics are therefore always also a series of ethical dilemmas. The twentieth century study of political economy under the title of Economics, despite its name, seems to have flourished when it ceased to observe different types of economy, and instead turned itself into an abstracted modelling process based on logical relationships between the entities it defines as economic forms. As such it no longer subsumes observation in political and ethical dilemmas, but instead adopts the epistemology of science, of hypotheses, laws and testing. The irony is that the more economics has eschewed direct observation of the world, the more it has grown pretensions to this scientific epistemology. Politics is thereby reduced to the means by which economic models can be realized in the world.

My point in this chapter is not merely to observe this contrast, but to try and explain it. Why should economics be so powerful and why should the alternative version of political economy now appear so antiquated and impotent, by comparison? But I also want to argue that we have no choice but to keep fighting against these trends; that students who at school learn the abstractions of economics and not the comparative study of economies that geography and history would have exposed to them, are actually less well educated and less able to be usefully involved in the political nexus that is termed civil society.

To explain as well as to observe the rise of modern Economics, we need to ground this shift within a larger sense of history as culture. I would argue that there are generalities that can be made which transform the accelerated rise of Economics over the last twenty years into simply one example in a larger set of changes that can be observed in many different phenomena. This will produce a paradox in the foundation of this chapter. In order to combat an increasingly abstract form of modelling, I will propose a rather abstract model. I would defend this in that my model arises out of a dialectical tradition (that began with the philosopher Hegel) in which a tendency to universalism is always matched by a commitment to particularity (for a recent account see Harvey, 1996). In a sense it is safer for the anthropologist to generalize, since we are committed to spending most of our time conducting ethnographic fieldwork in the most particular and specific examples of social action. The danger is abstract modelling that seems to have forgotten what it is that these were originally models of, that is social action, and led to the conceit that these models are reflections of abstract relations that are grounded in some imminent or transcendental economic laws, rather than the pluralism of culture.

The term 'virtualism' is proposed as the generalizing model for these historical shifts and five examples will be presented to make the case a more general one. These will be: structural adjustment, retail mergers, auditing, management consultancy and postmodern theory. Two basic principles link these five examples and these will therefore be proposed as fundamental attributes of contemporary virtualism. The first of these is that models which are thought to be descriptive of economic relations have become so powerful that they become in and of themselves the forces that determine economic relations. The second is that the sleight of hand described by Marx as commodity fetishism under which products no longer revealed the labour that created them, is replaced today by a consumer fetishism in which products no longer reveal the labour of consumption that could have appropriated them. The consumer is replaced by a virtual consumer.

I do not want to imply that all modern phenomena can be so described, but only that this is a useful set of generalities that help us account for some of the major changes we see around us. At its most ambitious the term virtualism would be proposed as either an extension of, or replacement for, the term 'Capitalism'. One problem with the term 'capitalism' is that it has now been used for so long and with respect to so many different phenomena that it may be losing any critical specificity that it once had. Virtualism, to be sure, is a pretty general term but it may help to bring up to date some of the implications that were once drawn attention to through the use of the term capitalism.

The relationship between the terms capitalism and virtualism may be drawn as follows. When Marx first proposed that the political economy of his time be subsumed under the term capitalism, he implied many things of which two stand out. The first was that the process by which capital reproduced had become abstracted from the ordinary working of society involved in economic relations to such a degree that, instead of reflecting the humanity of those that worked in the economy, it subsumed them as its instruments and became itself the principle imperative in economic relations. This was supported by observations on the industrial revolution where it seemed evident that workers had become mere pawns in the game of capital, reduced to the lowest expenditure possible for them still to survive and work for the interests of capital accumulation. At that time capital as an abstracted imperative could be linked to the interests of a particular class – the capitalists who tended to own and profit from the companies involved. Since the time of Marx we can see that even the humanizing involved in defining a class called capitalists may not be necessary, since many major corporations exist in the service of wage-based managers and shareholders who may not be so easily identified as a particular capitalist class, and yet the logic of abstract capital accumulation through profits still drives the corporation itself.

The second and equally important point was that capitalism operated through a sleight of hand – a cultural process that constituted a vast act of misrepresentation. This consisted of a device by which the worker whose labour was the foundation for all manufacture was separated from the products of his/her labour, which was no longer recognized as a product at all, but merely exchanged in the interests of capital accumulation. This was at the expense of the worker who had thereby invested him- or herself in the product as the concretized form of the labour and was then alienated from that objectified aspect of him- or herself. In effect instead of being developed by labour, in the way artists are developed by their art works, workers were diminished as their works were lost to them.

Marx began his academic work through equating observations on the conditions of the economy to principles borrowed mainly from the philosophical tradition of Hegel. Over time, however, this form of political economy became itself more abstract and by the time *Das Kapital* was written we can see a version of these ideas that lies much closer to what was going to develop in twentieth century economics as abstracted principles or even laws of economic relationships under the capitalist system. One of the reasons Marxism became a destructive science imposing its models on the world in the form of communism instead of simply a humanist critique of actual economies was that Marx himself moved in the direction of abstracted model building during the course of his life's work. All the more reason that a concept such as virtualism should not try to emulate these pretentions to general laws. My aim is to return to the more Hegelian roots of the younger Marx's ideas and critique. Hegel's attempt to write in the style of a grand narrative that is generalizing history as a kind of working out of a philosophical trajectory, is today deeply unpopular, but ironically I think with regard to our own time grand narrative is a great deal more appropriate than not. This tradition is being attacked at the very moment when history itself is as it were coming into line with its own story, since globalization is resulting in a much greater plausibility for any such claims to world history.

Hegel emphasized how at each historical cycle we reach a point when the forms and institutions we have created become so autonomous from us, their creators, so driven by their own logic that they become highly oppressive and dangerous to us. Marx exemplified these ideas but moved the focus from philosophy to a series of material changes in history which had indeed achieved new forms of abstraction on the one hand and particularity on the other. Marx focused on the growth of an autonomous logic to capitalism based on the original alienation of nature as private property. Marx did not suggest that capitalism was abstract because it was brutal or exploitative. It was brutal and exploitative because it was abstract, because it existed to follow a particular logic or rationality. It is equally misguided to see this as a triumph of the economic over the cultural, since capitalism was recognized by Marx to be that type of cultural phenomenon termed an ideology. One that dictated wages should not be increased because of a belief in the logical – i.e. rational – foundation of the market. The dominant belief system that served the dominant class.

Marx represented this not as immutable laws but as a stage in the story of history. It seems to me to follow that the one thing Marx could not have been if he were alive today would be a Marxist, since his sense of history was such that statements made in one century had to be superseded by a new understanding contemporary to the next historical moment. In short Marx today, if he were to be consistent, would have been searching not for capitalism but for some different but equivalent forces as relevant to the end of the twentieth century. He might have expected that, if communism was not to prove the end of history as he had hoped, then capitalism should have been superseded by some new even more abstract and alienating movement in history. In this I now want to suggest he would have been right.

Today we see a very different scenario from the kinds of capitalist society that Marx experienced and described. In most places our lives are dominated by many other forces than simply our niche within the circulation of capital. The diversity of society today is in complete repudiation of its homogenization under early capitalism. Today we have many different forms of capitalist societies

– unless of course you believe the postmodern mantra that all current diversity is simply superficiality. Ethnographers experience these differences as profound. The degree to which this has historically proved to be the case could hardly be clearer in that we live within the quite marvellous period within which the fastest growing version of capitalism is said to be chinese 'communist' capitalism (Smart, 1997). In Scandinavia, especially during the 1960s, a market system was made compatible with a highly welfare-orientated social democratic system, that could hardly have been more distinct from the conditions of the industrial revolution at the time of Marx. Another factor Marx ignored was the radical role that consumption would come to play in bringing back specificity in the form of a diversity of goods. The process of consumption often negated the alienation people felt from vast states and markets to a degree that their selection of specific goods could become a sign of the specificity of nationalism or the specificity of love. Elsewhere I have suggested that consumption could thereby be viewed as the negation of those general trends which were expressed as the abstraction and universalism of early capitalism (see Miller, 1987, 1995, 1997 and 1998a). The implication of these findings is that if it is the case that consumption and other forces have been used to counter the kinds of capitalism that Marx described, then I suspect Marx himself would be on the search for what might be called the negation of the negation. There should be some new force emergent that is based on the contradictions of these earlier forces – perhaps the contradictions of consumption. This would be some newly risen form of abstraction that will come to replace capitalism but must necessarily be more extreme than capitalism, more abstract and threaten to be still more dehumanizing until it in turn can be negated. Marx today would surely not be turning back to capitalism as its last predecessor but looking forward to this new target in its emergent formation.

That such a new chapter in the history of political economy might require writing is suggested by various changes in the last twenty years which seem to be putting into reverse many of the forces that had helped 'tame' capitalism during the previous century. The last twenty years seen from our local perspective (of western states) can only be regarded as regressive compared to the previous twenty years (i.e. 1959–1979). We see a decline in the movement towards greater equality, a decline in the faith in the welfare state, a decline in the sense of progressive potential that the 1960s more explicitly explored through an explosion of creativity in consumption itself – in music, clothing, lifestyles and so forth. This is conceded even by those most in favour of neo-classical economic perspectives. The journal *The Economist* which generally argues in favour of governments becoming more subservient to liberal economic ideologies, recently (20 December 1997) included a small note under the title 'Rising tide, falling boats'. Taking figures from the Centre on Budget and Policy Priorities it notes that in the United States from the late 1970s to today the family income of the richest fifth in that country has grown by 30% while that of the poorest fifth has declined by 21%, leaving the former with a 19.5 times higher income than the latter. Furthermore in all but one state the income of the middle fifth has also fallen during this period. It does not take a great deal of imagination to translate these figures into human experiences. There is a wealthy community for whom the additional increases merely add to monies that are almost beyond their ability to spend, but which front claims to the country as a whole becoming 'richer'. In the meantime those for whom every small shift represents a major constraint on their ability to realize their goals and values have seen a terrifying fall in their ability to participate meaningfully in their own society.

It is likely that similar figures would reflect a global scenario though data for the largest populations are unreliable. South Asia and the east coast of China have seen the rise of a formidable and evident middle class, but there is a suggestion that in the hinterlands of both areas vast populations have if anything faced a rise in inequality comparable to that of the United States. For any perspective with even a smidgen of concern for human welfare then, we live in regressive times. But why did this happen, what were the grounds for this turn around in our recent history, and how do they serve to negate the progressive potential of consumption in particular? I want to suggest that behind these figures lies the rise of virtualism, founded on the two defining characteristics described above. In order to give a more substantive sense of these two attributes of virtualism and how they appear in practice five examples follow which appear to suggest the generality of such trends.

## Case 1: structural adjustment and economics

The central player in the rise of virtualism is economics itself; that is, not economies, but the academic discipline of economics. This claim is, of itself, quite a radical departure from most studies of the contemporary political economy. In general it is assumed that it is transnational corporations, finance companies and the like which dominate the shifts in the economy, while economics is merely a discipline that describes and often gives moral support to such forces through its general ideological preference for free markets. My point is much more extreme. I want to suggest that under virtualism, the discipline of economics no longer merely describes the economy but rather the economy is increasingly forced to change itself in order to match the descriptions of abstracted models that are produced by academic economists.

The idea that this might be the case arose through a study of commerce in Trinidad (Miller, 1997). Trinidad is in a curious situation. In many respects the major capitalist institutions such as oil and grocery companies that had developed in order to exploit the island had been fought back – more through nationalism than socialism – and there had followed a degree of taming of the more negative effects of capitalist development, though more during the oil boom than today. Capitalism in contemporary Trinidad is surprisingly localized, including some powerful local transnationals that were starting to become incorporated within state-based welfare politics. Trinidad was no ideal society, but this was a clear step up from colonialism, let alone from slavery.

This progressive process is currently being undermined and I suspect negated because of the arrival of structural adjustment. This comprises a series of procedures and models that were devised by groups of economists working within some of the key institutions that were set up following the epochal meeting of Bretton Woods. These models fostered by the IMF, World Bank and their ilk are purely academic models, in the sense that they seem to pay no attention whatsoever to local context. Trinidad was being forced to adopt measures such as ending protectionism and abolishing currency control that would have been no different if this had been Nigeria or the Ukraine. These sometimes fitted the interests of capitalist corporations, but surprisingly often did not. I would argue this is because they are not the product of capitalism as an institutional practice of firms, nor of regional interests, though this has often been claimed. I would argue instead that they are simply idealized and abstract models that represent the university departments of economics engaged in academic modelling.

So while capitalism as a process by which firms seek to reproduce and increase capital through the manufacture and trade in commodities has become increasingly contextualized, another force has arisen which has become increasingly abstract. These are academics, paid for by states and given the freedom to rise above context to engage in highly speculative processes of modelling. While Marx had to tease out the abstract logic behind capitalism, today the still greater abstraction of academic economics is quite transparent and constantly confirmed by its very practitioners. We may not think of academics as particularly powerful, but then we are not economists. While capitalism was forced to engage with the world and was thus subject to the transformations of context, economics remains disengaged. Structural adjustment in Trinidad was not one iota Trinidadian, because economics has a form of power that again surpasses capitalism, that is the legitimate authority to transform the world into its own image. If we examine its details we find that all the changes that Trinidad was being asked to make were to remove what the economists call 'distortions'. For example, they must end subsidies to local companies because these distort our ability to see if that company is in fact competitive in the pure market. In short in every case where the existing world does not conform to the academic model, the onus is not on changing the model – i.e. testing it against the world – but on changing the world – i.e. testing it against the model. The very power of this new form of abstraction is that it can indeed act to eliminate the particularity of the world as a series of distortions which prevent the world from working as the model predicts it should (for other examples of structural adjustment see George and Sabelli, 1994; Mosely, Harrigan and Toye, 1991; McMichael, 1998).

So it is not that the principles of the market represent capitalism, but that capitalism is being instructed to transform itself into a better representation of a model of the market. If we look back home to the various forms of Thatcherism, Reaganomics and the degree to which we see them continued rather than rejected when their erstwhile opponents gain power as in Britain, then we see, as the Frankfurt school predicted, the rise of economics as the primary authority within politics. Just as capitalism had lost its ideological authority under a century of critique, it could be replaced by a safer more disengaged rhetoric. This process continues today. As I write (February 1999) the situation in Brazil has markedly deteriorated as the IMF attempts to resolve an economic crisis through what has become its classic strategy of making Brazil conform more closely to economic models. This includes massive privatization by which firms are purchased by outsiders and profits expatriated, a marked curb in social budgets and investment in schools, roads and other infrastructure. A prior attempt by government to give 200,000 people more land has reversed as 400,000 people have now lost land (*Guardian*, 1 February 1999, p. 19).

But in whose name is this being done, by what right do economists have this authority? All those institutions which proclaim the inevitability of the market, from economists and politicians downwards do so in the name of the consumer. The whole point of the market is that it is argued to be the sole process which might bring the best goods at the lowest prices to the consumer who thereby is the ultimate beneficiary from this process. Any force such as the state, the trade union or indeed the actual consumer that stands in the way of the realization of the economists' model constitutes a distortion preventing the realization of that ideal state in which everything works efficiently for the ultimate benefit of the consumer.

Looking as we always do with the hindsight of history, this seems almost inevitable. Since it was consumption as an expression of welfare that was the main instrument in cutting back the abstraction represented by previous forms of capitalism, the return of any still greater abstraction had to then overtrump consumption as human practice and put in its place a highly abstracted version of the consumer. The consumer of economic theory is not an actual flesh and blood consumer. Indeed, as practitioners of neo-classical economic theory constantly remind us, they make no claim to represent such flesh and blood consumers. They will tell us that their consumers are merely aggregate figures used in modelling. But the protestations of innocence are hollow, because these models that never claim to represent real consumers are the same models that are then used to force actual consumers to behave according to the predications of these virtual abstract figures. In short as though in some kind of global card trick an abstract virtual consumer simply steals the authority that had been accumulated for workers in their other role as consumers.

## Case 2: retail mergers and high finance

The second example arose out of an ethnographic study of shopping on a street in North London (Miller, 1998a; Miller *et al.*, 1998), and a subsequent attempt to link that study with the world of retail and the commerce of groceries (see Chap. 5 of Miller (in press) for a fuller description of this case study). The point of departure was the merger between the discount supermarket KwikSave, which was used by some of the most impoverished of the shoppers studied within the ethnography, and the Somerfield chain of supermarkets. At first this seemed to be a relatively straightforward story typical of mergers and consolidations within the world of business more generally. Seen from the point of view of strategies of supermarket stores KwikSave, having been for some time the most, indeed the only really successful discount grocery (Sparks, 1990) had from 1995 gone into decline. It attempted but failed to emulate the recent developments in rival stores such as the rise of own brands and value lines and was challenged by the rise of other discount supermarkets developed in continental Europe. Profits in 1995 of £125 million had fallen to £2.8 million by 1997. In the meantime Somerfield had survived a complex leveraged buy-out in 1989 which nearly destroyed it (in its earlier guise as Gateway) and the Asda chain as well, and was emerging as an important niche store, recolonizing high street locations and directing itself more to local communities. By 1996 it was strong enough for a return to market flotation. Given the general expectation that only the strongest chains would survive, a merger sounded a sensible measure at the time. At least this was the story generally to be found within the industry.

I would argue, however, that behind this simple tale lay another more complex story that goes right back to the Opium War in nineteenth century China. The change in fortune at KwikSave occurred when the original board of directors that had brought about its success was dominated by a new power representing the 29% stake held by a firm called Dairy Farm. This was in fact a supermarket chain which formed one branch of Jardine Matheson, a firm that owned much of the valuable down-town property of Hong Kong. This was the same firm that had been largely responsible for the Opium War and indeed the foundation of Hong Kong as a port protecting its original interests in exporting Indian grown opium into China. Not surprisingly it is not a firm much loved by the Chinese authorities. Under the looming threat of the return of Hong Kong to Chinese control Jardine Matheson had embarked on a number of ventures around the world in order to become less parochial. But just recently it found that it was under threat back in its heartland, including that of a possible takeover, and was therefore liquidating these ventures and bringing back the capital invested. So at the same time that it sold KwikSave (for a much larger sum, thanks to the merger with Somerfield, than would otherwise have been possible) it was also selling other European investments such as a Spanish supermarket chain which had undergone the opposite trajectory, from long-term loss making to a recent rise in profitability (see Miller, in press).

The implication of this particular study seemed to be that, in quite concrete events such as a particular merger, it may not be the logic of the particular type of commercial venture that dominated but rather the underlying dynamics of global finance. In which case my conclusions come close to the general findings of recent work on the British grocery trade by a group of what have come to be called 'new retail geographers', in particular, the work of Guy, Hallsworth, Marsden and Wrigley (see especially Wrigley and Lowe, 1996; Wrigley, 1998). Their starting point has been a general concentration in retail over the last three decades with a mere 39 retailers accounting for over half of all retail sales in Britain. In the grocery trade this was particularly dramatic. 'Between 1982 and 1990, the market share of the top five grocery retailers increased from under 25 per cent to 61 per cent of national sales' (Wrigley, 1993, p. 41). They are well informed about the more parochial factors such as the various technological innovations of just-in-time distribution and EPOS systems at the point of payment (see Hallsworth, 1992; Doel, 1996; Hughes, 1996), but here too it is larger exogamous factors that appear to dominate. For example, Wrigley (1996) noted that the huge investment in property at massive out-of-town sites that at first seemed to be a grounding of retail capital in assets became re-viewed as sunk costs, a more general problem arising from the fact that it can be easier to invest and expand in many areas than to accomplish 'market exit'. As a result of such factors the academics working on retail, just as the heads of retail corporations themselves, have to work in tension between financial markets that are only really interested in capital itself and how productive it is being as compared to where else it might be invested, and the specific considerations of the retail industry. It is when assets such as property valuations seem to shift in an unrelated sequence to the ordinary questions of profit from retail sales that the larger transcendent nature of business as merely an aspect of capitalism comes to the fore.

Hallsworth and Taylor (1996) provide the illustrative story of Asda, which started life as a fairly sensible retail chain with a regional base, funding its expansion largely from its own profits. The problem arose as it moved South and came within the vision of the City of London, where such forms of expansion are seen as an anathema. By contrast, the City reflects a financier's vision in which the driving force of good industry and efficient capital reproduction is debt, based either on stock or increasingly private forms of debt. The attraction of debt (and the fees paid to financial advisers) are the central imperatives behind the phenomenon of the leveraged buy-out which so nearly destroyed both Asda and Somerfield. This occurs when a company buys back all its own shares and then has to become 'leaner and fitter' in order to pay off the debt it has thereby accumulated. The LBO has become something of a fashion in grocery and in other trades in Britain and the US (see Burrough and Helyar, 1990, for the most spectacular examples of an LBO).

At this level finance itself tends to work to an agenda closer to that of economics where in some ways the more abstract the model and the more decontextualized the operation the higher the 'status' of the financier or economist concerned, many of whom seem to have a penchant for almost an aesthetic minimalism where idealized models of derivatives are preferred to the vulgarities of street nous and pragmatic or contextualized commerce such as retail. Followed upwards this leads ultimately to companies such as the absurdly named Long-Term Capital Management. This company is in fact all about the short-term gains of hedge funding, the use of 3 or 4 billion dollars of equity to invest more than 200 billion dollars in projects such as leverage buy-outs. It is the models of Nobel Prize winning economists that came to determine the shape and priorities of finance in practice. So if we follow the implications of supermarket mergers high enough we see behind them the first characteristic of virtualism, the determinant nature of abstract models over actual economic action.

The second defining feature, the fetishism of the consmer, has also been located by the new retail geographers. The primary growth of grocery retailing as one of the most dynamic sectors of British capitalism took place under the

Thatcher regime and as Wrigley argues it was the much more positive stance taken by the British state that accounts for much of the contrast with developments in the US where the state took a much more guarded view of the implications of such expansion (Wrigley, 1992). One of the contributory factors appears to have been the desire to delegate responsibility downwards for the whole arena of health regulation in foodstuffs, a topic which has become more and more important in terms of the public perception of risk and the potential for periodic crises of confidence (Flynn, Harrison and Marsden, 1998). But more generally the food retail sector has been prepared to take on a mediating role in the relationship between the state and the consumer. Marsden and Wrigley note that as a result the supermarkets took on a 'powerful role in structuring consumption around their own particular notions of the "consumer interest", and began increasingly to represent that "consumer interest" in their relations with government' (1996, p. 43). Thatcher never failed to describe herself as the Grocer's Daughter who rewarded the thrift of her population – in returning to what was presented as common sense values after its flirtation with left wing ideologies – by demonstrating that the market would deliver wealth and choice. It is very likely that the perceived high quality, high choice and good delivery of the major supermarkets was critical to the popularity of the political programme itself. It is in this sense that the supermarkets could be said to have delivered the citizen as consumer to the state and the promises of the state to the consumer. In short the supermarkets became the virtual consumer replacing the direct relationship between the state and its citizens as consumers.

## Case 3: auditors and the universities

To experience virtualism we do not need to look just to the esoteric world of high finance and economics. An example may be found just down the corridor. Over the same last two decades we have seen the expansion and democratization of higher education itself come under increasing threat. Most academics will happily bore anyone for hours about how the time they used to spend in teaching and research has become increasingly swallowed up in dealing with paperwork that has been generated by a variety of new institutional requirements that have in common the general term auditing; paperwork so formidable that what was once measured in pages is now measured by weight. These may be teaching (quality assurance) audits, HEFCE research audits (RAE), ESRC audits of courses that may receive grants. You can add your own favourite outrage – if you haven't yet had your Quake or Lemming like progress through the levels blasted by one of these demon audits – and lost a life or two – well, you will.

Ironically one result of that Thatcherite drive that I would suggest sincerely believed in limiting centralization and bureaucracy, was that it increased substantially the resources going into management and homogenizing bureaucracy. So this historical process cannot be understood as some simple expression of political will. It is rather an expression of contradiction.

Underlying these developments may be found once again the virtual consumer. All of these auditing procedures are justified on grounds that lead to the consumer, either as the taxpayer getting value for money or as the actual recipient of the services. Thus the clichés that those who were once students or patients are now understood as consumers of health or educational services. Indeed it is hard to imagine what political authority is left today that is not reducible to this hegemonic rhetoric of the consumer. Yet as any academic knows from experience the main effect on students is a loss of teaching time, since they are replaced by those who stand for them in aggregate form, that is the managers who carry out audits.

The argument is given firm grounding by Power (1994, see also 1997) in his analysis of *The Audit Explosion*. He starts by noting the sheer extent of growth in auditing now covering almost every branch of governance from medical, to criminal justice, from charities to corporate control. Some 8% of all graduates are now undertaking training within accountancy firms (ibid., p. 2). He also notes that auditing cannot be seen in isolation but is part of a growth of a more general culture of 'new public management' (ibid., p. 15).

As well as fetishing the consumer the audit explosion also conforms to the first characteristic of virtualism which is that it gives power and authority to abstract modelling. Power notes that 'One of the paradoxes of the audit explosion is that it does not correspond to more surveillance and more direct inspection. Instead, audits generally act indirectly upon systems of control' (1994, p. 19) so the national audit office is really the control of control, i.e. legitimacy now depends more on being seen to be audited than in actual outcomes of audit. 'By abstracting from local organisational diversity it has enabled audit to assume the status of an almost irresistible cultural logic' (ibid., p. 20). Instead of making organizations more transparent they tend to make them more opaque, adding another layer of control.

The other consequences are well known. Power notes that

> Scientists are changing research habits, and a whole menu of activities for which performance measures have not been devised have ceased to have official value. Editing books, organizing conferences and, paradoxically, reviewing and facilitating the publications efforts of others fall out of account. (1997, p. 100)

Strathern (1997) provides an anthropological perspective as participant observer of the audit process. She observes that under the strictures of audit, measures become reduced to targets (p. 308); research quality is conflated with research departments (p. 310); descriptions become prescriptions (p. 312); any sense of contradiction or ambiguity is replaced by canons of clarity and itemization (p. 315); and quality of teaching becomes reduced to a measure of usage of new information technologies (p. 318). Worst of all for a discipline whose strength lies in embedded knowledge and the study of context, terms such as transferable skills and the pressure to make intellectual work visible and measurable favours disembedded knowledge and attacks the very qualities of complexity and sensitivity that anthropologists are taught to value (p. 320).

# Case 4: *management consultancy and category management*

One of the arguments for relating all these examples to a larger trend termed virtualism is that this helps make more sense of each of them. Much of the writing on the rise of auditing assumes this to be a particular phenomenon of the public sector, which does not have to work under the same 'bottom line' constraints as the private sector. It is therefore highly instructive to note that almost exactly the same changes have been going on in the private sector at almost exactly the same time (though both now share an ideology of cost-benefit analysis: see contributions in Carrier, 1997). Just before it was taken over by Somerfield, KwikSave had secured the services of Andersen Consulting. This contract was ended early prior to the merger. One reason may have been that the £19 million anticipated bill together with the huge costs of carrying out the recommendations being made, and the fact that these recommendations were immediately disparaged by business journalists, suggest that far from being a potential cure this arrangement looked more like the straw that would break this particular camel's back. Andersen Consulting had started as a mere spin off from accountancy operations in 1989. By 1995 it had an annual income of US$4.2 billion, employed 44,000 people and had 152 offices in 47 countries. The scale and speed of this growth could almost be matched by other companies such as McKinsey and Figgie. Individual companies such as AT&T might spend half a billion US dollars alone on such consultants. O'Shea and Madigan (1997, see also surveys by *The Economist* over March 1997), who describe this transformation have used instructive sources, such as the accounts of court cases where management consultants have been sued, to shed considerable doubt as to what if anything companies gain from this vast expenditure on what they call 'Dangerous Company'. In particular commentators are starting to realize just how formulaic are the 'cures' that such companies tend to hawk from firm to firm, rather in the manner of nineteenth century patented medicines that will relieve all troublesome symptoms with sufficient doses of the right tonic.

Once again we can find the two defining characteristics of virtualism within this phenomenon. 'Cash-strapped' companies will spend vast sums on various forms of management consultancy, as part of what Thrift (1996) recently called soft capitalism, or Salaman (1997) called the new narrative of corporate culture. These are based on an equally abstracted discourse and are likely to be just as detrimental to the financial health of these willing victims. Thrift (1998) provides a detailed account of the long history behind what appears now as an explosion of virtual business knowledge and theory. Just as in the case of high finance and academic economics, a firm such as Andersen Consulting will take bright new recruits from any number of disciplines and train them in the formulaic procedures of management consulting, which has a strong tendency to prescribe a cure including very expensive changes in informational technology which can, of course, be provided at a high price from the likes of Andersen Consulting.

Just how formulaic these procedures turn out to be and how far therefore they accord with the two definitional characteristics of virtualism may be seen if we move the focus in on to the detail of what may occur. This is possible thanks to some recent research by Cook, Crang and Thorpe (forthcoming) on the operation of 'category management'. This phrase has become a mantra which is supposed to almost guarantee an increase in profits to whichever line of groceries it is applied to. It is not hard to find the details of this fashion, since the Institute of Grocery Distribution provides its own handy guide to the practice (McGrath, 1997), without any of the costs of the management consultants that usually supply it. It involves the reordering of goods within supermarkets to match a new set of categories that are supposed to more clearly reflect the consumer, e.g. all breakfast goods together. In practice, as Cook, Crang and Thorpe have argued, there are grave doubts as to whether any real benefits accrue, in large measure because, as with so many of these fashions, they are applied across the board with little concern for the local knowledges that many of the grocery firms' own employees have garnered from years of service. What is particularly apropos here, however, is that category management is merely one part of a larger argument that management consultancy uses to justify the claims to its importance for grocery retail. As such it is part of a larger master plan which is called 'efficient consumer response'. This, in turn, comes with a whole jargon of key categories such as 'sleepers' and 'supply chain management'. Throughout this exercise then the legitimation of management consultants has to be that somehow they represent the consumer, or the way the companies can come closer to the consumer as shopper, even though it is generally evident that the opposite would be the case. The supermarket is probably much closer to the shopper through their everyday business than the consultants who promise to deliver the shopper to them. What management consultants and category management provide to retail is once again the virtual consumer, where real consumers are replaced by their representation in aggregate form within formulaic models.

## Case 5: postmodernism and the virtual consumer

So far my story has found villains that are easily identified as villainous, but for my final example I want to pick a less readily acceptable target, that is ourselves. After all economists are generated by academic constructions, and we are all also academics. During the same period that economists have been gaining in power, and auditing has expanded, social scientists such as human geographers and sociologists have been increasingly concerned with related phenomena. In particular a literature on consumption that was once mostly conspicuous by its absence is now almost impossible to keep track of (see Miller, 1995). The key term that developed over these same two decades to characterize the new style and content of such writing was postmodernism. This has many connotations but I refer to those ideas about our contemporary world that roughly follow Jameson's (1984) original *New Left Review* article. Consumption was certainly central to the analysis of this new age, and Baudrillard, a key figure in this literature, began his work in the analysis of consumption. But once again appearances may be deceptive. What was produced was not an empathetic understanding of consumption as complex human practice and the struggle for human welfare within capitalism, but a rhetoric based on the homogenizing term 'consumer society' or 'consumer culture', that simply stood for the consumer, whether historically as in citing Benjamin's arcades project or as contemporary symbol of superficial difference. In this literature people as consumers become reduced to two dimensions. Rather than describing actual consumption, we find once again an academic model of the virtual consumer that comes to stand for and to supplant actual consumers engaged in consumption as practice. One can see a similar movement in the way the original ethnographic foundations of Stuart Hall's cultural studies at Birmingham quickly became its opposite, as cultural studies was transformed into the vast reading of texts allowing academics to read off the nature of the consumer society according to what have become rather set models rather than undertaking the original task of an empathetic experience of moments of consumption.

There are today plenty of studies of consumers and shoppers based on detailed ethnographic and other studies (see Miller, 1995), but these lend no support whatsoever to the general theorizing of the consumer society as postmodern. For example, in my own work on shopping (Miller, 1998a) I have argued that where others see de-contextualized processes that merely echo the imperative of commerce, I have found a developing technology for the expression of love in families that is principally concerned to negate the interests and alienation caused by both the market and the state. So in the story of the rise of virtualism I am afraid we academics of the humanities and social sciences are not the grand heroes and heroines who have stood up against the rise of economics and their virtual consumers, but simply a less powerful but parallel academic movement that reflects the very same trajectory that the economists have followed; equally expressive of the movement from capitalism to virtualism within which academics have generally formed the vanguard. The only reason that this case study does not fit quite as well as the previous four is that although there is a clear fetishism of the consumer at the heart of the theory of postmodernism, the advocates of this theory tend to have rather less power in the world to transform society into becoming merely a better fit to their models. Although within academia itself one can see precisely this phenomenon. That is to say if one was to look for an example of postmodernism that is a highly superficial genre of endlessly circulating citations which increasingly bears little relationship of signification to any external world then the obvious example would be the literature on postmodernism which is generated by just such a process of circulating citations. So the very human geographers who proclaimed most stridently that they were attacking economists through the cultural turn were, I would argue, merely creating the equivalent movement within their own discipline (for evidence behind this assertion see Miller, 1998b).

## Conclusion

Dialectical stories are epigenetic – that is, they absorb rather than supplant their previous chapters – so that other symptoms of virtualism are easier to see as continuity with capitalism. They appear as simply more abstract forms of macro-capitalism. An example would be the increase in abstract or virtual versions of capital itself in the extraordinary new forms of highly esoteric finance capital that have been studied by geographers such as Nigel Thrift and David Harvey. Of course different regions also remain at different stages in this story. We co-exist with some plantation-based societies where Marxism remains the much more relevant discourse for understanding the present predicament than would virtualism be. Nor do I wish to suggest that virtualism is as all encompassing a perspective as some uses of the term capitalism aimed to be. A conspicuous absence from my list has been the rise of new informational technologies and the internet which gave rise to the term virtual in the sense that I have used it. I would suggest in a final irony that this is the one area where things are actually much less virtual, since the capacity for ordinary consumers to appropriate the internet and turn it back into high plural and meaningful worlds is becoming ever more evident.

My story began with capitalism and then told the various ways capitalism as a force for abstraction from humanity had been tamed – a chapter within which consumption comes to play a conspicuous role in creating new particularity and localisms. It went on to suggest that new forms of abstraction have arisen. We do not discern these at first since they use the very medium that had transformed capitalism as their vehicle for replacing capitalism. In all five examples that I provided the terms consumer and consumer culture remain central to their authority. But in each case these involve substituting a virtual consumer for the presence of actual consumers. Also in all but the final case they demonstrated the power to change that which they were supposed to be describing into a better fit to the models they proclaimed as descriptive but which were in practice prescriptive.

It follows from a dialectical perspective that all new forms of abstraction contain positive potentials for some time in the future. Such a highly abstract argument as that delivered in this chapter cannot claim opposition to abstraction *per se*. I prefer to follow Simmel's (1978) perspective on money in welcoming all new forms of abstraction in their potential contribution towards understanding these same processes, including therefore the potential of economics – modern governance depends upon modern economists, though it could do more to defend other forms of political ideology. The stances I have taken relate not to intrinsic properties of institutions but historical moments in a story. The narrative only suggests that, at this particular moment of history, we should be fearful that if these forces continue to be allowed to manifest their internal logic towards abstraction with increasing power to mould the world to these abstract shapes, then the result is already evidently regressive, leading to a negation of a series of progressive movements and forces and replacing these with what may become increasingly oppressive and vast rhetorics and practices in the image of the economists' model of the market. In as much as we could better understand these current trends by looking forward and seeing them as incipient developments rather than trying to fit them into earlier critiques then I have suggested the term virtualism to summarize this phase in history as a replacement for the critique of capitalism.

To conclude: I have no argument with a cultural turn in that a concern with the minutiae of cultural life as, for example, the world of consumption needed to be developed. But this should have been seen as complementary rather than in opposition to the grand narrative that attempted to create a larger understanding of the political economy. Rather I would argue that both are part and parcel of a continued commitment to political economy, precisely because the political economy is itself not simply an economic force, but a historical movement in culture. I am not opposed to abstraction, though personally as an ethnographer I do not particularly enjoy this eagle eyed perching in order to determine the changing landscape of history and discern the shape of the woods. I am happier wandering amongst the trees with all the sensual delight and satisfaction to be found in ethnographic stumbling around the leaves and flowers of the undergrowth, and most of my work is I hope firmly located in the mud and mire, for example, of shopping. But as I have tried to suggest in Miller (in press) a dialectic perspective calls for the constant re-immersion of the particular in the universal and the universal within the particular since, if the problem of modernity is the degree to which abstracted forms fly off out of reach of human welfare, then our task is to continually re-forge the link between the intimate worlds uncovered in ethnographic encounters and the empathy with ordinary human welfare and the macro knowledge exemplified in grand narrative.

## Note

1. This chapter represents a precis and an amalgamation of the arguments and examples to be found at greater length in Miller (1998b) and Chap. 5 of Miller (in press). Acknowledgements will be found in Miller (in press).

## References

Burrough, B. and Helyar, J. 1990 *Barbarians at the Gate*. London: Arrow.

Carrier, J. 1997 *Meanings of the Market*. Oxford: Berg.

Carrier, J. 1998 'Introduction.' In J. Carrier and D. Miller Eds. *Virtualism: A new political economy*. Oxford: Berg, 1–24.

Cook, I., Crang, P. and Thorpe, M. (forthcoming) 'Constructing the consumer: category management and circuits of cultures in the UK food business.' *Geoforum*.

Doel, C. 1996 'Market development and organizational change: the case of the food industry.' In N. Wrigley and M. Lowe Eds. *Retailing Consumption and Capital*. London: Longman, 48–67.

Flynn, A., Harrison, M. and Marsden, T. 1998 'Regulation rights and the structuring of food choices.' In A. Murcott Ed. *The Nation's Diet*. London: Longman, 152–67.

George, S. and Sabelli, F. 1994 *Faith and Credit*. London: Penguin.

Hallsworth, A. 1992 *The New Geography of Consumer Spending: A political economy approach*. London: Bellhaven.

Hallsworth, A. and Taylor, M. 1996 'Buying power: interpreting retail change in a circuits of power framework.' *Environment and Planning* A: 28: 2125–37.

Harvey, D. 1996 *Justice, Nature and the Geography of Difference*. Oxford: Blackwell.

Hughes, A. 1996 'Forging new cultures of retailer – manufacturer relations?' In N. Wrigley and M. Lowe Eds. *Retailing Consumption and Capital*. London: Longman, 90–115.

Jameson, F. 1984 'Postmodernism, or the cultural logic of late capitalism.' *New Left Review* 146, 53–92.

McGrath, M. 1997 *Category Management*. Institute of Grocery Distributors.

MacLennan, C. 1997 'Democracy under the influence: cost-benefit analysis in the United States.' In J. Carrier Ed. *Meanings of the Market*. Oxford: Berg, 195–224.

McMichael, P. 1998 'Development and structural adjustment.' In J. Carrier and D. Miller Eds. *Virtualism: A new political economy*. Oxford: Berg, 96–116.

Marsden, T. and Wrigley, N. 1996 'Retailing, the food system, and the regulatory state' 33–47 in Wrigley, N. and Lowe, M. (eds) *Retailing, Consumption and Capital: Towards the New Retail Geography*. Harlow: Longman.

Miller, D. 1987 *Material Culture and Mass Consumption*. Oxford: Blackwell.

Miller, D. 1995 Ed. 'Introduction.' In D. Miller Ed. *Acknowledging Consumption*. London: Routledge.

Miller, D. 1997 *Capitalism: An Ethnographic Approach*. Oxford: Berg.

Miller, D. 1998a *A Theory of Shopping*. Cambridge: Polity Press.

Miller, D. 1998b 'A theory of virtualism.' In J. Carrier and D. Miller Eds. *Virtualism: A new political economy*. Oxford: Berg, 187–215.

Miller, D. (forthcoming) *The Dialectics of Shopping*. Chicago: University of Chicago Press.

Miller, D., Jackson, P., Thrift, N., Holbrook, B. and Rowlands, M. 1998 *Shopping, Place and Identity*. London: Routledge.

Mosely, P., Harrigan, J. and Toye, J. 1991 *Aid and Power: the World Bank and Policy Based Lending*. London: Routledge.

O'Shea, J. and Madigan, C. 1997 *Dangerous Company. The Consulting Powerhouses and the Businesses they Save and Ruin*. London: Nicholas Brearley.

Power, M. 1994 *The Audit Explosion*. London: Demos.

Power, M. 1997 *The Audit Society*. Oxford: Oxford University Press.

Salaman, G. 1997 'Culturing Production.' In P. DuGay Ed. *Production of Culture/ Cultures of Production*. London: Sage, 235–72.

Simmel, G. 1978 *The Philosophy of Money*. London: Routledge and Kegan Paul.

Smart, A. 1997 'Oriental despotism and sugar coated bullets: representations of the market in China.' In J. Carrier Ed. *Meanings of the Market*. Oxford: Berg, 195–224.

Sparks, L. 1990 'Spatial structural relations in the retail corporate sector.' *Service Industries Journal* 10: 25–84.

Strathern, M. 1997 ' "Improving ratings": audit in the British University System.' *European Review* 5: 305–21.

Thrift, N. 1996 'Soft Capitalism'. *Cultural Values* vol 1, 27–57.

Thrift, N. 1998 'Virtual capitalism: the globalisation of reflexive business knowledge.' In J. Carrier and D. Miller Eds. *Virtualism: A new political economy*. Oxford: Berg, 161–86.

Wrigley, N. 1992 'Antitrust regulation and the restructuring of grocery retailing in Britain and the USA.' *Environment and Planning A*: 24: 727–49.

Wrigley, N. 1993 'Retail concentration and the internationalization of British grocery retailing.' In R. Bromley and C. Thomas Eds. *Retail Change*. London: UCL Press, 41–68.

Wrigley, N. 1996 'Sunk costs and corporate restructuring: British food retailing and the property crisis.' In N. Wrigley and M. Lowe Eds. *Retailing Consumption and Capital*. London: Longman, 116–36.

Wrigley, N. 1998 Ed. 'Theme issue: retail development.' *Environment and Planning A*: 30: 13–66.

Wrigley, N. and Lowe, M. Eds. 1996 *Retailing Consumption and Capital*. London: Longman.

# Cultivating ambivalence

## The unhinging of culture and economy

*David B. Clarke and Marcus A. Doel*

> The 'hard law of value', the 'law set in stone' – when it abandons us, what sadness, what panic!
>
> (Baudrillard, 1994, p. 156)

> value has itself . . . chosen a fatal strategy . . . by liquidating everything and charging on regardless
>
> (Baudrillard, 1998, p. 3)

On finding themselves hopelessly lost, a couple stop to ask directions: 'Say! How do we get to Leatherhead?' After lengthy consideration, the stranger replies: 'Well, if I were you, I wouldn't start from here.' A somewhat feeble joke, perhaps; but nonetheless, a perfect illustration of the problems faced when trying to get beyond the impasse that has been produced by the long-standing separation of 'culture' and 'economy'. Needless to say, many are already somewhere else. The majority, however, are stuck in a rut, desperately attempting to conjugate culture and economy, through such well-worn notions as the 'commodification of culture' and the 'postindustrialization of economy', or through newfangled notions like 'infantile', 'soft', or 'virtual' capitalism. Where it was once possible to feign a separation of cultural and economic spheres, such a conjuring trick has become increasingly difficult to pull off. At the very least, economy and culture depend upon one another, bend to one another, and penetrate one another. Such is the co-implication and interlacing of one with the other that attempts to distil the essence of each seem increasingly anachronistic and doomed to failure. For it is not so much that culture and economy affect and determine one another, or that each is contextualized, actualized and embedded within the vicinity of the other, but rather that the differentiation of each from the other simply does not hold. For us, there is neither economy nor culture nor a conjugation of the two. Something else is in play alongside such an unproductive exchange of terms: a pointless circulation of signs. Imagine: what if culture and economy alike hinged upon a useless exchange of signs?

Take any old thing, such as a festive edition of the British Broadcasting Corporation's illustrious *Radio Times* magazine. Our copy is advertising a set of *Stars Wars* trilogy calendars for only £14.99. We wonder what one would do with a *set* of calendars. We suspect that they should not be split: one for me; one for him; one for you. After all, a trilogy is not a one plus one plus one. It does not amount to three. No, a set is *a* set: it is a quality rather than a quantity. A set is more than the sum of its parts. Or rather, it exists *alongside* its parts, with a presence all of its own. (And we wonder if one set will suffice: for what? – precisely.) So, should we leave the set of calendars in the original wrapping, or should we nail them up in a line, recording each red-letter day in triplicate like an old-style bureaucrat? And what difference would it make to live one's life in four dimensions: real-time and calendar-times? Or would that be four-by-four dimensions – or even more? Our useless speculation is brought to an untimely end when we notice that for a mere £5.99 we could buy the *Changing Rooms* 1999 calendar. Great value! It claims to be 'the official calendar'. We infer that it wants to be obeyed and adored. A sign of the times, perhaps. Like the television series on which it is based, the calendar depicts homely make-overs, featuring 'before and after shots as drab and ordinary rooms become colourful and exciting living spaces'. It promises to 'give *you* creative inspiration' in your endeavour to make a house into your home. With *Changing Rooms*, economy again becomes a matter of household management, just as culture again becomes inhabitation and worship. But what if the house in question were a haunted house? What if this useless do-it-yourself plasticization of commodities and signs could barely suffice to paper over the 'desert of the real'? And what if, when all was said and done, no one and no-thing were at home amidst these 'signs taken for wonders'?

In our mind's eye we see the pleasing prospect of walls shedding their pictures, trinkets and crockery; casting off their paint and wallpaper; and filling up with calendars of every conceivable complexion: one wall fills with land-scapes; another with figures; another with film stars; another with cats. The hall is becoming monochrome in a series of hundreds of calendars that graduates from white to black. The kitchen swells with square calendars offset against round ones. The utility rooms are giving themselves over to flesh-toned Januarys, lush-green Augusts, and azure Decembers. The stairs take on a rainbow effect shot through with discarded spiral binders that give an impression of heavy rain, while the bathroom undulates with water-colour calendars from the early 1980s. The loft has become a refuge for boy bands. Bedrooms are in turn social realist, abstract expressionist, surrealist, minimalist, primitivist, etcetera. The garage has become a focus for soft porn; the shed an outpost of hardcore. Garden calendars cover the gardens; car calendars cover the cars; calendars with misprints or tears fill the bins. Meanwhile, the street is overflowing with humor-ous calendars. Delivery vans are swept away on the crest of calendar waves, while still more calendars tumble from their flung-open doors. The world is now one mad profusion of calendars, stretching as far as the eye can see. Even the sky has succumbed, as billions upon billions of digits float by in sublime cloud formations. And as the Op Art sun finally sinks from view something dawns on us: *the self-evidence of something's use-value is an illusory effect of scarcity, constriction and conscription.* This is why 'use' is an order-word to which one is expected to submit. It carries the force of a social, structural and functional imperative. By contrast, *excess puts everything into suspense.* Consequently, 'a' is a pass-word since one can no longer know in advance what to do with *a* calendar, *a* letter, *a* table, etcetera. Each event is always already and forever 'up in the air', so to speak. Everything is irreducibly untimely and undecidable. Everything trembles as a result of this excess. For excess, like a set, is not a quantity, but a quality. It does not mean 'great number', but trembling suspense: to become disoriented (in a manner akin to existential nausea: existence without essence). Similarly, scarcity does not mean 'scant number', but numb finality: to resign oneself to what appears given. Scarcity pacifies, constrains and restricts; excess sweeps one away. However, it is not that scarcity and excess are opposed, just that the former is a certain disavowal of the latter; which is to say that numb finality is a way of enduring, of living with, perpetual suspense. One should recall that the word 'scarce' derives from the French *(e)scars* and *eschars*, which in turn come from the Latin *excerpere*, to pluck out of. So, faced with an excess of reality, wouldn't you want it to mean something, to be worth something, to be useful for something or other? Wouldn't you want a reason for existence? But we suspect that the world's 'mere' existence will exact its revenge against all gestures of onto-theology (cf. Baudrillard, 1996). We suspect an immanent return of the repressed: an outpouring of useless, meaningless and valueless excess! Such is the cancerization of culture and economy, of sign-value and use-value: function and finality derailed and deranged. Such is the useless exchange of signs that subtends each act of making sense and having an effect. The structure that constricts is perpetually breaking down. The long and the short of it is that each is given in excess. One begins with excess, with the excessiveness of mere existence, with 'existence before essence', as Sartre aptly phrased it.

The *Radio Times* calendars are available by credit-card hotline and will arrive in time for Christmas. There are dozens to choose from, and they would make perfect gifts for all those who would like to conjugate economy and culture, commodities and signs. We thought that the use-value of a calendar was unproblematic: now we are not so sure. We believed that they should be used one at a time, and that in their rolling over from year to year one could rely upon a regular rhythm and invariant structure to one's life. For example, birthdays, anniversaries, and holidays appeared to offer an endless return of the same: as do Mondays and Thursdays; Mays and Junes; and 2:30s and 11:15s. 'Have you not done tormenting me with your accursed time? It's abominable. When! When!' says Pozzo to Vladimir [*suddenly furious*] in Beckett's *Waiting for Godot* (1986, p. 82). 'One day, is that not enough for you, one day like any other, one day he went dumb, one day I went blind, one day we'll go deaf, one day we were born, one day we'll die, the same day, the same second, is that not enough for you?' Such are the differences that produce repetition: again and again and again. But what should one *do* with many calendars, with many *sets* of calendars? Should one abandon all but one of them? And what about the numerous diaries and year planners that come one's way, not to mention the memo-pads, reminder-notes, etcetera? Should all of the permutations and combinations be sacrificed for just one? Amid the swarm of calendars and pads, each moment seems to proliferate wildly. From the off, each seems to 'go off' in a myriad of incommensurable ways. One plays out one's life in innumerable dimensions. Such a 'pullulation of time' neatly expresses the unsuturable disjointure and interminable stuttering of difference-producing repetition. We remain eternally disoriented. Such is an immanent and impersonal life.

With an excess of calendars already in the offing from our trawl of *Radio Times* magazine, our attention is drawn to the nine letters of the just-for-fun 'Trackword' that are arranged in a three-by-three table. One is invited to string together words of three letters or more. It seems to be a timely distraction. *Challenge*: to fashion seemingly meaningful events amid an indifferent and arbitrary structure; to make meaning out of non-meaning. *Brainbox*: 51 words. *Average*: 30 words. D-O-E-L is in the table. So are B-U-N, D-O-U-B-L-E and L-O-O-T. *Average* seems a long way off. Our stock of cultural capital feels somewhat depleted. We are not very good in the pointless-stringing-together-of-signs stakes, in endlessly exchanging them and substituting them in rounds of interminable circulation. It is too formal, too abstract, too structural. We are not fond of producing merely for the sake of producing. It is too – processive. There's B-L-O-T, D-O-T, T-O-O-L and T-O-O-L-E-D. But is anything of any real value actually being produced in this seemingly useless enchaining of signs? Or are we merely caught up in a simulacrum of meaningful production, whose only significant out-turn is the endless investment and reinvestment in the code itself? And will we ever exhaust and overcome this little set of structural relations that can be reproduced and extended to infinity? For there is something cancerous about the structural play of signs, about this simulacrum of movement: something that introduces disorder into communication and reproduction from the off. Perhaps we should move away from mere word-play and turn towards something more solid: a material commodity, perhaps.

'A commodity', writes Marx (1954, p. 76), 'appears, at first sight, a very trivial thing, and easily understood.' However, 'Its analysis shows that it is, in reality, a very queer thing.' Take wooden tables, which may, like calendars, turn out to be useful for someone or other, perhaps even useful as tables – or calendars: Dr d. woz ere in '93.

> The form of wood . . . is altered, by making a table out of it. Yet, for all that, the table continues to be that common, every-day thing, wood. But, so soon as it steps forth as a commodity, it is changed into something transcendent. It not only stands with its feet on the ground, but, in relation to all other commodities, it stands on its head, and evolves out of its wooden brain grotesque ideas, far more wonderful than 'table-turning' ever was. (Marx, 1954, p. 76)

This is what the dialectical materialist tells us. The wooden table has its feet *and* its head on the ground. It is not that the table is really a wooden thing and that its facing up to other commodities is illusory. Its discourse with other commodities is as real and as intense as its woodenness. This folding table bends towards the material and the immaterial, to the useful flesh of the earth that it is (it is wooden through and through) and to the grotesque ideas that it spins off (it is conversing with others, with other folded things, which may or may not have been – Once Upon A Time, Not So Long Ago – wooden through and through). For a commodity is not a thing. It is an embodiment of differential relations ('the interval takes all. The interval is substance' (Deleuze and Guattari, 1988, p. 478)). Like a contortionist, the commodity holds the tension between the concrete (its wooden substance) and the abstract (its relation to other commodities) within itself, and this is why the spring-loaded fold of real contradiction lends itself to dialectical relief. The tension that articulates a commodity awaits release. Its differential calculus begets reintegration. Or does it?

We prefer to affirm, with Baudrillard (1993, p. 8), 'The end of labour. The end of production. The end of political economy.' But we should stress at once that this is not the end in the sense of a point that would terminate a line, still less a moment of dialectical sublation would lift a contradiction to a higher dimension. We are at the end in the sense of withdrawal and ex-termination: not surpassing but perpetually falling short. This is how one may finally have a chance of getting out of the dead-end traced by the aporetic pincer movement of culture and economy. To initiate such an act of disappearance we will begin by refolding the fourfold nature of the commodity as conventionally depicted in political economy: the folding of labour around the hinge of concrete labour and abstract labour; and the folding of value around the hinge of use-value and exchange-value. In so doing, we will accompany Baudrillard in unfolding a political economy of the sign.

## Value controversies

> There is a mirror, and the commodity form is also this mirror, but since all of a sudden it no longer plays its role, since it does not reflect back the expected image, those who are looking for themselves can no longer find themselves in it. Men no longer recognize in it the *social* character of their *own* labor. It is as if they are becoming ghosts in their turn. . . . Now that is what happens with the *commerce* of the commodities *among themselves*. (Derrida, 1994, pp. 155–6)

In the wake of the so-called 'value controversy' (Mandel and Freeman, 1984; Steedman *et al.*, 1981), which focuses on the aporetic nature of the value–price transformation problem, we can no longer trust in and speculate on the alchemical transmogrification of labour into value. The 'value controversy' pertains to the working of matter into useful objects, such that these variegated objects thereby embody – or, as Marx prefers, 'crystallize' – a common (social) substance: namely *human labour*, both concretely (in its heterogeneous specificity as a singular quality) and abstractly (in its homogeneous generality as a quantitative portion of social labour). When 'commodities are looked at as crystals of this social substance, common to them all, they are – Values' (Marx, 1954, p. 46). Whilst this distinction between concrete and abstract labour represents Marx's point of departure from Ricardo's 'labour theory of value', the fact that labour is always already 'crystallizable' undermines any attempt to depict labour as innocent, pure and full, while characterizing capital alone as corrupt, impure and parasitic: as if there were 'proper' socially useful labour until the advent of 'alienated' and 'exploited' labour under capitalism. Marx, moreover, reflexively applied this revised labour theory of value to the labour market itself, thereby revealing the creation and expropriation of surplus-value as the essence of capital-in-process. Value, 'while constantly assuming the form in turn of money and commodities, it at the same time changes in magnitude, differentiates itself by throwing off surplus-value from itself' (Marx, *Capital I*, cited in Harvey, 1982, p. 20).

Marx lay considerable emphasis on this 'double being of the commodity', on the

> twofold character of the labour whose product it is: the *useful* labour, i.e. the concrete modes of the labours which create use-values; and the *abstract* labour, labour as the expenditure of labour power, whatever the 'useful' mode in which it is expended (on which depends the later representation of the production process). (Quoted in Althusser and Balibar, 1979, p. 79)

The value of a commodity is no longer 'given' by the actual expenditure of concrete labour, but will have been *retroactively* determined by the labour deemed *socially necessary* to produce that commodity. Such is the disjointure in the temporalities of political economy, capitalism and labour. Inasmuch as social necessity can only be retroactively determined through 'realization' on the market, value is lodged in the untimely temporality of the future perfect (*post modo*). Like sense, value *will have been*. Value is *spectral*. Thus, 'labour can create no value unless it creates social use values – use values for others. . . . value has to be created in production and realized through exchange and consumption if it is to remain value' (Harvey, 1982, p. 16). Value is an *untimely* category; it is 'presupposed' that it will have been, prior to any act of exchange, which means that it is not a concept *abstracted from* the empirical diversity of the world but *projected on to it*. Value is delayed and relayed from the off. Like a calendar, it is summoned from the future.

Under capitalism, then, human labour is expended not in order to create useful *objects* but in order to produce *commodities*: material embodiments of abstract, 'crystallized' labour, whose expenditure is always already given over to the differential calculus of exchange, and thence the realization of surplus-value. A commodity is not so much a thing as a real abstraction, like clock-time or the number one. It is an articulation of structural relations: a concrete abstraction that manifests the knotting of differential relations. This knot is what political economists call 'value'. Consequently, capital, as value *in process*, never materializes as such, and is never given as such. It only exists insofar as it circulates through the untimely moments of various manifestations. It is possessed by a phantom objectivity and a real ideality. But capital is not merely spooky. It is a vampiric and wraithlike form. Dead labour feeds off living labour. The more labour that capital sucks, the more capital appears to live; and the more that labour is sucked by capital, the more labour appears to become undead and alien to itself. The distinction between the *object-form*, where circulation is confined to the 'exchange' of useful objects created through incommensurable concrete labour, and the *commodity-form* is therefore vital. For strictly speaking the commodity is not an object: it phantomalizes itself. The material through which the commodity passes, such as the wood of a wooden table, is a haunted house: occultation of culture and economy. Like (counterfeit) money, capital

> is not a thing like any other, precisely, in the strictly determined sense of thing; it is 'something' like a sign, and even a false sign, or rather a true sign with a false value, a sign whose signified seems . . . finally not to correspond or be equivalent to anything, a fictive sign without *secure* signification, a simulacrum, the double of a sign or a signifier. (Derrida, 1992, p. 93)

The commodity-form hinges on a threefold splaying out or ghosting of value. It projects use-values on to the object-form. It occludes concrete labour through a 'crystallizing out' of abstract labour, and it plays on the social intercourse of commodities amongst themselves. Since value can only be established retroactively as a phantom objectivity in the continuous circulation of capital, its magnitude can most effectively be determined by way of a 'reflex, thrown upon a single commodity, of the value relations of all the rest' (Marx, *Capital I*, cited in Harvey, 1982, p. 11). Yet this reflex commodity – paradigmatically, the money-form – exemplifies the paradox of value under capitalism. It is only by expressing all commodities through the medium of money that objects may become valued. The actuality of value is achieved only in and through price. Yet it is this very process which actually conceals, instead of disclosing, the social character of labour. In this way, the commodity-form in general, and the money-form in particular, yield a *false mirror* that fails to reflect value as a *social* relation. Hereinafter, producers and consumers relate to each other through commodities and as commodities, often misrecognized yet again as mere things. Such is our estrangement and alienation. Worse, these 'things' are seconded a speech and a will. Commodities appear to have business, commerce and socio-sexual relations amongst themselves. In commodity fetishism 'the relations connecting the labour of one individual with that of the rest appear, not as direct social relations between individuals at work, but as what they really are, material relations between persons and social relations between things' (Marx, 1954, p. 78). Fetishism

> is not a subjective mystification, but the mode of appearance of reality. . . .
> In the capitalist mode of production it takes the form of *the fetishism of*

*commodities*, i.e., the personification of certain things (money-capital) and the 'reification' of a certain relationship (labour). It does not consist of a *general* 'reification' of *all* relationships, as some humanist interpretations of Marx argue. (Brewster, 1979, p. 313)

Folded double, with their feet on the ground and standing on their heads, commodities really do have intercourse amongst themselves (Doel and Clarke, 1999). In this way, 'The system . . . reproduces capital according to its most rigorous definition, as the *form of social relations*, rather than in its vulgar sense as money, profits and the economic system' (Baudrillard, 1993, p. 28). It is no longer tenable to believe in the modern myth that 'We live in a society where there is a sharp distinction . . . between persons and things', as Mauss (1990, p. 47) once put it. And where does that leave use-value?

## Use-value: the alibi of the commodity-form

To say that the same thing, the wooden table for example, *comes on stage* as commodity *after* having been but an ordinary thing in its use-value is to grant an origin to the ghostly moment. Its use-value, Marx seems to imply, was intact. It was what it was, use-value, identical to itself. The phantasmagoria, like capital, would begin with exchange-value and the commodity-form. It is only then that the ghost 'comes on stage'. (Derrida, 1994, p. 159)

all illusions converge . . . on use value, idealized by its opposition to exchange value, when it [is] in fact only the latter's naturalized form. (Baudrillard, 1981, p. 139)

For Marx (1973, p. 881), a use-value, the 'material side' of the commodity, is 'the object of the satisfaction of any system whatever of human needs'. As such, its 'examination . . . lies beyond political economy'. However, 'use value falls within the realm of political economy as soon as it becomes modified by the modern relations of production, or as it, in turn, intervenes to modify them.' Given the evident subtlety of the appreciation of use-value already developed in the *Grundrisse*, Marx (1976, p. 215) feels entitled to proclaim that 'Only an obscurantist who has not understood a word of *Capital* can conclude [that] use value plays no role in [the] work.' Nonetheless, the role of use-value in Marx is ambiguous. Despite his claim that, in his hands, the issue received its full and due attention, there can be little doubt that the real potency of Marx's analysis lies elsewhere: in the analysis of exchange-value. Whatever the strength of Marx's distinction between use-value and exchange-value, virtually all efforts to reconstruct the Marxian theory of the commodity soon find their way into the mysteries of exchange-value, without pausing for long over use-value. It is here that the critique of political economy is at its shakiest. For it assumes a difference in kind between use-value and exchange-value, insofar as the latter is given over to a calculus of comparison and equivalence, while the former expresses an irreducible incommensurability. An exchange-value is set in relation to others. An exchange-value never comes alone; it always comes as part of a pack. By contrast, a use-value has only itself. It is always a singular quality; one that retains its specificity even when it is herded together with other such use-values. This is what Marx asks us to believe. However, use-values do indeed depend on a logic of comparison and equivalence. Exchange-value is itself only able to appear as a system governed by a logic of equivalence, in apparent contradistinction to 'incomparable' use-values, as a result of a parallel process of abstract formalization and reduction in the sphere of use-value: 'Considered as useful values, all goods are already comparable among themselves, because they are assigned to the same rational-functional common denominator, the same abstract determination' (Baudrillard, 1981, pp. 131–2). Function, utility, and use serve not as marks of qualitative singularity, but as general equivalents that cloak the world in the anthropocentric form of use-value. In neglecting this,

> Marxist analysis has contributed to the mythology (a veritable rationalist mystique) that allows the relation of the individual to objects conceived as use values to pass for a concrete and objective – in sum, 'natural' – relation between man's needs and the function proper to the object. (Baudrillard, 1981, p. 134)

Given its embodiment of a universality based on the abstract formalization of *functionality*, use-value is, like exchange-value, a social relation. It relates to 'needs' only insofar as it is the concrete abstraction of a *system* of needs. Use-value cannot, in any sense, be taken as 'the object of the satisfaction of any system whatever of human needs' (Marx, 1973, p. 881). To the contrary, use-value is the abstract determination of a system of needs intrinsic to capitalism which, as its effect, engenders precisely the 'objectivity' of any given object-form, and this does not depend on who uses it and what purpose it serves. (Indeed, Marx states as much when he insists that use-values are necessarily *social* use-values.) As a social relation predicated upon a logic of equivalence paralleling that determining exchange-value, the system from which use-value emerges is 'founded on the mere adequation of an object to its (useful) end' (Baudrillard, 1981, p. 132). In short, use-value cannot be taken as the innate quality of an object that would hold fast irrespective of the social formation. As Marx (1954, p. 87) implicitly recognized: 'Could commodities themselves speak, they would say: Our use-value may be a thing that interests men. It is no part of us as objects.' However, Marx failed to push this insight to its logical conclusion: that it is the abstract equivalence of use-values that defines objects *qua* use-values, in terms of the rational–functional calculus of the *object-form*. The object-form supplements the commodity-form. For use-value is nothing other than the naturalized form of exchange-value, which is thereby set to serve as its alibi, inasmuch as the system of needs, of which it is the abstract determination, is 'cloaked in the false evidence of a concrete destination and purpose, an intrinsic finality of goods and products' (Baudrillard, 1981, p. 131). This is apparent in the social imperative to consume. Indeed, one's duty as a citizen of the consumer society is to consume. Consumption manifests itself as a deontology, underwritten by the naturalized law of needs, wants and desires. By a strange twist of logic, use has become an obligation and an ethic – 'Conform! Obey! Consume!' (Graffiti, Leeds, 1999) –, and woe betide anyone who misuses and abuses objects, commodities and signs! Such is the moral geography of use-value – although it is a purely *formal* morality: not *how* to consume; just *to consume*.

Use-value, then, is neither the 'beyond' nor the 'before' of exchange-value. It is the satellite and alibi of exchange-value. Use-values and the object-form enable the completion of the commodity-form of value, and not its surpassing. The commodity-form affects *in advance* the use-value of a thing. Hence, 'for its first presumed owner, the man who takes it to market as use-value meant *for others*, the first use-value is an exchange-value' (Derrida, 1994, p. 161). But that is not all,

> Since any use-value is marked by this possibility of being used by the other or being used another time, this alterity or iterability projects it a priori onto the market of equivalences. . . . In its originary iterability, a use-value is in advance promised, promised to exchange and beyond exchange. It is in advance thrown onto the market of equivalences. (Derrida, 1994, p. 162)

In short, the commodity-form is folded double: use-value bends to the rhythm of exchange-value; and exchange-value bends to the beat of use-value. However,

> if use value has no autonomy . . . then it is no longer possible to posit use value as an alternative to exchange value. Nor, therefore, is it possible to posit the 'restitution' of use value, at the end of political economy, under the sign of the 'liberation of needs' and the 'administration of things' as a revolutionary perspective. (Baudrillard, 1981, p. 139)

We can no longer suspend our disbelief and hold faith with the 'idealized' relations of equivalence that the notions of use-value and utility imply. Hereinafter, Marxism's revolutionary refrain – 'from each according to ability; to each according to need' – sloganizes not the overcoming of political economy, but its completion; not our emancipation from the tyranny of capital, but our total submission.

## *The political economy of the sign*

Bad thing, to signify – y'hear me? (Kerouac, 1976, p. 256)

Even signs must burn (Baudrillard, 1981, p. 163)

If one of the most common reactions to the difficulties presented by the commodity and value has been to flee the arduous terrain of political economy for the relatively autonomous safety of the 'cultural', such a move fails to recognize that the same phantom logic haunts the supposed superstructure as shakes the trembling foundations of the economic base. This is particularly evident with respect to Saussure's (1974, p. 16) project of '*A science that studies the life of signs within society . . . semiology* (from the Greek *semeion*, "sign").' As we shall see, the mirroring of political economy in semiology is uncanny.

For Saussure, signification was only graspable in its *synchronic* dimension, which constitutes the structural system of *langue*. The terms of this system, signs, are predicated on their exchangeability along two dimensions. First, signs are able to signify – refer to or designate – 'real-world' referents. Second, signs are able to function in this way by means of the structural system of *langue*, whereby the terms of the system engage one another *diacritically* through a process of negative differentiation (each term refers to every other in such a way that it is the sum of what it is not). This is the sense in which sign-value is a structural relational that results from a differential calculus. Of these two dimensions, the latter amounts to the structural dimension of signification, whereas the former acts as the functional (referential) dimension. When the structural and functional dimensions are synchronized, designation appears as the finality of signification (its essence and telos): the structural play of signs appears to be motivated by the need to express a fragment of reality. Thus, meaning emerges from the system of *langue*, possessing the apparent capability of functioning referentially. However, the implication of this structural understanding of signs is that they are necessarily marked by an arbitrary and conventional quality, rather than an essential motivation issuing from the direction of the referent. The structural dimension of the sign-system is prerequisite to the possibility of reference to the world. In this way, the sign takes its place in the system of differential relations – a sign amongst signs – as a purely *mediating* presence between subject and object, its role being the designation in discourse of the referent as a fragment of the real. However, although the conception of the conventionality of the sign discloses meaning as issuing from an *articulation* of reality, it nonetheless fails to acknowledge the nature and status of the medium the sign constitutes. This medium can be figured as an articulation of two aspects. First, the *signifier*, which is the material support of signification, such as sounds, marks and images. One could call the signifier a sign-vehicle or the substance of expression. Second, the *signified*, which is the immaterial content of expression, such as meanings, values and concepts. And given the structural constitution of signifiers and signifieds, each of which is irreducibly caught up in the play of differential relations, a sign crystallizes out from the articulation of a signifier (Sr) and a signified (Sd).

Since Lacan, the sign has been understood not merely as an assemblage of signifier and signified, but as a form that prioritizes the former over the latter. This may be expressed in the formula Sr/Sd, the bar of which undermines the Saussurean assumption that each pairing of signifier and signified composes a discrete sign, expressed as an exchange of equivalents (a signifier for each signified, and vice versa). The dissolution of this assumption destroys the commonsensical notion that systems of signification find their true form, function and motivation in the re-presentation of an anterior reality comprising given referents. Lacan (1977), however, was concerned to demonstrate the primacy of the signifier in relation to the process of subjectivation, in the sense that one becomes what one will have been through the play of signifiers; that the meaning and value of one's life is a special and retroactive effect of articulating signs. In this way, signification and subjectivation experience the ceaseless *sliding* of the signified under the signifier: sign-value is an occult (folded) form of the play of signifiers. For example, consider the semantic slippage of 'the' signified of the signifier 'city' in the signs 'city', 'garden-city', 'the City', and 'Leicester City'. Sign-value can only be achieved through a forced and temporary stabilization of the syntagmatic chain of signifiers; through a blockage of semantic dissemination. Values are 'anchoring points' (*points de capiton*) that never stop drifting. For there is no value that is not worked over by differential relations and dissemination. Consequently, the commonplace overshadowing of form by content, of supposing that signifiers submit to given signifieds and referents, causes Baudrillard to speak of the 'political economy of the sign'. For this moral privileging of contents, such as use-values, signifieds and referents, only serves to conceal the more decisive privileging of forms: exchange-values and signifiers. Accordingly, the political economy of the sign needs to undergo a double invagination that would demonstrate that content is a special effect of form, and that the 'beyond' of exchange-value and sign-value is neither use-value nor reality, but something else, which Baudrillard dubs 'symbolic exchange': the destruction of value in all of its manifestations and determinations. Or rather, the suspension of value, signification, subjectivation, designation and determination in what Derrida might call the 'realm of the undecidable'. In a way that is almost Kafkaesque, the surmounting of political economy may be found not in the slogan 'We will be satisfied!', but in the refrain 'We will have been disoriented.' Our world is composed of events that are incompossible.

## The false guarantee of the real

> The crucial thing is to see that the separation of the sign and the world is a fiction, and leads to a science fiction. (Baudrillard, 1981, p. 152)

> The sign is a discriminant: it structures itself through exclusion. (Baudrillard, 1981, p. 149)

As we have seen, a fundamental problem with respect to the analysis of the commodity-form derives from the misrecognition of the anthropological horizon of the system of political economy as the possibility of its transcendence. This is repeated in the attempt to rescue the signified and referent from the terrorism of the signifier. Indeed, the desire to criticize the signifier on behalf of the signified and referent, to treat 'reality' as the perfect embodiment of the alternative to the abstract play of signs, is of a piece with the idealism of use-value. It is only by analyzing the structural relations that articulate the commodity and the sign that this situation reveals itself. Paralleling the commodity-form, sign-value is an exchange-value determined by the differential network of signifiers (*langue*), while the signified, especially in its referential function, appears essentially as a use-value:

> just as use value, the 'literal' and ideal finality of the object, resurges continually from the system of exchange value, the effect of concreteness, reality and denotation results from the complex play of interference of networks and codes – just as white light results from the interference of the colors of the spectrum. So the white light of denotation is only the play of the spectrum – the chromatic ghost – of connotations. (Baudrillard, 1981, p. 158)

Accordingly, the logic of the sign cannot be completed by pinning it down to the real, which is always already of its own making. A sign cannot re-present a portion of the world supposedly 'out there' since the 'world' 'evoked' by the sign is nothing but the effect of the play of signs. Thus, 'the referent does not constitute an autonomous concrete reality at all; it is only the extrapolation of the excision [*decoupage*] established by the logic of the sign onto the world of things' (Baudrillard, 1981, p. 155). What we are objecting to is the *a priori* separation of 'world' and 'signs', a separation that would ensure that mediation and re-presentation would take centre stage. It is not that the 'world' beyond is ungraspable and unpresentable – except in and through signs (the sorrows of being 'laden' with language) – but that such a 'beyonding' is illusory. Signs world. 'Reality is the phantasm by means of which the sign is indefinitely preserved from the symbolic deconstruction that haunts it' (Baudrillard, 1981, p. 156). If signs do not re-present it is because they are transformers. They do not merely participate in the world; they participate of the world. Signs are events: intensities, affects, redistributions of energy. In this way, representation is returned to the simulacrum: representation is merely a special effect; and signification is simply the simulation of meaning. What becomes apparent here is the structural complicity of commodity and sign in the evoking of a functional and meaningful world that would be 'beyond' our calculi of identity and difference. This naturalizing reality-effect depends on the disavowal of ambivalence, undecidability, and excess. Here, then, lies a further glimpse of the direction in which we are heading: 'Only ambivalence (as a *rupture* of value, of another side or beyond of sign value, and as the *emergence of the symbolic*) sustains a challenge to the legibility, the false transparency of the sign' (Baudrillard, 1981, p. 150).

The 'false transparency' of the sign and the commodity – that is to say, their illusory 'beyonding' of reality and simulated adherence to re-presentation – proceeds from the same structural law of value. Such an homology between the commodity-form and sign-form may be schematized as:

$$\frac{EV}{UV} = \frac{Sr}{Sd}$$

In its vertical dimension, this suggests a correlation between the internal structure of the commodity-form and sign-form – exchange-value (EV) is to use-value

(UV) as signifier (Sr) is to signified (Sd) – while in its horizontal dimension, exchange-value is to signifier as use-value is to signified. The commodity- and sign-forms thereby express a double reciprocity and a double structural relation of equivalence. In short, this formula neatly encapsulates the structural, formal and abstract field of 'general political economy'. It privileges form, function and process over and against singular content, and it does so in the name of a symbolic cancellation or ex-termination that results in indeterminacy, incommensurability, and undecidability. Hereinafter, '*the logic of the commodity and of political economy is at the heart of the sign*' since 'signs can function as exchange-value (the discourse of communication) and as use-value (rational decoding and distinctive social use)'. Conversely, '*the structure of the sign is at the very heart of the commodity form* [since] the commodity can take on, immediately, the effect of signification' (Baudrillard, 1981, p. 146). Thus, the object of a *general* political economy is neither commodity nor sign, but a blending of each into the other. Both are liquidated as specific determinations, but not as forms. General political economy reveals the extent to which sign and commodity are traversed by the same form and governed by the same logic. In the case of the commodity-form, use-value is seen to be just as much a social relation and an abstract determination of capitalism as is exchange-value. Moreover, this social relation reveals itself through the *detour* of content, as a kind of naturalized, specular completion of, and alibi for, exchange-value. Likewise, in the case of the sign-form, the signified and referent act as the alibis and guarantors of the signifier. Both the sign and the commodity find their coherence in codes that ceaselessly manage the circulation and exchange of values. In both cases, 'It is in the "materiality" of content that form consumes its abstraction and reproduces itself as form' (Baudrillard, 1981, p. 145).

## Ex-terminus: *forget value*

> an outline of social relations emerges, based on the extermination of value. (Baudrillard, 1993, p. 1)

> Against the differential play of value, the dual play of form: reversibility and metamorphosis. . . . For if all values do indeed seem to be disappearing as part of an irresistible process, forms, for their part, seem indestructible – at least in dreams. (Baudrillard, 1998, p. 4)

As we once witnessed the passage from the classical law of value (the beneficence of nature) to the commodity law of value (the rationality of production), we must now bear witness to the spectral play of the structural law of value – and thence its cancerous and viral degeneration. Under the classical law of value, the real proclaimed itself in the form of use-value. Both political economy *and* its critique appropriated the natural law of value as their imaginary system of reference: one must satisfy individual and collective 'needs' (as well as 'wants' and 'desires') by matching them with objective 'uses'. Whilst presenting themselves as 'beyond' and 'before' exchange-value and the sign, use-value and the functional naturalism of the referent are effects of, and alibis for, the hegemonic and despotic reign of the code. Value is neither given nor motivated by a referent. It issues forth from a generalized system of 'total relativity, general commutation, combination and simulation – simulation in the sense that, from now on, signs are exchanged against each other rather than against the real' (Baudrillard, 1993, p. 7). Needs are not articulated around the demand of a desiring subject. They derive their coherence from elsewhere; from a generalized system of needs that is to desire what the system of exchange-value is to concrete labour as the source of value. Hereinafter, the referential and functional character of language becomes one aspect in the structural play of signs, just as realism and perspectivism in the cinema are two contingent modes of cinematography, rather than its 'natural' or 'given' essence. Henceforth, denotation is a special effect of connotation.

Now, if one can speak of a 'structural revolution of value', it is because the two dimensions of value that Marx and Saussure mistakenly took as 'eternally bound as if by a natural law' (Baudrillard, 1993, p. 6) have once again become unhinged and left in suspense. Whereas the classical law of value operated 'simultaneously in every instance (language, production, etc.), despite these latter remaining distinct according to their sphere of reference', the structural law of value effects the indeterminacy of every sphere in relation to every other, marking a passage from the determinate realm of signs to the de-differentiated and indeterminate space of the code. Similarly, 'the entire apparatus of the commodity law of value is absorbed and recycled in the larger apparatus of the structural law of value. . . . Political economy is thus assured a *second life* . . . maintain[ing] an effective presence as a system of reference for simulation' (Baudrillard, 1993, p. 2). We are witness to a potentially infinite process of 'gearing down', whereby each phase of value adsorbs its prior stage on to itself as its alibi and simulated reference; as its own reality-principle.

In the wake of this 'gearing down', through which the purported 'outside' functions as the structural alibi of the 'inside', it should be clear why such things as use-value, utility and need cannot offer an escape from the tyranny of the code. For when all is said and done, use-values, as abstract determinations of the system of objects and the system of needs – above all as *values* – are the cancellation of the reciprocity and ambivalence of symbolic exchange. They ensure that '*All ambivalence is reduced by equivalence*' (Baudrillard, 1981, p. 135). And to the extent that the 'critique' of political economy invests in the cultivation of use-values it only serves to exacerbate such a reduction of ambivalence, undecidability and suspense. Guattari (1992, p. 46) puts it beautifully: 'Capital, Energy, Information, the Signifier are so many categories which would have us believe in the ontological homogeneity of referents (biological, ethological, economic, phonological, scriptural, musical, etc.).' Or again: 'Structuralists have been content to erect the Signifier as a category unifying all expressive economies: language, icon, gesture, urbanism or the cinema, etc. They have postulated a general signifying translatability for all forms of discursivity' (Guattari, 1992, p. 37).

While the commodity destroys ambivalence through a principle of equivalence, the sign achieves the same effect through its arbitrary discretion. The destructive character of the sign comes from the assumption of a perfect correlation between a discrete signifier and an equally discrete signified-cum-referent (its pointillism fails to do justice to the general relativity of the structural play of signs). Its arbitrariness comes from the positing of an equivalence between such and such a signifier and such and such a signified–referent. There can only be equivalence, correlation, equilibrium, balance, and harmony on the basis of discrete signifiers, discrete signifieds, and discrete referents. Identity, and the differences between identities, are therefore the result of a moral topology that insists on individuation and frowns upon interpenetration. (Parenthetically, this suggests that the discretion and pointillism of the political economy of the sign is aligned with an orthodox ontology of phallocentric sexual difference: both commodity and sign are conventionally depicted in terms of the metaphysics of presence.) Moreover, it is by way of such arbitrary discretion that the sign's status as, precisely, a *value* lies. Discretion underwrites the reality-principle of the sign. One can only maintain the reality-effect of discrete signs by disavowing everything that overflows the calculus of equivalence and signification. The sign 'functions as the agent of abstraction and universal reduction of all potentialities and qualities and meanings [*sens*] that do not depend on or derive from the respective framing, equivalence, and specular relation of a signifier and a signified' (Baudrillard, 1981, p. 149). What cannot be translated into the calculus of discretion is simply discounted, overlooked or destroyed.

Insofar as commodities and signs rest upon discretion we can discern their structural homology and functional complicity in the disavowal of ambivalence. The commodity seems to function as exchange-value in order better to hide the fact that it functions as a sign: not to denote and accumulate value (e.g. values translated into prices and stored in money), but to reproduce the code (the calculus of discretion and digitality). Hereinafter, the structural revolution of value operates with political economy as its simulated model. The end of political economy announces the era of generalized reproduction, in which

> all values are rehabilitated and indiscriminately permutated, . . . but what we do not know how to re-create is the electricity generated by their contradiction. . . . We have rehabilitated them not in their dialectical tension [but] as endangered masterpieces. Emptied of this negative tension, they become equivalent, substitutable. Each shows through the other. . . . Each one squints through the other. A generalized strabismus of value. (Baudrillard, 1998, p. 2)

Such is the passage from the 'transvaluation of all values' to the viral 'involution of all values'. At the heart of the commodity resides not the 'real contradiction' between use-value and exchange-value, or between concrete labour and abstract labour, that would ceaselessly threaten to blow the commodity law of value apart, but the structural, spectral and simulacral play of sign-value. Strictly speaking, then, there is nothing *in* the commodity that could be released and discharged. This 'is announced everywhere by the commutability of formerly contradictory or dialectically opposed terms' (Baudrillard, 1993, p. 8). And the characteristic effect of the dominion of the code is precisely this: in spite of itself, 'Everything becomes undecidable' (Baudrillard, 1993, p. 9).

Such is the end of dialectical materialism: not as finality, resolution or completion; but as declination and peregrination. Any thoroughgoing materialism must strive towards an *aleatory materialism*; a 'materialism of the encounter' (cf. Callari and Ruccio, 1996; Derrida, 1994; Elliott, 1998). Structure is no longer an enduring isomorphism that holds fast between contents, such as the structural relations of capital and sexual difference that are repeatedly expressed and materialized in particular manifestations of capitalist and patriarchal social formations. Structure is transversal, to be sure, but it is always already set in motion: the structuralist's imposition of a 'general signifying translatability' gives way to an asignifying driftwork in continuous variation and polymorphous perversion (Deleuze and Guattari, 1984; Lyotard, 1992). Translation as a formal equalization of two heterogeneous and incompossible contents gives way to translation as a difference-producing repetition. Signs are not representatives in another medium or dimension. They are conductors, and therefore transformers of intensities and affects. Such is one's entry into the unstable and undecidable world of poststructuralist solicitations (Doel, 1999).

Value *as such* has always been dedicated to the annihilation of the symbolic: equivalence pitched against the reciprocity of symbolic exchange. It rests on the principle of opposition; and opposition presupposes the full presence of a *given* term – discretely isolated from its flawed double, and expressive of the desire for a final, dialectical resolution. Following Mauss (1990), however, we recognize that what is given must also be *returned*. The reversibility of symbolic exchange proceeds on the basis of the non-opposition of terms; on a formal antagonism that admits no reconciliation, sublation or resolution. In subjecting what is given to the irreversibility of the code, and hence by denying the reversibility of symbolic exchange, value can result only in undecidability and ex-termination. Just as Bataille (1988) recognized the necessity of sacrifice – of a useless expenditure of energy – we must recognize that the irreversibility of value rests on the denial of the 'accursed share' and the disavowal of the symbolic. Hence, if the irreversibility of value is to be done away with, this can only be achieved by invoking that which is other than the code: 'We must . . . displace everything into the sphere of the symbolic, where challenge, reversal and overbidding are the law' (Baudrillard, 1993, p. 36).

Political economy and its critique work on terms and terminals. At bottom, commodities and signs are engaged in acts of discrimination and discretion: they work by creating identities through structural differentiation. A commodity or sign takes its place amongst others, and in so doing is apportioned value, attributed meaning, and lent sense. Neither commodity nor sign is given before such a structural embedment. Each is an assemblage, an articulation, or still better, an unfolding of differential relations. In contrast to the discrimination of commodities and signs, symbolic exchange rests on reversibility, exemplified in the reversibility of the gift in the counter-gift, of the giving in the taking:

symbolic exchange, where the law is that something is given to you and you have to give it back and, if possible, give back *more* – the *surplus-value* of symbolic exchange. The position is this, then, that the world is given to us, and given to us as unintelligible: we have to render it even more unintelligible. (Baudrillard *et al.*, 1995, p. 82)

Faced with the play of the code, it becomes necessary to '*Ex-terminate* every *term*, abolish value in the term's revolution against itself: that is the only symbolic violence equivalent to and triumphant over the structural violence of the code' (Baudrillard, 1993, p. 5). It is this heightening of suspense – the unfolding of a dissipative, asignifying and metamorphic driftwork – that renders symbolic exchange adialectical.

If commodity exchange is based on discrete terms and their equalization-cum-extermination, then it should be evident that symbolic exchange is an 'impossible exchange'; a modality of exchange that is incompossible with and impossible for any system of terms and terminals, of poles and polarities. Reversibility takes time, demands more and is duplicitous. While commodity exchange depends upon circulatory systems of *value*, which are inevitably perishable and deconstructable, symbolic exchange is a play of plastic and mutable forms. And unlike terms, codes and values, such play is reversible, undecidable and indestructible. Such is the (s)playing out of restricted into general economy: 'there is no residue, nothing is left over, since everything is exhausted in the reciprocity of terms, in the reversibility of terms' (Baudrillard *et al.*, 1995, p. 87). The risk that Baudrillard takes 'beyond' the event horizon of terms, codes and values is that one might recover *suspense* – or what he prefers to call 'indifference' and 'objective indifference' – from the decomposition of metamorphic *forms* (i.e. cycles of transmogrification, such as the transformation of reality into signs or labour into commodities) into floating *formulæ* (e.g. EV/UV = Sr/Sd). This is why Baudrillard claims that there are effectively three possibilities: *symbolic exchange*, marked by a reversibility of forms; *dialectical exchange*, marked by the opposition and thence sublation of terms; and *indeterminacy*, which issues from the short-circuiting, scrambling, blurring, and extermination (finalization) of terms: the world of the anomalous, the chaotic, the viral, and the cancerous. The return to symbolic exchange is not, then, a primitivist nostalgia; it is the affirmation of the reversibility in play *alongside* irreversibility, of undecidability *alongside* indeterminacy, and of ex-termination *alongside* extermination. We are drawn by what is ex-terminus, by what is beyond the end, or rather, by what declines – in the sense of bending away from – the end. So, to the discriminant one sets in play the clandestine, the imperceptible and the indiscernible. One turns away from finalization, resolution and termination. Hereinafter, one is obligated to endure the suspense of a world without ends, of a world that has swerved away from terms. For ours is a world that is cast adrift in perpetual (s)play, peregrination and suspended animation. Our strategy, then, is not to partake of the cult of sign-values, exchange-values and use-values, least of all is it to invest in the cult of labour, needs and motivations. We advocate only this: the cultivation of ambivalence, disorientation and suspense.

## Acknowledgements

This work was presented to the 'Cultural Turns/Geographical Turns' conference, 16–18 September 1997, University of Oxford. Dave Clarke acknowledges the support of ESRC Award No. H52427002294.

## References

Althusser, L. and Balibar, E. (1979) *Reading Capital* Verso, London.

Bataille, G. (1988) *The Accursed Share: An Essay in General Economy. Volume One: Consumption* Zone, New York.

Baudrillard, J. (1981) *For a Critique of the Political Economy of the Sign* Telos, St Louis.

Baudrillard, J. (1993) *Symbolic Exchange and Death* Sage, London.

Baudrillard, J. (1994) *Simulacra and Simulation* University of Michigan, Ann Arbor.

Baudrillard, J. (1996) *The Perfect Crime* Verso, London.

Baudrillard, J. (1998) *Paroxysm* Verso, London.

Baudrillard, J., Boyne, R. and Lash, S. (1995) 'Symbolic exchange: taking theory seriously. An interview with Jean Baudrillard' *Theory, Culture and Society* 12: 79–95.

Beckett, S. (1986) 'Waiting for Godot' in *Samuel Beckett: The Complete Dramatic Works* Faber and Faber, London pp. 7–87.

Brewster, B. (1979) 'Glossary' in Althusser, L. and Balibar, E. *Reading Capital* Verso, London, pp. 309–24.

Callari, A. and Ruccio, D. (eds) (1996) *Postmodern Materialism and the Future of Marxist Thought* Wesleyan University Press, London.

Deleuze, G. and Guattari, F. (1984) *Anti-Oedipus: Capitalism and Schizophrenia* Athlone, London.

Deleuze, G. and Guattari, F. (1988) *A Thousand Plateaus: Capitalism and Schizophrenia* Athlone, London.

Derrida, J. (1992) *Given Time: 1 Counterfeit Money* University of Chicago Press, Chicago.

Derrida, J. (1994) *Specters of Marx: The State of the Debt, the Work of Mourning, and the New International* Routledge, London.

Doel, M.A. (1999) *Poststructuralist Geographies: The Diabolical Art of Spatial Science* Edinburgh University Press, Edinburgh.

Doel, M.A. and Clarke, D.B. (1999) 'Dark Panopticon. Or, Attack of the Killer Tomatoes' *Environment and Planning D: Society and Space* Vol 17, No 4, pp. 427–50.

Elliott, G. (1998) 'Ghostlier demarcations: on the posthumous edition of Althusser's writings' *Radical Philosophy* 90, 20–32.

Guattari, F. (1992) *Chaosmosis: An Ethico-Aesthetic Paradigm* Power Publications, Sydney.

Harvey, D. (1982) *The Limits to Capital* Blackwell, Oxford.

Kerouac, J. (1976) *On the Road* Penguin, Harmondsworth.

Lacan, J. (1977) *Ecrits: A Selection* Tavistock, London.

Lyotard, J.-F. (1992) *Libidinal Economy* Athlone, London.

Mandel, E. and Freeman, A. (eds) (1984) *Ricardo, Marx, Sraffa: The Langston Memorial Volume* Verso, London.

Marx, K. (1954) *Capital: A Critique of Political Economy* Volume 1. Lawrence & Wishart, London.

Marx, K. (1973) *Grundrisse* Penguin, Harmondsworth.

Marx, K. (1976) *Value: Studies by Marx* (ed. A. Dragstedt) New Park Publications, London.

Mauss, M. (1990) *The Gift: The Form and Reason for Exchange in Archaic Societies* Routledge, London.

de Saussure, F. (1974) *Course in General Linguistics* Fontana, London.

Steedman, I., Sweezy, P., Wright, E.O., Hodgson, G., Bandyopadhyay, P., Itoh, M., De Vroey, M., Cohen, G.A., Himmelweit, S., Mohun, S., Shaikh, A. (1981) *The Value Controversy* Verso, London.

# Imagined regional communities

## Undecidable geographies

*James D. Sidaway*

## Introductions to the plot and cast

What is economy? Among its irreducible predicates or semantic values, economy no doubt includes the values of law (*nomos*) and of home (*oikos*, home, property, family, the hearth, the fire indoors). *Nomos* does not only signify the law in general, but also the law of distribution (*nemein*), the law of sharing or partition [*partage*], the law as partition (*moira*), the given or assigned part, participation. Another sort of tautology already implies the economic within the nomic as such. As soon as there is law, there is partition: as soon as there is *nomy*, there is economy. Besides the values of law and home, of distribution and partition, economy implies the idea of exchange, of circulation, of return. The figure of the circle is obviously *at the center*, if that can still be said of a circle. It stands at the center of any problematic of *oikonomia*, as it does of any economic field: circular exchange, circulation of goods, products, monetary signs of merchandise, amortization of expenditures, revenues, substitution of use values and exchange values. This motif of circulation can lead one to think that the law of economy is the-circular-return to the point of departure.

(Derrida, 1992, pp. 6–7)

Society, economy, culture: each of these 'areas', now tagged by a concept, is a comparatively recent historical formulation. 'Society' was active fellowship, company, 'common doing', before it became the description of a general system or order. 'Economy' was the management of a household and then the management of a community before it became the description of a perceived system of production, distribution, and exchange. 'Culture', before these transitions, was the growth and tending of crops and animals, and by extension the growth and tending of human faculties. In their modern development the three concepts did not move in step, but each at a critical point, was affected by the movement of the others. At least this is how we may now see their history. But in the run of real changes what was being put into the new ideas, and to some extent fixed in them, was an always complex and largely unprecedented experience. 'Society' with its received emphasis on immediate relationships was a conscious alternative to the formal rigidities of an inherited, then seen as imposed order: a 'state'. 'Economy', with its received emphasis on management, was a conscious attempt to understand and control a body of activities which had been taken not only as necessary but as given.

(Williams, 1977, pp. 11–12)

Comparing, contrasting and moving between political science, international relations, economics, cultural studies and geography, this chapter provides a case study of how circles of 'cultural', 'economic' and 'political' (all of which are also 'social') powers intersect and (re)produce each other and how the state emerges out of the trace marked by these interplays. The chapter is interested in what the jargon of philosophy calls the 'ontological status' of the social, cultural, economic and political, and the state. That is, what are they like? For example, is the state to be conceptualized as a thing in itself, or as a result of (more fundamental?) 'economic' or 'cultural' expressions? In turn, how do these relate to each other? These are big questions and for many years the idea of a general theory that resolves once and for all such issues has been recognized as problematic or impossible. Indeed, as Raymond Williams (cited above) explains, these terms have complex histories and have signified different things in different times and places. So there can be no question of a perfect and complete theory in general of how, say, 'culture' relates to the 'economy'. Instead we are forced to study what forms such relations have taken in specified historical and geographical contexts. This is done here through reference to the trajectory of formal regional associations of states. Of these the (currently) 15 member European Union is probably the best known. But a huge variety of other regional communities of states have been established in recent years. These are often described as responses to putative 'globalization' or as a feature of the post-Cold War world.

However, such narratives are comparatively novel, and there are other longer established theoretical approaches to understanding regional communities. Indeed, since the middle of the twentieth century, regional communities (including the European Coal and Steel Community, which was established in 1953 and which later grew to be the European Union) have been subject to a corpus of theoretical treatments of which the bulk have been conducted in branches of economics or political science. These will be briefly described here, together with some contributions from international relations and history.

The most significant accounts have come from the discipline of economics. The significance of the abstract models of customs unions and trade blocs that have been formulated, debated and refined by economists arises from the way that economics has come to occupy a privileged position as a source of policy. The chapter will describe later how this has operated through a series of institutions that regulate global 'free trade' that were set up under American leadership in the 1940s and which have developed normative models of 'the market' into which regional communities are required to fit. Although they largely side-step questions of power, subjectivity and agency, such models have therefore proven powerful. In other words, the 'science' of economics abstracts from the world a highly rarefied vision of how the world operates, usually described by a set of equations or graphs. But, in turn, these abstractions become reimposed as the world is forced to become more like them through policy. As critical commentators (for a selection, see Carrier, 1997; Carrier and Miller, 1998; and Miller, this volume) have argued, this abstraction and reimposition in the form of policy is a deeply political process with often profound social consequences – but is disguised as the technocratic, scientific and 'neutral' scientific jargon of economics.

Political science is not quite so influential as economics. But the impact of political science theories of integration should not be discounted. Elaborated in part as a way of understanding (and therefore providing a formula for accelerating) European integration, so-called 'functionalist' theories described how new regional communities would develop following a logic of 'spill-over', whereby economic, social and political functions would be ceded by states to a higher 'supranational' authority. The ceding of functions in one domain (say an aspect of economic policy) was expected to create demands and pressures for a similar assignment of others. Functionalism was particularly influential amongst many of those involved in establishing European integration in the 1950s and 1960s, and despite problems with demonstrating when or exactly by what means 'spill-over' really operated, functionalism has continued to exercise a certain influence.

However, it is to the discipline of international relations that we may turn for some more subtle (and culturally grounded) theoretical treatments of regional communities. Such subtlety is not always evident, for sometimes international relations theory (backed up by historical accounts) has simply reduced regional communities to the interaction of sovereign state interests. But such approaches (which are known as 'realist') have come in for considerable criticism for taking for granted and reifying the identity and power of the states, almost as if the latter were features of the natural, not social world. The sense of accelerating 'globalization' has reinforced such criticism. Indeed, the end of the big picture of the Cold War (*The Fifty Years War* as one author recently called it[1]) combined the sense of disorientations and juxtapositions which go under the sign of 'globalization' allowing many narratives of global politics, including debates about culture, values and scales and objects of analysis. Traditional realist approaches remain but, with so much going on, they have become less credible.

In such contexts, a certain 'cultural turn' might even be detected in the minds of some senior civil servants at the British Foreign and Commonwealth Office (FCO). When men such as Robert Cooper, the director for Asia and the Pacific at the FCO, tells readers of monthly *Prospect* magazine (which calls itself 'Britain's Intelligent Conversation') that: 'It is time we had a new theory of international relations', that 'an existentialist/post-modern perspective' might be 'right for the spirit of the times?', and that: 'International relations today is about values and identity at least as much as it is about ideas of balance or plans to engineer peace or romantic theories of progress' (1998, p. 58), one senses that something is afoot.[2]

Whilst it could credibly be argued that 'values' and 'identity' are in fact exactly what traditional FCO discourse was about, a call for existentialism or postmodernism is rather novel. I think that it is fair to say that Foreign Office mandarins are not known for their embrace of postmodernism or continental theories and that the FCO has hardly been a major centre for innovative theory. The FCO's most significant 'theoretical' contributions were in the days of British colonial empire. Together with a few anthropologists and geographers, employees of the British Foreign and Colonial Office developed elaborate notions of racial difference, masculine heroism (the boy scout, Kiplingesque world of adventure) and colonial administration, to be applied in Africa and Asia with a mixture of 'civilizing mission' (an imperial mission to bring 'civilization' or as it was later called 'modernization' to the 'natives') and brute force.

But in what he judges is now a 'post-heroic, post-imperial, post-modern society', Robert Cooper is interested in thinking (with the help of texts such as that by former US secretary of state Christopher Coker, whose book on the *Twilight of the West* Cooper reviews for *Prospect*): 'about the imagination of *international* communities, and about the re-imagining of the countries that make them up' (1998, p. 58).

Wondering about how they might also apply for 'international communities', Cooper (1998, p. 58) mentions Benedict Anderson's idea of nations being *Imagined Communities*:

> Identity has long been important for the internal cohesion of nations. Nations, as Benedict Anderson taught us, are imagined communities. A nation must first exist in the mind of its citizens before it can hold together on the battlefield or football terraces; or before its parliament can exact loyalty and taxes.

Cooper seems to read Anderson's (first published in 1983, but revised and expanded in 1991) *Imagined Communities* rather narrowly here. For as Tønnesson and Antlöv (1996, p. 7) explain:

> Anderson argues that the nation is a cultural construct, not in the sense of building on historical tradition but in that of being collectively imagined by all those going to the same kinds of school, viewing or listening to the same media, sharing the same mental map of the nation and its surrounding world, or visiting the same museums. There is thus nothing immanent or original about the nation: it is a construct, similar everywhere, only using different symbols, but it always considers itself as antique: it creates its own narrative, imagining itself as 'awakening from sleep'.

Anderson's understanding of the rise of the nation is therefore of a certain political economy of culture, a *cultural economy*, based around many things, but with a special emphasis on the rise of the media and the circulation of texts that declare and regulate nationhood:

> What, in a positive sense, made the new [national] communities imaginable was a half-fortuitous, but explosive interaction between a system of production and productive relations (capitalism), a technology of communications (print), and the fatality of human linguistic diversity. (Anderson, 1991, pp. 42–3)

This economy of circulation was at first mostly restricted to the Americas and Europe. But:

> since the end of the eighteenth century nationalism has undergone a process of modulation and adaptation, according to different eras, political regimes, economies and social structures. The 'imagined community' has, as a result, spread out to every conceivable contemporary society. (Anderson, 1991, p. 157)

Although his rendition of Anderson's book might be an oversimplification, Cooper is also suggestive in so far as he appreciates that something in Anderson's thesis might be applied to thinking about certain international communities, such as the (re)invention of 'Europe' in the form of the European Union. This has not escaped other observers. Expressing it simply, Andrew Hurrell (1995, p. 41) says that: 'As with nations, so regions can be seen as imagined communities which rest on mental maps whose lines highlight some features whilst ignoring others.'

In similar terms, in an original analysis of 'region building in Northern Europe', Iver Neumann (1994) insists that formal regional communities of an international form, such as the EU or the Nordic Countries are constituted in part out of a discourse whereby spokespersons for the community: 'as part of some political project imagine a certain spatial and chronological identity for a region and disseminate their imagined identity to others' (Neumann, 1994, p. 58).

More widely, the presence of regional communities has been extensively noted in recent years (and thereby rendered more real and important) in a variety of media and academic discourses. For example, Walter Rostow (1990) detected the signs of 'The coming age of regionalism'. Back in the 1960s, Rostow had been assistant national security adviser to the Kennedy Administration and codifier of highly influential theories of (western-led) modernization for the Third World. More recently though for Rostow (1990), writing in *Encounter* (a conservative American policy journal previously funded by the CIA), regionalism in general could be 'a metaphor for our time'.

Other influential commentators would have it that, in an age of supposed 'globalization', to talk of nations and states (let alone nation-states) is not always enough. For example, during his first official visit to Portugal in July 1995, Fernando Henrique Cardoso, the then recently elected Brazilian president and one time academic sociologist (who had written rebuttals of Rostow's version of 'development') discussed links between the seven independent countries with Portuguese as an official language. President Cardoso commented that, today: 'We can't reason only in terms of countries – there are regional blocs' (cited in de Vasconcelos, 1995, p. 9, my translation).

With President Cardoso's stark declaration and Walter Rostow's divinations in mind, following the suggestions by Robert Cooper, Iver Neumann and others, and wishing to take into account the circles of 'economy', 'politics' and 'culture' mentioned in the opening words, the bulk of what follows reconsiders some (cultural) aspects of the imagination of regional communities.[3] This is done through attention to some of the metaphors and images that are invoked when regional communities are spoken of. In turn, this allows a reflection on the way that imagination of regional communities is also reimagination of the nations or states that constitute them. I will suggest that the 'regional community' is therefore always something through which the state (and the *nation*, which the state claims at once to arise from and to build) are also able to appear as 'real'. The chapter will indicate that it might be impossible to decide if the constituent states 'produce' the regional community or if the latter (together with other diplomatic fora) serve to reproduce the state. In turn, this allows some broader reflections about the interlocked circles of 'culture', 'economy' and 'politics'. Prior to this, however, some basic contextualization of contemporary regional communities is in order.

## Contextualizing regional communities: the story

> As Marx said, every child knows that a social formation which did not reproduce the conditions of production at the same time as it produced would not last a year. (Althusser, 1984, p. 1)

> Order is, at one and the same time, that which is given in things as their inner law, the hidden network that determines the way they confront one another, and also that which has no existence except in the grid created by a glance, an examination, a language; and it is only in the blank spaces of this grid that order manifests itself in depth as though already there, waiting in silence for the moment of its expression. (Michel Foucault, 1970, p. xx)

There are many regional communities – with distinct histories and contexts, making generalization at once difficult and problematic. The idea that one theory (or set of theories) may provide appropriate understandings of them all has long been criticized. This has not stopped some commentators from producing universal models of (political and/or economic) integration, or projecting theories developed in highly specific contexts (usually European integration) on to other quite different communities and contexts (such as Asia, Africa or South America). Many of the 'models' and theories from economics and political science have suffered from this. I shall not be directly concerned much more with these formal theories of integration, for which many commentaries are anyway already available.[4] As has been noted, such theories (functionalism, federalism, economic 'models' of customs unions and so on) have their *aficionados*, impacts and insights. But my interest in irreducible and intertwined circles of (economic, cultural and political) power requires that I ask questions of regionalism that such theories largely take for granted. This allows a few claims in the final notes which I hope will prove suggestive.

A good way to begin our select(ive) story here is with a subsequent commentary on President Cardoso's words, in which a Portuguese journalist decided that they implied that:

> The Portuguese-speaking community is not an alternative to the insertion of the countries that constitute it into their own regional spaces . . . For example, the case of Portugal and Brazil is not different to that of Angola and Mozambique in relation to Southern Africa. When the links of the countries that compose the respective regional spaces are stronger, the international clout of the Portuguese-speaking community is greater. (de Vasconcelos, 1995, p. 9, my translation)

The 'Portuguese-speaking community' that the journalist mentioned is a formal organization, established in 1995 which 'brings together' the five states in Africa or off Africa's Atlantic coast which have Portuguese as an official language (Angola, Cape Verde, Guinea Bissau, Mozambique, São Tome and Principe) plus Brazil and Portugal, in a loose cultural and political association. Until the mid-1970s all of these African and Atlantic island territories were ruled from Portugal. Brazil had also been a Portuguese colony until 1822. Indeed the Portuguese-speaking community represented a project by the Portuguese government to set up the style of organization (bringing together the metropole and former colonies) that Paris (via the Communauté Francophone) and London (via the Commonwealth) had long sought to utilize to project certain versions of French and British 'culture' in the world. However, despite substantial economic, cultural and political links that endure from the age of empire, these postcolonial clubs have not precluded members (including the former colonial powers themselves) participating in various sorts of regional communities on a continental or sub-continental basis with neighbouring countries. The putative complementarity or synergy between continental or regional communities and linguistic/cultural associations of places previously united by old imperial links is what our Portuguese journalist was referring to in his commentary on Cardoso's words.

As has been noted, the best known of these regional, continental or transcontinental groupings is the European Union (née European Community 1992–1994, European Economic Community 1957–1992, European Coal and Steel Community, 1953–1957). But there are many others: for example, Asia-Pacific Economic Co-operation (APEC), the Association of South-East Asian Nations (ASEAN), the Economic Community of West African States (ECOWAS), the Caribbean Community and Common Market (Caricom), the Southern African Development Community (SADC), the North American Free Trade Area (NAFTA), and the Mercado Común del Sur (Mercosur, comprising Argentina, Brazil, Paraguay and Uruguay). And whilst these (mapped in Figure 1) are some

**Figure 1** Regional communities mentioned in the text

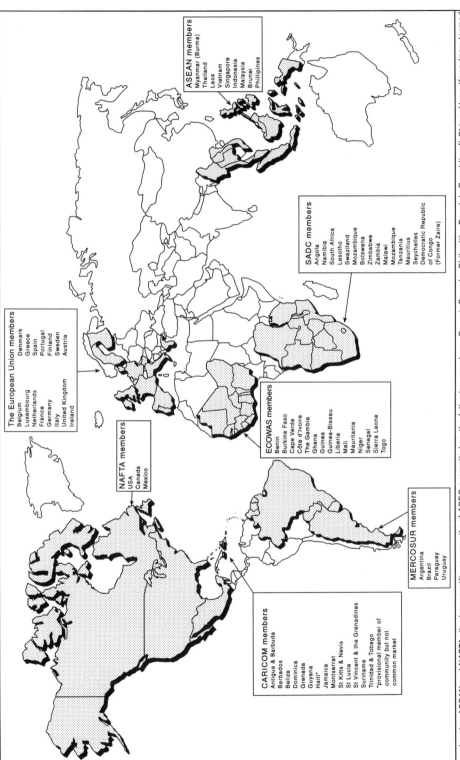

The European Union members
Belgium          Denmark
Luxembourg       Greece
Netherlands      Spain
France           Portugal
Germany          Finland
Italy            Sweden
United Kingdom   Austria
Ireland

ASEAN members
Myanmar (Burma)
Thailand
Laos
Vietnam
Singapore
Indonesia
Malaysia
Brunei
Philippines

SADC members
Angola
Namibia
South Africa
Lesotho
Swaziland
Mozambique
Botswana
Zimbabwe
Zambia
Malawi
Mozambique
Tanzania
Mauritius
Seychelles
Democratic Republic
of Congo
(Former Zaire)

ECOWAS members
Benin
Burkina Faso
Cape Verde
Côte d'Ivoire
The Gambia
Ghana
Guinea
Guinea-Bissau
Liberia
Mali
Mauritania
Niger
Senegal
Sierra Leone
Togo

NAFTA members
USA
Canada
Mexico

CARICOM members
Antigua & Barbuda
Barbados
Belize
Dominica
Grenada
Guyana
Haiti*
Jamaica
Montserrat
St Kitts & Nevis
St Lucia
St Vincent & the Grenadines
Suriname
Trinidad & Tobago
*provisional member of
community but not
common market

MERCOSUR members
Argentina
Brazil
Paraguay
Uruguay

Overlapping ASEAN and NAFTA, the transpacific community of APEC currently has the following members: Brunei; Canada; Chile; (the People's Republic of) China; Hong Kong (now termed Hong Kong, China in APEC documentation); Indonesia; Japan; (South) Korea; Malaysia; Mexico; Philippines; Russia; Singapore; Taiwan (always termed Chinese Taipei in APEC documents so as not to offend Beijing); Thailand; and the USA.

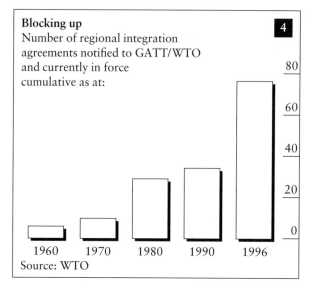

**Figure 2** The number of regional agreements notified to the GATT/WTO and currently in force
Source: Adapted from the *Economist* (London), 7 December 1996

of the best known and most active communities, there have been many more. The World Trade Organization (WTO) noted in 1995 that in the 47 year life of its forerunner (the General Agreement on Tariffs and Trade or GATT, which operated from 1947 to 1994) a total of 108 bilateral or regional agreements had been notified to it. Within these, over 70 notifications of regional agreements (or amendments to expand them) had been registered. Furthermore, as the *Economist* magazine detailed in a 1996 cover story on regional economic communities, of the: '76 free trade areas or customs unions set up or modified since 1948 . . . more than half have come in the 1990s' (Anon, 1996, p. 27) (see Figure 2).

Many of these, notably those in the 'Third World' were postcolonial projects, bringing together states and societies which had been divided by colonialisms. In a rather different, but parallel sense, the EU was also given impetus by the perception amongst the old imperial powers that the world 'led' by European imperialism had been supplanted by continental superpowers (the USA and the USSR).

At the end of the Second World War, most of the European overseas colonial empires were still intact (though terminally weakened). Britain, Portugal, the Netherlands and France sought to regain and reconsolidate empires in what had become known during the war as South-East Asia.[5] Britain was also still a colonial power in what would soon usually be designated as South Asia as well as in Africa (with France, Portugal, Spain and Belgium), in what was now called the Middle East (with France) and in the Caribbean (with France and the

Netherlands). All of these imperial projects operated various kinds of preferential trade. Indeed, protected markets, imperial preferences and so on, had been put in place by and formed part of the rationale for colonialism (itself, at once, economic, political and cultural).

Yet by 1945 the colonial projects were all decisively weakened (even if this was not always obvious to all of the imperialists at the time). It had already been clearly demonstrated that for all the racist nonsense about the superiority of white ('European'[6]) civilization or 'races', western colonial power was *not* invulnerable nor eternal. The speed of its collapse when faced by the military challenge of Japanese imperialism in 'South-East Asia' in 1941 and the rise of formal anti-imperialism[7] in the form of US and Soviet (super)power provided the global backdrop to the rise of national liberation movements which would contest the old colonial powers. The ensuing collision of national liberation struggles, superpower politics and dying colonialisms is vastly complex, and our story really ought to become fractured here.[8]

But amidst all the collisions and complexities was a strong American commitment to global 'free trade'. This was embodied in American aid (always also a trade policy) and more directly in the regulatory regimes established under American leadership in the 1940s. The latter regimes, the GATT, the World Bank and the IMF (together with the machinery of the UN), had universalist aspirations, but were profoundly shaped by American power. What went with them, and which they were supposed to ensure, was the construction of a relatively open trading order. All the power of the discipline of economics was also invoked and harnessed here, in part via analyses of the way that regional communities could best complement and operate within the logic of 'the Market'. All this not only suited US transnational corporations, but was an expression of a new US strategic commitment to what it scripted as the non-communist 'Free World'. Such commitment was expressed too in a series of military alliances with the US at their core. These alliances and the trade regimes which were associated with them, combined to produce a certain space of flows (at first, with the exception of raw materials, overwhelmingly *from* the US) of capital, commodities, diplomatic exchange, cultural artefacts (such as music, cinema and fashion) and military forces. After 1945, the Soviet bloc was partly constituted and imagined as a space of resistance to these flows. Revisionist histories of the origins of the Cold War in Europe have accordingly stressed how it emerged in part from Soviet reaction to the potential extension of this space of flows into Soviet-occupied eastern and central Europe.[9] As Stalin recognized, this would have undermined Soviet hegemony in the region. Since the latter had been established at such great cost and was seen as a vital buffer against further German or western aggression, the logic of resistance left the Soviets little alternative but to seek to close off this space (at least partially) and to constitute an alternative.

The presence of a Soviet-Communist other just across the Iron Curtain and the possibility of its spread through revolutionary upheavals in the Third World, further embedded the American-led space of flows in the west and ensured that the US and its local allies would entangle it with other military alliances and systems of 'containment'. These were extended across the Pacific to Japan (formalized in a treaty signed in 1951), into Korea, Indochina and South-East Asia, the Middle East (where they were complicated by a close relationship with Israel) and southwards into the American continents and surrounding seas.

I have condensed and retold this (familiar) story here in order to note how the logics within it both allowed and shaped the formats for regional communities. Leaving aside the predominately transatlantic and transpacific military communities, such as the North Atlantic Treaty Organization (NATO), it is notable that the US had *contradictory* stances with regard to the emergence of potential economic and political regional communities. On the one hand, Washington was disposed to see them as ways of consolidating capitalism and (at least in western Europe) forms of liberal democracy, that would keep communists and radical forces outside the main power structures. On the other hand, America (along with the regulatory institutions) was wary of the possibility that regional communities could become protectionist blocs and disrupt the space of flows in which the USA was represented as having a great stake.

This is clearest in the case of the European Union. When it was first established as the six member EEC in 1957, it broke GATT rules, but was allowed to pass in part because the USA was supportive of it as (in summary) a means to promote capitalist and democratic stability in western Europe and bind together American allies on the European continent (see Lundestad, 1998). Meanwhile Soviet international relations 'experts' and politicians were scripting the EEC as one of a range of economic, political-ideological or military institutions, designed to 'integrate' or 'fuse' what they termed the aggressive bloc of western (monopoly) capitalist states supported or led by the US (see Neumann, 1989).

## A *host of metaphors*

– But Prime Minister, we don't really belong to the region do we? We're not at the *centre* of it are we?

– Well, it's a basin – the Pacific Basin. There's no-one at the centre. We're at the edges of it and so is everybody else. ([Australian Prime Minister] Paul Keating, interviewed by Paul Lyneham *The 7.30 Report* 5 March 1992). (Cited in Gibson (1994, p. 83))

For Britain to get the best out of the EU we must be players on the pitch, not commentators from the stands. (British Prime Minister, Tony Blair, 2 February 1999). (Cited in White (1999, p. 2))

Since its 1950s foundations, the European Economic Community has evolved into the European Union, now with a single currency (at least for most of its members) which may come to challenge the dollar as the global reserve currency. All this, plus the end of the Cold War (which had served to allow something of a *common* western identity to be scripted against a Soviet other) and periodic trade disputes between the EU, Japan and US have allowed a geo-economic script to circulate about the prospect of (re)newed global 'tripartite splits' between 'blocs' constructed around each of these. The establishment of a North American Free Trade Area (NAFTA) and Japan's more assertive role during the early 1990s (after which severe recession hampered it) are also part of what grants such stories greater power and credibility.

The monthly *Time* magazine has frequently served as a forum for codifying American strategy to the 'educated' citizen (see Sidaway, 1998a). The 15 June 1992 issue duly reminded its readers of George Orwell's (1949) dystopian vision of three 'regions' (which he called Eurasia, Eastasia and Oceania), perpetually at war with each other. Although nominally writing about the future, Orwell was undoubtedly reflecting on the pre-1945 inter-imperialist competition and the Nazi geopolitical discourse of 'panregions' (see O'Loughlin and van der Wusten, 1990). With this dreadful backdrop in vision, *Time* would have it that unless care is taken to preserve American-led global 'free trade' from its many potential opponents, then the world could 'split' into three competitive regions. Likewise, for the UK-based weekly *Economist* magazine, in the mid-1990s, the greatest threat to global 'free trade' was in the proliferation of regional communities. On its front cover of 7 December 1996, the *Economist* featured a vision of the proverb of 'too many cooks spoil the broth'. The 'cooks' in question were regional communities and the 'broth' world trade (see Figure 3).

Such metaphors, whilst seemingly trivial or banal, are in fact the very stuff of both elite and popular forms of discourse about regionalisms and geopolitics more generally. In addition to the languages of imperial and Cold War geo-politics – 'heartlands', 'rimlands', 'dominoes', 'blocs', and so on (see Ó Tuathail, 1996; Ó Tuathail, Dalby and Routledge, 1998), territorial metaphors of centre and margin or *inside* and *outside* provide the often taken for granted structure to virtually all elite, academic and popular discourses about (inter)national[10] relations (see Agnew, 1994; Agnew and Corbridge, 1994; Walker, 1993).

In the case of regional communities, the notion of them as containers with varying degrees of 'openness' or 'closure' has been important in debates about the threats that they might pose to global 'free trade'. These have acquired particular currency across the Pacific. Speaking in the name of APEC, member state politicians have reiterated its openness and complementarity with global free trade and capital mobility. At the same time, however (as with Europe), there is no consensus on the issue. Whilst APEC's official discourse stresses 'open regionalism', other versions of what it is and might become are articulated.

In part, this is because APEC represents a response by Australia, New Zealand, the US and Canada to versions of Pacific or East Asian co-operation which might exclude them. This is not the place for lengthy comment on the contested geopolitics of APEC, about which a great deal has been written to which interested readers are referred.[11] Instead, I will briefly note how, particularly for Australia, New Zealand (and to an extent for Canada and the USA), a sense of rising Asian confidence and the threats that this might pose, produced

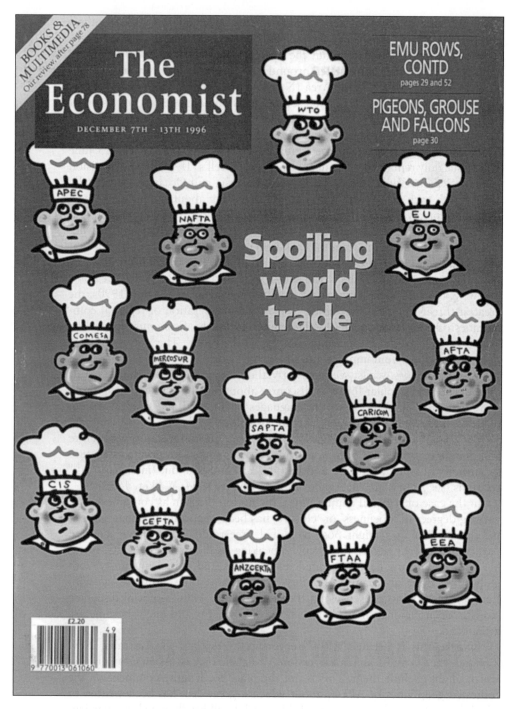

**Figure 3** 'Spoiling World Trade'
Source: *The Economist* (London), 7 December 1996

**Figure 4** 'APEC Asunder'
Source: *Asiaweek* (Hong Kong), 4 December 1998

the possibility of responding to it or engaging with it through projection of a new Pacific Age. Seeking a sense of renewal, the old colonial-settler states are held (also) to belong to such a bright future, along with dynamic postcolonial Asia.[12] The 'Pacific Rim' thereby becomes a signifier filled with a whole set of ('white') fantasies, fears and senses of possibility (Dirlik, 1993). But for many of the Asian countries a rather different notion of a (shared) Asian modernity, which is distinct from European or American cultural and political forms (or, as they are sometimes called, 'civilizations') is registered in discourses about and metaphors of an 'Asian way' projected through APEC and other regional communities such as ASEAN.[13]

Another set of metaphors stress forward movement or travel – sometimes a difficult journey. This cropped up following the contentious 1998 APEC summit, when the Hong Kong based *Asiaweek* magazine published an editorial calling for a rethink of APEC's rationale. With political differences between the US and host country Malaysia particularly evident, the (metaphorical) notion of the integrative vehicle of APEC, with its members as passengers being shaken by the rocky road of 'politics' was portrayed in a cartoon that headed the *Asiaweek* editorial (see Figure 4).

At other times a metaphor of flight provided a more elegant (and appropriately, of Japanese origin) visualization of East Asian or Pacific 'integration'. A pre-war Japanese economist, Kaname Akamatsu, compared the destiny of East Asia with the flying pattern of wild geese (*ganko keitai*). In this, the pilot bird of Japan leads the flock and other 'birds' (i.e. countries) which are aided in the slip-stream of Japanese energy. The notion of *ganko keitai* was recycled in the 1970s and 1980s by influential Japanese economists and one time foreign

**Figure 5** 'Southern Africa here we come!'
Source: *Financial Gazette* (Johannesburg), 21 October 1993

minister Okita Saburo. In the 1990s it enjoyed a new life within discourses about integration. But whenever it has been restated, *ganko keitai* has complex historical resonances with Japanese imperial geopolitical discourses[14] and permits a clear naturalization of or at least an assertion of potential Japanese leadership of Asian or Pacific-Asian integration.

Moving to another regional community, the third cartoon reprinted here, this time from a South African newspaper (Figure 5), is superficially different from that of Figure 4 in so far as this time the collective vehicle portrayed is the state (of South Africa) whose post-apartheid government is keen to set out on the path to integration (following the road marked by the sign pointing to the regional community of the SADC), but whose vehicle of state lacks the required motive capacity (the wheels!). The new government is in the driving seat (it occupies the government offices), but cannot translate this formal political power into the requisite energy to propel South Africa in the specified direction. A decade earlier, just after the launch of what was to become the SADC, a Malawian minister had invoked 'roads' and 'battles':

> There are countless pot-holes on the road to economic liberation and what SADCC [the Southern African Development Coordination Conference, the predecessor to the SADC] has managed to achieve so far is only a very short distance on the long road to economic independence. But united we are, and

with the support received from many governments and organisations, we are confident that we will, with good intentions win the battle. (L. Chakakala Chaziya, Malawian Minister of Finance, cited in SADCC, 1981, p. 7)

Leaving the 'trajectory' of the SADC aside,[15] we should note that such a metaphor of a path, road or railway to 'development' and integration, 'driven' by the 'motor' of the regional community are very common. They have cropped up in most of the (albeit limited) range of examples that I have drawn upon here and their apparent frequency is so much the case that they start to form part of the basic maps of meaning, the discursive horizons, involved whenever 'integration' is discussed. That the associated notion of forward progress is also more or less universal within the global discourse of development would suggest that a deeper, common element of discourses of modernity[16] is finding expression.

We have not exhausted the metaphorical repertoire here. For example, in the case of debates about and discourse of the EU, Chris Shore (1997, p. 139) notes how an architectural metaphor of 'building' Europe has (like 'building communism' in the old Soviet Union), 'become part of the ideological formula'. He goes on to note that:

> Other metaphors of Europe included containers ('entering the ERM', 'going into Europe'), gastronomy (France says 'no to à la carte Europe'), weddings ('the ERM marriage'), mathematics ('variable geometry', 'concentric circles'), illness ('healing Europe's wounds') and rites of passage (the 'death of the ERM', the 'birth of the new currency', the 'christening of the Euro'). To this list John Major added 'sport' when he summed up the 253 pages of the Maastricht Treaty signed in February 1992 as 'Game, set and match to Britain'. . . . What is significant about these metaphors is the way they are used to set the terms for debate and provide discursive models for promoting alternative scenarios. In each case, use of these metaphors can be linked to rival and often conflicting visions of Europe, and reflect attempts by different political agents to exert hegemony over the European debate. (p. 140)

With such a diversity of metaphors in mind, in the space that remains I want to turn to what else is present within (albeit sometimes obscured or unacknowledged) and what unifies *diverse* discourses of integration.

## Final notes: undecidable geographies[17]

> Communities are to be distinguished, not by their falsity/genuineness, but by the style in which they are imagined. (Benedict Anderson, 1991, p. 6)

Although sometimes it is less than immediately evident, the idea or claim of integration expressed through a regional community presupposes the existence of states. We might say that rather than the state simply preceding and constituting (together with other such 'sovereign actors') the community, the latter allows the state to be invoked and made to seem more real. The community is 'other' to, an extension of, or exterior to the state. But once it is constituted, once a treaty to establish it is signed by representatives of the member states (as in Figure 6 which reproduces the signatures from the treaty which established the SADC), then the community also serves to remind anyone who would interact with it or come into contact with a representation of it, that the states of which it is formed apparently *exist* as sovereign, tangible, real things (rather than, for example, as contested, simulated and ramshackle sets of social relations masquerading as states).

At first, all this is simply in the form of signatures on a piece of paper which together invoke and create a community and *at the same time* presume a state by and through those who *sign*, in place of, in the name of, or as we might say, *representing* the latter. Later, meetings, declarations, secretariats and commissions all come into force to reconfirm this. Whilst the regional community is in place we are stuck with a certain circularity. We can read and hear about how the community overcomes, pools or perhaps is hostage to the sovereign states of which it is made. These states are therefore *confirmed* as having a real existing presence. That such 'presence' is itself in part the result of such signatory acts (and similar declarations, proclamations and constitutions) is taken for granted and the state is naturalized, made to seem simply a thing, in and of itself, with its own secure proper identity and history.

An example may serve to indicate something of how all this operates. Turning again to Figure 6, we might note that in the signature on the document establishing the SADC, Angolan President Eduardo dos Santos appears as a representative of (standing in for) Angolan sovereignty. The president signs in the name of his state and the people subject to it. Leaving aside the fact that the signature (t)here is a reproduction of a reproduction (through electronic and mechanical means of photocopying, printing and so on), and that the original is itself an *imitation*, by dos Santos himself, of every other signature by him (that is, no two signatures by the same person can ever be identical, unless they are copies and not 'originals') it nevertheless stands (t)here as an uncontested representative of Angolan sovereignty – in the presence of his other cosignatories.

Yet at the time that this signature was made, the Angolan government was not in basic administrative control of all of its war-torn 'national' territory. For example, swathes of the centre and south of Angola were controlled by the rebel guerrilla forces of the União Nacional para a Independência Total de Angola (UNITA). Therefore, as I have noted elsewhere, when signing documents at a summit meeting the SADC:

> Angolan ministers are uncontested embodiments of the Angolan government's sovereignty. But within the geopolitical space called Angola, government forces, the insurgent movements of UNITA and FLEC, foreign oil companies,

**Figure 6** Extract from the Treaty of the Southern African Development Community
Source: Declaration Treaty and Protocol of the Southern African Development Community (SADC, Gaborone)

## ARTICLE 41

### ENTRY INTO FORCE

This Treaty shall enter into force thirty (30) days after the deposit of the instruments of ratification by two thirds of the States listed in the Preamble.

## ARTICLE 42

### ACCESSION

This Treaty shall remain open for accession by any state subject to Article 8 of this Treaty.

## ARTICLE 43

### DEPOSITARY

1. The original texts of this Treaty and Protocols and all instruments of ratification and accession shall be deposited with the Executive Secretary of SADC, who shall transmit certified copies to all Member States.

2. The Executive Secretary shall register this Treaty with the Secretariats of the United Nations Organisation and the Organisation of African Unity.

## CHAPTER EIGHTEEN

### TERMINATION OF THE MEMORANDUM OF UNDERSTANDING

### ARTICLE 44

This Treaty replaces the Memorandum of Understanding on the Institutions of the Southern African Development Coordination Conference dated 20th July, 1981.

28

IN WITNESS WHEREOF, WE, the Heads of State or Government have signed this Treaty.

DONE AT Windhoek, on 17th ... Day of August, 1992 in two (2) original texts in the English and Portuguese languages, both texts being equally authentic.

THE PEOPLE'S REPUBLIC OF ANGOLA

KINGDOM OF LESOTHO

REPUBLIC OF MOZAMBIQUE

KINGDOM OF SWAZILAND

REPUBLIC OF ZAMBIA

REPUBLIC OF BOTSWANA

REPUBLIC OF MALAWI

REPUBLIC OF NAMIBIA

UNITED REPUBLIC OF TANZANIA

REPUBLIC OF ZIMBABWE

29

mercenaries, and at various times, Cuban, South African and United Nations peacekeeping troops have all invoked *sovereign* authority. Angola represents something of an extreme example (certainly in the scale of dislocation and human suffering). Yet at the same time it forces us to recognise what is taken for granted when sovereignty is invoked and discussed. (Sidaway, 1998b, p. 570)

A parallel point is made by Montecinos (1996, p. 120) in respect to what she terms 'ceremonial regionalism' in Latin America:

The organisations created to support regional integration should not be conceptualised as only a technical response to the problems of economic development. They have also served a ceremonial display. Regional agencies were intended to facilitate the obtention of resources, acceptance, and legitimacy from significant 'outsiders.' To the extent that the ceremonial functions of integration were successful, failures in effectiveness were continuously overlooked.

A kind of taken for grantedness or overlooking is an important part of what allows a circle of recognition, presence and simulation. It forms a condition of possibility for the claiming of sovereignty. Regional communities are not the only such basis to establish such a 'circle'. For once states are recognized in and participate in other fora, notably and most significantly, the United Nations, so another circle of discourse, of simulation, of sovereign presence-absence-difference is invoked. Spotting this, Tim Luke (1993) argues that the United Nations exists in part to demonstrate that all the member states are utterly 'real'. Exceptions, those states such as Switzerland which has never joined, or Taiwan which lost its seat to mainland China, are rare and anyway not magically outside circles of presence-absence-recognition.

For most countries of the world, Luke notes that their member-statehood is renarrated at the UN whenever a representative of that state is allowed to take a seat, make a statement or hoist a flag outside its institutions. Blowing in the New York wind, those flags continually reiterate the fixity, presence and reality of the member states. And with them, circulating around the manifestation, duplication or spectacle of the state, are conceptions of 'national' economies and 'national' cultures. Never mind what is going on elsewhere, stand in front of those flagpoles and the presence of all those countries seems perfectly real. Yet, since both the appearance of presence (in a territory and at the UN) and 'representation' (at the UN and in a territory) are required, who can decide exactly which is producing what? Usually therefore, one cannot decide if the delegation simply represents *or* if it produces the state. It turns out that this ambiguity, this *undecidability* is in fact part of what is required to make the state seem real. This is an important source of the proper identity of the state and ultimately also that of the 'imagined regional community', whose ambiguous confirmations of authority are akin to those of the UN.

More widely, it is in such repeated performances, inscriptions and claims that the state, the international or regional community, and for that matter, culture, economy and politics are constituted. None of these should be seen as stable fixed components that simply come together in some mixture or other to form a regulatory nexus. It is more helpful to read the state, the community, culture, the economy and so on as expressions of power whose meaning is itself constructed through the nexus. It is hard, indeed impossible, to decide what precedes what. If all this tautology sounds confusing, I can only plea that the confusion is not entirely of my own making. To simplify things and bearing in mind the figure of the circle that was invoked in the opening quotation to this chapter, we might say that the story is always *circular*, so the apparent end of one is simply the beginning of another – and the beginning of another merely a sequel to that which has gone before. And so on, and so forth, until the plot and cast start to seem thoroughly real.

## Acknowledgements

This chapter draws upon research on the SADC funded by an ESRC Research Fellowship. A visit to Singapore and the APEC secretariat was funded by the 20th International Geographical Congress Fund. I am grateful to both the ESRC and the IGC Fund for this support. The chapter was written whilst I was affiliated to the University of Seville funded by an EU Training and Mobility of Researchers Grant. I would also like to thank Ian Cook for bearing with me and for his comments on earlier drafts.

## Notes

1. Richard Crockatt (1995).
2. Cooper had previously written a longer essay on 'The post-modern state and the world order' which was published by the 'think-tank' Demos. For a laudatory review of this, see the *Economist*, whose anonymous reviewer entitles his article about Cooper's pamphlet (under the rubric of 'foreign policy') 'Not quite a new world order, more a three-way split' (Anon, 1997).
3. This chapter by no means exhausts these. A fuller treatment would require consideration of debates about 'culture' as a mode of constructing difference/identity of regional communities. For a survey of such debates with reference to integration in Europe and Pacific Asia, see Wesley (1997).
4. The introductory chapters to Gibb and Michalak (1994), Fawcett and Hurrell (1995) and Mansfield and Milner (1997) may usefully be consulted for guides. Two undergraduate geography texts have good sections on regionalism which touch on some of the theoretical literatures, see the chapter by Brook (1995) and associated readings in a text produced for an Open University course, and at a more basic level, Chap. 11 in Knox and Agnew (1998). A more detailed survey may be found in the recently published four volume set on *International Economic Integration* edited by Jovanovic (1998) which runs to 2966 pages!
5. Although the geopolitical designations of 'South-East Asia' and the 'Middle East' may appear common sense and deeply rooted, they are recent inventions. On the former, Benedict Anderson (1998, p. 3) notes that:

> As a meaningful imaginary, it has had a very short life, shorter than my own. Not surprisingly, its naming came from outside, and even today very few among the

almost 500 million inhabiting its roughly 1,750,000 square miles of land (to say nothing of water), ever think of themselves as 'Southeast Asians.' The older Chinese concept *nan-yang* referred vaguely to a 'southern' region to be reached by sea . . . Its later Japanese derivation, *nampó*, stretched out broadly and elastically into what the Americans would call the Southwest Pacific. Southeast Asia, as such, emerged as a significant political term only in the summer of 1943 with the creation of Louis Mountbatten's South-East Asia Command, an offshoot of the more traditional India Command. But this command was based in Kandy, and its territorial responsibilities included both Ceylon and the Raj's Northeast Frontier (neither in 'Southeast Asia' today) and excluded the Netherlands Indies (till July 1945), as well as the Philippines. Yet the naming was clearly a response to the fact that for the first time in history a single power – that of Hirohito's armies – effectively controlled the entire stretch between British Burma and the Hispano-American Philippines.

The term 'Middle East' has a similar vintage. As I have noted elsewhere: 'though it may be traced back as a label used by the British India Office since the middle of the 19th century . . . it entered popular discourse more recently via the writings of the American geopolitician Alfred Mahan; who in his 1902 geopolitical text on seapower, scripted the region around the Gulf as neither 'Near East' nor 'Far East' (Sidaway, 1994, 357).

Blake and Drysdale (1985, p. 11) describe how the term became much more familiar in the United States and Western Europe during the Second World War: 'when both the British and the Allied headquarters in Cairo – known as H.Q. Middle East – covered large parts of northern and eastern Africa as well as Iran, Turkey, and all the Arab states east of the Suez canal.'

Today, although few see themselves foremost as South-East Asians or of the Middle East, the terms do circulate in the designated regions (as well, of course, much more widely) and in Asia have entered state discourse as in the ASEAN.

'Latin America' as a designation has a longer vintage and has entered discourse and patterns of self-identification there. However, as a coherent object of study, it has been shaped by the US geopolitical imagination, particularly since 1945 (see Berger, 1995).

6. It is notable that the term 'European' seems to have been first most commonly used as a designation of white settlers or administrators in the colonial empires. Whilst not identical with the post-1945 (or indeed earlier) notions of European identity and citizenship as an expression of continental unity, the latter has been indelibly marked by the former.

7. Both superpowers claimed to be anti-imperialist, asserting their difference from the old European powers, whilst accusing each other of practising imperialism. It is not difficult to deconstruct these discourses, indicating that both were based on dubious assertions and debatable claims. Nevertheless at a certain rhetorical and material level superpower antipathy towards the old European imperialisms was a decisive feature of world politics after 1945.

8. See Stephanson (1997) for a rich argument that resists the tendency to reduce the story of the Cold War to a single narrative. As he notes:

> Typically, moreover, the obvious end is retrospectively inscribed in the beginning and in the whole nature of the period so as to allow its history to be rewritten as an 'explanation' of the obvious. Meanwhile other possible periodizations are barred or simply subsumed, periodizations, say, in terms of 'decolonization,' the economic

rise of Japan and Germany,' or 'the universalization of the European model of the nation-state'. . . . Yet the picture to be completed always seems to expand and indeed always will. There is no final or pristine cold war in the archives, or anywhere else for that matter, waiting to be discovered or uncovered. (pp. 62–3)

9.   For a short rendition of 'orthodox' versus 'revisionist' histories of the Cold War, see Chap. 4 in Richard Crockatt (1995). This may usefully be compared with Chap. 2 of Fred Halliday (1986).

10.  And not just (inter)national relations. Michael Herzfeld (1992, p. 109) explains how: 'All other bureaucratic classifications are ultimately calibrated to the state's ability to distinguish between insiders and outsiders. Thus ... one can see in bureaucratic encounters a ritualistic enactment of the fundamental principles upon which the very apparatus of state rests.'

11.  Richard Higgott (1997, p. 165) notes how:

> The study of Asian regionalism is extremely fashionable nowadays. In addition to the vast and growing body of monograph literature on Asia Pacific regionalism of both a scholarly and a policy oriented nature, the pages of the specialist regional journals and specialist journals of international economics and international political economy abound with analyses of regional economic growth at all levels in the region.

With this in mind, I will make no attempt to summarize the literatures here, save noting that I have found Beeson and Jayasuriya (1998) and Terada (1998), to be useful starting points. Higgott and Stubbs (1995) is also a stimulating primer on the different conceptions of East Asian/Asia-Pacific regionalism and what is at stake in the articulation and circulation of these differences.

12.  On the Australian version of this reinscription, and Canberra's role in promoting APEC, see Bell (1997).

13.  ASEAN has a fascinating history. Beginning as a late 1960s Cold War scheme to associate the often divided anti-communist states of South-East Asia, ASEAN has slowly developed a wider economic function. Although for a long time and with some justification mostly described as a Cold War project, ASEAN's celebrated diplomatic style relying on 'consensus' needs to be analysed in its local cultural specificity. Thambipillai and Saravanamuttu (1985) is a primer on this.

14.  For a study which both demonstrates this and specifies the differences between contemporary Japanese discourse about Asia-Pacific co-operation and imperial Japanese geopolitics, see Koschmann (1997).

15.  For a study of the SADC and suggestions for further reading, see Sidaway (1998b). SADC's forerunner is compared and contrasted with ASEAN in Curry (1991).

16.  Two of the best known accounts of discourses of modernity are Habermas (1987) and Berman (1982). Whilst both are excellent, they are also both marked by drawing mostly upon 'First World' experiences and a fuller account would want to acknowledge the diversity of expressions of modernity (within which some might accommodate postmodernity). Escobar (1995), Miller (1994) and Rist (1997) may be consulted on how modernity has been incorporated into discourses of development.

17.  In addition to Tim Luke's always stimulating writings (which are cited in the main text), this section owes a good deal to the interpretations of Bartelson (1998), Constantinou (1996), Derrida (1986, 1988), Dillon and Everard (1992), and Mitchell (1991).

# References

Agnew J. (1994) 'The territorial trap: the geographical assumptions of international relations theory', *Review of International Political Economy*, 1, 1, 53–80.

Agnew J. and Corbridge S. (1994) *Mastering Space: Hegemony, territory and international political economy* (Routledge, London and New York).

Althusser L. (1984) *Essays on Ideology* (Verso, London).

Anderson B. (1991) *Imagined Communities: Reflections on the origin and spread of nationalism* (Second Edition) (Verso, London).

Anderson B. (1998) *Spectres of Comparison: Nationalism, Southeast Asia and the World* (Verso, London).

Anon (1996) 'All free traders now?' *Economist*, 7 December, 25–7.

Anon (1997) 'Not quite a new world order, more a three-way split', *Economist*, 20 December, 51–5.

Bartelson J. (1998) 'Second Natures: is the state identical with itself?' *European Journal of International Relations*, 4, 3, 295–326.

Beeson M. and Jayasuriya K. (1998) 'The political rationalities of regionalism: APEC and the EU in comparative perspective', *The Pacific Review*, 11, 3, 311–36.

Bell R. (1997) 'Anticipating the Pacific Century? Australian responses to realignments in the Asia-Pacific', in M.T. Berger and D.A. Borer (eds) *The Rise of East Asia: Critical visions of the Pacific Century* (Routledge, London and New York) 193–218.

Berger M.T. (1995) *Under Northern Eyes: Latin American studies and US hegemony in the Americas, 1898–1990* (Simon and Schuster, Bloomington, Berman).

Berman M. (1982) *All that is solid melts into air* (New York).

Blake G.H. and Drysdale A. (1985) *The Middle East and North Africa: A political geography* (Oxford University Press, Oxford).

Brook C. (1995) 'The drive to global regions?' In J. Anderson, C. Brook and A. Cochrane (eds) *A Global World?: Re-ordering political space* (The Open University and Oxford University Press) 114–65.

Carrier J.G. (1997) (ed.) *Meanings of the Market: The free market in western culture* (Berg, Oxford).

Carrier J. and Miller D.G. (1998) (eds) *Virtualism: A new political economy* (Berg, Oxford).

Constantinou C.M. (1996) *On the Way to Diplomacy* (University of Minnesota Press, Minneapolis and London).

Cooper R. (1998) 'Irony and foreign policy', *Prospect*, December, 58–60.

Crockatt R. (1995) *The Fifty Years War: The United States and the Soviet Union in world politics, 1941–1991* (Routledge, London and New York).

Curry R.L. (1991) 'Regional economic cooperation in Southern Africa and Southeast Asia', ASEAN *Economic Bulletin*, 8, 1, 15–28.

Derrida J. (1986) 'Declarations of independence', *New Political Science*, 15, 7–15.

Derrida J. (1988) *Limited Inc* (Northwestern University Press, Evanston, IL).

Derrida J. (1992) *Given Time: I counterfeit money* (University of Chicago Press, Chicago and London).

de Vasconcelos A. (1995) 'Sobre Fernando Henrique Cardoso', *Público*, 22 July, 9.

Dillon G.M. and Everard J. (1992) 'Stat(e)ing Australia: squid jigging and the masque of state', *Alternatives*, 17, 281–312.

Dirlik A. (1993) (ed.) *What is in a Rim? Critical Perspectives on the Pacific Rim Idea* (Westview Press, Boulder, CO).

Escobar A. (1995) *Encountering Development: The making and unmaking of the Third World* (Princeton University Press, Princeton, NJ).

Fawcett L. and Hurrell A. (1995) (eds) *Regionalism in World Politics: Regional organisations and international order* (Oxford University Press, Oxford).

Foucault M. (1970) *The Order of Things: An archaeology of the human sciences* (Tavistock, London).

Gibb R. and Michalak W. (1994) (eds) *Continental Trading Blocs: The growth of regionalism in the world economy* (John Wiley, Chichester).

Gibson, M. (1994) 'A centre of flux: Japan in the Australian business press', *Continuum: the Australian journal of media and culture*, 8, 2, 83–102

Habermas J. (1987) *The Philosophical Discourse of Modernity: Twelve lectures* (Polity Press, Cambridge).

Halliday F. (1986) *The Making of the Second Cold War* (Verso, London).

Herzfeld M. (1992) *The Social Production of Indifference: Exploring the symbolic roots of Western bureaucracy* (The University of Chicago Press, Chicago and London).

Higgott R. (1997) '*De Facto* and *de jure* regionalism: the double discourse of regionalism in the Asia Pacific', *Global Society*, 11, 2, 165–85.

Higgott R. and Stubbs R. (1995) 'Competing conceptions of economic regionalism: APEC versus EAEC in the Asia Pacific', *Review of International Political Economy*, 2, 3, 516–35.

Hurrell A. (1995) 'Regionalism in theoretical perspective', in L. Fawcett and A. Hurrell (eds) *Regionalism in World Politics: Regional organisations and international order* (Oxford University Press, Oxford) 37–73.

Jovanovic M.N. (1998) *International Economic Integration: Critical perspectives on the world economy* (four volumes) (Routledge, London and New York).

Knox P. and Agnew J. (1998) *The Geography of the World Economy* (Arnold, London and Wiley, New York).

Koschmann J.V. (1997) 'Asianism's ambivalent legacy', in P.J. Katzenstein and T. Shiraishi (eds) *Network Power: Japan and the new Asia* (Cornell University Press, Ithaca, NY) 83–110.

Luke T. (1993) 'Discourses of disintegration, texts of transformation: re-reading realism in the New World Order', *Alternatives*, 18, 229–58.

Lundestad G. (1998) *'Empire' by Integration: The United States and European integration, 1945–1997* (Oxford University Press, Oxford).

Mansfield E.D. and Milner H.V. (1997) (eds) *The Political Economy of Regionalism* (Columbia University Press, New York).

Miller D. (1994) *Modernity: An Ethnographic Approach. Dualism and Mass Consumption in Trinidad* (Berg, Oxford).

Mitchell T. (1991) 'The limits of the state: beyond statist approaches and their critics', *American Political Science Review*, 85, 77–96.

Montecinos V. (1996) 'Ceremonial regionalism, institutions and integration in the Americas', *Studies in Comparative International Development*, 31, 2, 110–23.

Neumann I.B. (1989) *Soviet Perceptions of the European Community, 1950–1998* (Norsk Utenrikspolitisk Institut, Oslo).

Neumann I.B. (1994) 'A region-building approach to Northern Europe', *Review of International Studies*, 20, 53–74.

O'Loughlin J. and van der Wusten H. (1990) 'The political geography of panregions', *Geographical Review*, 80, 1, 1–20.

Orwell G. (1949) *Nineteen Eighty-Four* (Secker and Warburg, London).

Ó Tuathail G. (1996) *Critical Geopolitics: The politics of writing global space* (Routledge, London and New York).

Ó Tuathail G., Dalby S. and Routledge P. (1998) (eds) *The Geopolitics Reader* (Routledge, London and New York).

Rist G. (1997) *The History of Development: From Western origins to global faith* (Zed Books, London and New York).

Rostow W.W. (1990) 'The coming age of regionalism', *Encounter*, 74/5, 3–7.

SADCC (1981) *Southern African Development Co-ordination: From Dependence and Poverty toward Economic Liberation* (SADCC, Gaborone).

Shore C. (1997) 'Metaphors of Europe: integration and the politics of language', in S. Nugent and C. Shore (eds) *Anthropology and Cultural Studies* (Pluto Press, London and Chicago, IL) 126–59.

Sidaway J.D. (1994) 'Geopolitics, geography and "terrorism" in the Middle East', *Environment and Planning D: Society and Space*, 12, 357–72.

Sidaway J.D. (1998a) 'What is in a Gulf? From the "arc of crisis" to the Gulf war', in G. Ó Tuathail and S. Dalby (eds) *Rethinking Geopolitics* (Routledge, London and New York) 224–39.

Sidaway J.D. (1998b) 'The (geo)politics of regional integration: the example of the Southern African Development Community', *Environment and Planning D: Society and Space*, 16, 549–76.

Stephanson A. (1997) 'Fourteen notes on the very concept of the cold war', in G. Ó Tuathail and S. Dalby (eds) *Rethinking Geopolitics* (Routledge, London and New York) 62–85.

Talbott S. (1992) 'Beware of the three-way split', *Time*, 15 June, 39.

Terada T. (1998) 'The origins of Japan's APEC policy: Foreign Minister Takeo Miki's Asia-Pacific policy and current implications', *The Pacific Review*, 11, 3, 337–63.

Thambipillai P. and Saravanamuttu J. (1985) *ASEAN negotiations: Two insights* (Institute for SouthEast Asian Studies, Singapore).

Tønnesson S. and Antlöv H. (1996) 'Asia in theories of nationalism and national identity', in S. Tønnesson and H. Antlöv (eds) *Asian Forms of the Nation* (Curzon Press, Richmond) 1–39.

Walker R.B.J. (1993) *Inside/outside: International relations as political theory* (Cambridge University Press, Cambridge).

Wesley M. (1997) 'The politics of exclusion: Australia, Turkey and definitions of regionalism', *The Pacific Review*, 10, 4, 523–55.

White M. (1999) 'Want to find out what's really going on in British politics? Read a women's magazine', *The Guardian Europe 2*, 3 February, 2.

Williams R. (1977) *Marxism and Literature* (Oxford University Press, Oxford).

World Trade Organization (1995) *Press Release 18 April 1995. No evidence of polarisation of World Trade among three 'blocs' and no clash between world trade systems – says new WTO report* (World Trade Organization, Geneva).

# PART IV

# Nature and society

Westbay, a small seaside town in Dorset. 1997 © Martin Parr/Magnum Photos

# Introduction

## Simon Naylor

Investigating the relationships between the natural and social realms has been an abiding preoccupation for geographers. If, for argument's sake, we took G.P. Marsh's (1864) *Man and Nature* as a marker for the introduction of modern environmentalism into the geographical discipline, we could state that a significant part of geography's history had been consumed by discussions of nature and the environment and their relationships with humans and society. Richard Grove (1997) suggests that the rise of western environmentalism in the 1860s was pre-dated by policies enacted in colonial contexts in the eighteenth century, and as the nascent geographical discipline was thoroughly embroiled in the imperial endeavour so we might argue that its interest in nature–society interactions extends back accordingly. However we look at it, geography has a long history of investigating human (social)–nature relations (see Livingstone, 1992; Macnaughten and Urry, 1998).

The fact that as a discipline geography spans the natural and human sciences, might enable us to conclude that it is well placed to provide further insights into the human–nature interface, unattainable from other disciplinary standpoints. However, for those working within its boundaries this synthesis of ideas, theories, methods and practices has patently not taken place. Indeed, the schism between 'physical' geography, 'human' geography and their myriad practitioners has only *increased* since the discipline's inception in the early years of the nineteenth century. This is in large part due to their following quite different epistemological and methodological trajectories: physical geography has been happy to position itself within the scientific camp, whilst human geography has utilized a range of theories and techniques from the social sciences and the humanities (as well as sometimes harbouring distinctly 'scientific' approaches to its subjects, of course). However, human geography's adoption of ideas and methods from cultural, social and feminist theory over the last few decades and their deployment in the development of new geographies of nature and the nature–social interface has, I think, marked one real rupture between the two subdisciplines. The venom directed at human geographers for suggesting that nature might be as much discursive as real has only been matched by the scorn of human geographers when they hear one of their scientific counterparts suggesting that their microscopes, quadrants and test-tubes have yielded better and more accurate insights on natural processes and the natural world.[1]

Although the minor antagonisms that often imbue geography departments over the opinions of its faculty members concerning the status of nature and human–non-human relations have not really been reflected in geographical literatures (possibly because of a tacit acknowledgement that we/they have to share corridors, meeting rooms and teaching loads!),[2] they have emerged in other realms. This antagonism has probably been most bloody in the aptly named (although arguably over-conflated) 'science wars', where luminaries in the sciences and the social sciences and humanities have engaged in open combat over the others' very belief systems.

Of course, arguments concerning the 'nature' of nature do not simply range over the frontiers of the human and natural scientific divide (with geography lying spread-eagled across the very border). Academic practitioners within the human sciences have been just as factious in their debates over nature and the environment. These, as Sarah Whatmore explains in her paper in this section, have often been centred around an (environmental) political issue: researchers worried about the destruction of threatened natures and habitats have voiced concern that 'discursive' or 'constructivist' treatments of nature disenable the protection, and sometimes even *facilitate* the destruction of the natural world (see also Soper, 1996 for a discussion of the differences between 'ecological' and 'postmodernist' approaches to nature). Whilst it is probably fair to say that some of the work on nature that has emerged out of the cultural turn suffers from an overt obsession with the 'textualization' of our worlds, it is blatantly unjust to label all social scientific work on nature that is informed by cultural theory as apolitical, averse to discussions of environmental policy or shy of materiality. A case in point is Whatmore's paper.

Whatmore urges the development of new understandings of life that don't simply support entrenched nature–society, human–non-human divisions, but which seek to map the heterogeneous geographies of life. This new cartography of hybridity – which recognizes nature as 'an always already *inhabited* achievement of heterogeneous social encounters' (p. 270) – not only images the non-human realm as complexly interwoven with the actions of its human counterpart; it also invokes what Whatmore refers to elsewhere as a 'relational ethics' (1997). This articulates a position where we humans can no longer hide behind the binary categories that have so successfully shielded us from forms of corporeal responsibility to the non-human realm, but forces us to face up to a suddenly enlarged community that is no longer 'other' (ibid.); a constituency which is very much bound up in the shaping of the 'business of [our] everyday living' (this volume, p. 270).

Jacquelin Burgess also confronts some of the misconceptions about culturally informed work on social–natural relations. Her paper is directly concerned with the development of new approaches to the design of environmental policy.

While she is wary of the excesses of 'high' and heavy-handed cultural theory she also rejects the reductive cost-benefit analyses that have characterized many attempts to measure the 'value' of nature. But that is not to say that she gives up in her attempt to develop new approaches to managing the local environment. Rather, Burgess outlines a project completed with the Environment Agency that was and is based on new discursive, collaborative and reflexive strategies – new forms of inclusive and consensual decision making – for the fashioning of environmental policy. Burgess's paper not only forces a rejection of some of the criticisms levelled at cultural theorists of nature over their disregard for environmental destruction, it also illustrates the ways in which environmental policy makers are *actively adopting* elements of cultural theory in their work.

Judith Gerber also takes a relational approach to the study of nature–social relations. In her paper on royal and plantation forests she demonstrates the mutable qualities of entities often projected as enduring and unchanging. Specifically, Gerber indexes the production of forests to historical and political-economic changes. The royal forest, she argues, was very much a product of medieval feudalism and the Crown's sovereignty over land. However, the plantation forests that emerged in the early twentieth century were closely associated with capitalism, the state, and a new rational, scientific planning discourse. These natural forms embodied social relations through and through, whilst their attributes changed according to the demise and rise of shifts in these relations. And, like their social counterparts, these natures were never left unchallenged; these 'natural' landscapes were open to the same forms of criticism as were the political-economic regimes that supported them.

Although all the papers in Part IV in some ways make the point that nature and the social are complexly interpellated, the paper by Sheila Hones usefully shifts the lens somewhat, away from a focus on the socialization of nature to an engagement with the *naturalization* of social life. Hones's paper in particular engages very directly with the ways in which *social* relations are often constructed upon particular natural(izing) metaphors. Analysing the writings in the American journal the *Atlantic Monthly* from 1880 to 1884, Hones discusses the solutions posited to deal with the 'pauper problem' in the US. She points out that in the alleviation of poverty and homelessness, private charity was normalized over public charity by articulating the former as natural and moral, whilst the latter was described as artificial, unhealthy and immoral. Thus social control was kept in the hands of the wealthy classes and the 'natural' social orders maintained.

Each of these papers deals with quite different empirical issues. However, as we have seen, common threads or assumptions interweave them all. All of the authors in this section start from the assumption that nature and society are complexly intertwined entities; and that their seeming separation is attributable not to their distinct ontologies, but to the great deal of purificatory work that makes them appear so (see Latour, 1993). As such, all of the papers are at least nominally supportive of a culturally informed approach to the study of nature–social relations. Although that is perhaps unsurprising considering the subject of this volume, it should be noted that there are some differences in approach too. For instance, Hones's overtly textual approach to the naturalization of social orders is quite different from Burgess's policy concerns, while Gerber's dialectic approach to the production of nature is subtly variant from Whatmore's network analysis of human–non-human relations. However, these tensions are not, I believe, the marker of a nascent field of enquiry that is unsure of itself or of how to use the cultural tools at its disposal. Rather it signifies a fast-growing and dynamic area of research which has benefited greatly from the introduction of cultural and spatial theory.

## Notes

1.  This obviously paints too simple a picture and fails to account for physical geographers who have engaged with 'post-normal' science and complexity theory, or for human geographers who harbour a view of the world that is as pre-eminently classificatory and objectifying as their scientific colleagues.
2.  Although I agree with Crang's dismissal of the environment as a potential peacemaker for the geographical discipline's two factions, and applaud his call for more sustained studies of the 'making' of physical geographies through 'field, laboratory and conference' (1998, p. 1973), I withhold my judgement over his assertion that human geographers' study of the practices and theories of physical geography might 'bring the two tribes of geography into dialogue' (ibid.).

## References

Crang, M. (1998) 'Places of practice, and the practice of science', *Environment and Planning A*, 30 (1), pp. 1971–4.

Grove, R. (1997) *Ecology, Climate and Empire: colonialism and global environmental history 1400–1940*, Cambridge, White Horse.

Latour, B. (1993) *We Have Never Been Modern*, London, Harvester Wheatsheaf.

Livingstone, D. (1992) *The Geographical Tradition*, Oxford, Blackwell.

Macnaughten, P. and Urry, J. (1998) *Contested Natures*, London, Sage.

Marsh, G.P. (1864) *Man and Nature: or physical geography as transformed by human action*, New York, Scribner.

Soper, K. (1996) 'Nature/"nature"', in Robertson, G. *et al. FutureNatural. Nature, science, culture*, London, Routledge.

Whatmore, S. (1997) 'Dissecting the autonomous self: hybrid cartographies for a relational ethics', *Environment and Planning D: Society and Space*, 15, pp. 37–53.

# Heterogeneous geographies

## Reimagining the spaces of N/nature

*Sarah Whatmore*

## Introduction

Human geography finds itself at an important juncture in its critical engagement with the question of nature in which neither the 'bracketing off' of an environmental sub-field common in other disciplines, nor the threadbare promise of a reintegration of physical and human geography, will suffice. Recent debate has congealed into a standoff between versions of 'social constructionism', in which Nature is treated as an inescapably mediated artefact of the social imagination, and versions of 'natural realism', in which 'nature' is the bedrock of a 'real' world of substantive entities and objective forces (Soper, 1995).[1] A vivid illustration of the mutual, and often hostile, incomprehension that passes for conversation between the two can be found in a collection of conference papers called *Reinventing nature. Responses to postmodern deconstruction* (1995). The editors Michael Soulé and Gary Lease, Professor of Conservation Biology and Dean of Humanities, respectively, at the University of California, Santa Cruz, describe the book as being 'about a clash of intellectual cultures' between

> certain radical forms of 'postmodern deconstruction that question the concepts of nature and wilderness, sometimes in order to justify further exploitative tinkering with what little remains of wildness' . . . [and] 'the opposing view [which] . . . assumes that the world, including its living components, really does exist apart from humanity's perceptions and beliefs about it'. (1995, p. xv)

By the end of the book, those misguided intellectuals who find themselves tagged 'postmodern' have been neatly aligned with other 'city people' (p. 162) who, it is implied, can have neither the expert knowledge nor the moral commitment necessary to securing a future for this dwindling natural world. The venom invested in those small words 'city people' spits off the page and is directed at all those, including human geographers, who have drunk from the poisoned chalice of the so-called 'cultural turn'. Doubtless we have played our parts in engendering this dialogue of the deaf, choosing the comfort of cabalistic networks and arcane codes as readily as scientists and laying ourselves open to the charge that our concerns with the cultural freight of the designation 'nature' have overshadowed those for the well-being of the creatures who inhabit it. But the irony is that for all their loudly declared enmity, the analytic encampments which pass for constructionism and realism have more in common than their protagonists would like to admit. For both sides of the Nature/nature debate premise their arguments on the acceptance, however unrecognised, of an a priori purification of the things of the world according to the magnetic poles of the 'natural' and the 'cultural'; the 'real' and the 're-presented'. In different ways, both return us – to use Bill Cronon's phrase (1995) – to 'the wrong nature'.

This binary impulse reverberates through more everyday geographical imaginations and environmental sensibilities rehearsed in pervasive distinctions between 'built environments' (the cultural pole) and 'natural environments' (the natural pole), with hierarchies of human 'settlement' in between marking inverse gradations of social/natural presence and absence. From the conventions of cartographic colour coding to the protocols of land use planning or of environmental designation, numerous professional and policy practices impress this binary imaginary upon the fabric of the world. With its celebration of 'wild(er)ness', configured as both species and places marked out precisely by their distance from humankind, much environmentalist rhetoric is also complicit in this purification of the spaces of 'society' and 'nature'. As the anthropologist Tim Ingold has observed 'Something . . . must be wrong somewhere, if the only way to understand our own creative involvement in the world is by [first] taking ourselves out of it' (1995a, p. 58).

My purpose in this chapter then is to refuse these terms of engagement with the question of nature and to begin instead, following Donna Haraway's lead, to work towards understandings of being in the world whose 'geometries, paradigms and logics breakout of binaries . . . and nature/culture modes of any kind' (1991, p. 129). This, of course, is easier said than done but has crystallized for me in an ongoing effort to rethink the humanist assumptions of 'human geography' and to join others in exploring ways of recognizing and accommodating the presence of non-humans in the worlds we inhabit (see, for example, Wolch and Emel, 1998). Such a 'hybrid' enterprise, as I have called it elsewhere (Whatmore, 1997), is concerned with the living fabrics rather than abstract spaces of social life; relational configurations spun between the capacities and effects of organic beings, technological devices and discursive codes within which people are differently and plurally articulated. At their most basic, the heterogeneous geographies that I want to sketch here imply a radically different understanding of 'who' (what) constitutes the worlds 'we' inhabit.

Such an understanding decentres social agency from the unitary figure of individual intent, recognizing it instead as a relational achievement, and it decouples social agency from the logocentric assumptions that restrict the capacity to act or to have effects to human beings,[2] admitting other players to the networks of social life. In a longer paper on which this chapter draws, I have elaborated three key theoretical manoeuvres drawn principally, but not exclusively, from science studies which underpin this relational understanding of social life under the headings hybridity, collectivity and corporeality (Whatmore, 1999). All three manoeuvres I suggest are directed, in different ways, against the lexical cast of the cultural turn and are associated with various theories of practice, or what Nigel Thrift has called 'non-representational theories of the social' (1996), which have very different philosophical lineages from those of deconstruction.

I do not want to spend time here rehearsing these manoeuvres but to focus instead on their geographical implications in the belief that theoretical ventures in this vein cannot succeed without disrupting the purified spaces of 'nature' and 'society' which litter contemporary environmental thinking and practice. Re-cognizing nature not as 'a physical place to which one can go' (Haraway, 1992, p. 66) but as an active, changeable presence that is always already in our midst challenges spatial, as well as social predispositions.

## Geographies in/of motion

> If [nature] can stop being (just) out there and start being (also) in here, . . .
> then perhaps we can get on with the unending task of struggling to live rightly
> in the world – not just in the garden, not just in the wilderness, but in the
> home that encompasses them both. (Cronon, 1995, p. 90)

In a controversial intervention questioning the North American reverence for wilderness, the environmental historian William Cronon signals the importance of geographical imaginations both to keeping 'nature' and 'society' in their proper place and to liberating them from this binary world. The coincidence between 'wild' plants and animals (species) and the 'wild' spaces they inhabit (habitats) pervades western environmental sensibilities (Whatmore and Thorne, 1998). It is powerfully evoked, for example, in the protocols of 'global environmental management' which police the place of nature by means of territorial archetypes – like biodiversity reserves – that enact a scientific blueprint of who and what should live there (McNeely et al., 1990). But it is a coincidence that is no less resonant in the political dramatics of radical environmental groups like EarthFirst!, or Greenpeace, which reinforce the place of nature by means of iconographic landscapes – like the rainforest – that are framed by/as their televised sites of struggle (see Baldwin et al., 1994). Such imagined spaces all too readily become flesh as heterogeneous communities are purified in their name through the sometimes violent removal of people, animals and plants who find themselves on the wrong side of the wire. The ethnic minority Karen people in southern Burma are even now being forcibly ejected by the military government from their traditional lands to make way for the million hectare Myinmoletkat 'Biosphere' Reserve.[3] In Britain, the ruddy duck recently found itself the target of a bizarre alliance of ornithological and nature conservation agencies intent on culling (i.e. killing) its insurgent population here in order to preserve the genetic purity and species integrity of the 'indigenous' European whiteheaded duck from its 'aggressive' mating habits (Lawson, 1997).

Are these examples of the 'living world that exists apart from humanity's perceptions of it' invoked by conservation biologists like Michael Soulé, to fortify their authority to speak and act for this exiled nature? The expert knowledges and moral choices threaded through such bodies, communities and places may be better disguised than is the case for celebrated creatures like Dolly the cloned sheep, but they are nonetheless crafted for that. The scientific manipulation of animals in the name of wildlife conservation – from the reproductive technologies of captive breeding, to those of electronic monitoring and population management in the 'wild' – is a ghostly presence in the landscapes of wildlife that occupy our TV screens, camera lenses and policy discourses. Figure 1, for example, illustrates the use of cross-species embryo transfer to reproduce endangered animal species, in this case with an adult female eland giving birth to a bongo calf at Cincinnati Zoo (Kaufman and Mallory, 1993).

This is a wildlife in which the sensuous, social and creative creatures who inhabit that designation are all too readily reduced to gutless units in an indifferent census of species and genetic diversity as biological resources in which scientists, however disingenuously, have been instrumental.[4] Accommodating non-humans in the fabric of social life requires more intimate, lively and promiscuous geographies than these quarantined fragments of a too precious nature.

The heterogeneous or hybrid geographies that I have in mind unsettle this glib coincidence of the things/spaces of nature fixed somewhere, always at a distance, and alert us to a world in commotion in which wildlife emerges within the routine, interweavings of people, organisms, elements and machines as these configure the partial, plural and sometimes overlapping time/spaces of everyday living. These humdrum spaces include, amongst others, the mutable flesh of embodiment, the ordinary motions of inhabitation and the mediating devices that make 'us' present even in 'our' absence. In place of the rigid contours of the flat maps and species inventories of conservation science, or the objectifying gaze of landscape studies, a topology of wildlife is a much more fluid beast in at least three senses.

The first of these senses concerns the spaces of embodiment. The mutability of organisms (including humans), in terms of their intrinsic organization and morphological plasticity (Goodwin, 1988), has been somewhat overshadowed by the heady talk of their malleability in the socio-technical networks of genetic engineering; organ transplantation and the like. Yet at the very heart of these artefactual worlds we are reminded by the proliferation of changling viruses, mutant cells (dis)figuring corporeal stability, or the startling appearance of pink and purple frogs in suburban garden ponds, that we are not the only agents in their fabrication.

The second sense of fluidity has to do with the spaces of motion. Animals (including humans) and, rather less obviously, plants lead mobile lives – on scales that vary from the Lilliputian travels of a dung beetle to the global navigations

**Figure 1** A bongo calf born to an eland mother as a result of embryo transfer. Such cross-species transfers indicate how rapidly this kind of reproductive technology is advancing. Photograph courtesy of Cincinnati Zoo. Image and text as printed in Figure 5.7 in Kaufman and Mallory (1987).

of migrating whales and birds. Their mobilities are relational achievements –
plant seeds journeying in the bellies of animals; the learning of spatial markers
and seasonal routines within creature communities. Moreover, plants and
animals have been caught up in socio-technical networks with 'humans' for well
over 30,000 years, before 'we' recognized ourselves as *Homo sapiens sapiens*
(Ingold, 1995b), unsettling the categorical boundaries between the wild and the
cultivated that we now insist on long before the unravelling of DNA. Efforts
like the UN Convention on Biological Diversity to fix their place in the world
as 'indigenous species' within 'natural habitats' is a no less political regulation
of mobile lives than the paraphernalia of passports and border controls.

The third sense in which a topological rendition of wildlife is more fluid
concerns the spaces of relation. 'Wild' animals and plants whose designation
depends on their being forever somewhere else find their place in the world less
than secure. The radio collars and tags which adorn the remotest parts of the
animal kingdom, no less than their daily exhibition in the wildlife document-
aries that occupy TV screens in millions of homes around the world, disturb the
geometry of distance and proximity. In place of a straight line from here to
there, or a relation rooted in the same spot, the wild and the domestic get swept
up in the volatile eddies and flows of socio-material networks that bring
people, living organisms and machines together in varied and particular ways.

## Conclusions

I have tried to suggest that disrupting the binary terms in which the question of
nature has been posed implies a radical re-cognition of the intimate, sensible
and hectic bonds through which people, organisms, machines and elements
make and hold their shape in relation to each other in the business of everyday
living. This upheaval implicates geographical imaginations and practices both
in the purifying logic which, like 'ethnic cleansing', fragments living fabrics of
association and designates the proper places of 'nature' and 'society', and in the
promise of its refusal. That refusal does not lead to a world in which the properties and things ascribed to nature have been comprehensively extinguished by,
or absorbed within, the compass of those ascribed to human society.

Rather, it is a refusal which challenges the expert and moral authority of
environmental science to keep 'nature' at a distance, beyond everyday understandings and sensibilities. Refusing the purified spaces of nature and society
requires an acceptance of the world as it is – an always already inhabited
achievement of heterogeneous social encounters where, as Donna Haraway
reminds us, 'all of the actors are not human and all of the humans are not "us"
however defined' (1992, p. 67).

## Acknowledgements

This chapter, and the conference presentation on which it is based, condense and reframe
arguments developed elsewhere (see Whatmore, 1999).

# Notes

1.  In human geography these battle lines are variously rehearsed and interrogated in papers by Demeritt (1996), from a broadly poststructuralist perspective, and Gandy (1996), from a critical realist perspective.

2.  Of course the cognitive and linguistic competences that conventionally define the fully-fledged subject and social actor are patriarchal constructs from which various categories of 'humans' have been historically, and continue to be, excluded. Moreover, their status as the distinguishing mark of 'humanity' is troubled by the comparable skills of other classes of animals (notably, primates and cetaceous mammals) and broader reassessments of animal cognition (see Ingold, 1988; Noske, 1989).

3.  Leading international conservation agencies, including the WWF-UK, the Wildlife Conservation Society in New York and the Washington-based Smithsonian Institute, lent their scientific expertise and credentials to the designation of the Myinmoletkat Reserve and continue to pursue research and conservation programmes there on 'endangered' species like the Sumatran rhinoceros and tiger (*The Observer*, 23 March 1997).

4.  I am not suggesting that conservation biology, let alone science, are unitary knowledge projects or social institutions and, hence, that scientists are uniformly committed to this kind of reductionism. But I am suggesting that countervailing currents in the ethics and practice of biological research have been marginalized in terms of the institutional configuration of global conservation programmes like the Biodiversity Programme.

# References

Baldwin A., J. DeLuce and C. Pletsch (eds), 1994. *Beyond Preservation*. University of Minnesota Press, Minneapolis.

Cronon W., 1995. 'The trouble with wilderness: or getting back to the wrong nature'. In Cronon W. (ed.), *Uncommon Ground: Toward reinventing nature*. W.W. Norton & Co., New York, 69–90.

Demeritt D., 1996. 'Social theory and the reconstruction of science and geography'. *Transactions of the Institute of British Geographers*, 21/3: 484–503.

Gandy M., 1996. 'Crumbling land: the postmodernity debate and the analysis of environmental problems.' *Progress in Human Geography*, 20/1: 23–40.

Goodwin B., 1988. 'Organisms and minds: the dialectics of the animal–human interface in biology.' In Ingold T. (ed.), *What is an animal?*: 100–9. Allen and Unwin, London.

Haraway D., 1991. 'Situated knowledges: the science question in feminism and the privilege of partial perspective.' In *Simians, cyborgs and women. The reinvention of nature*: 183–202. Free Association Books, San Francisco.

Haraway D., 1992. 'Otherworldly conversations; terrain topics; local terms.' *Science as Culture*, 3/1: 64–98.

Ingold T. (ed.), 1988. *What is an animal?* Allen and Unwin, London.

Ingold T., 1995a. 'Building, dwelling, living. How animals and people make themselves at home in the world.' In Strathern M. (ed.), *Shifting contexts. Transformations in anthropological knowledge*: 57–80. Routledge, London.

Ingold T., 1995b. ' "People like us": the concept of the anatomically modern human.' *Cultural Dynamics*, 7/2: 187–214.

Kaufman L. and K. Mallory (eds), 1993. *The last extinction*. MIT Press, Cambridge (Mass.). Second edition 1987.

Lawson T., 1997. 'Brent duck.' *Ecos*, 17: 27–34.

McNeely J., K. Miller, W. Reid, R. Mittermeier and T. Werner, 1990. *Conserving the world's biological diversity*. IUCN, Geneva.

Noske B., 1989. *Humans and other animals. Beyond the boundaries of anthropology*. Pluto, London.

Soper K., 1995. *What is nature?* Basil Blackwell, Oxford.

Soulé M. and G. Lease (eds), 1995. *Reinventing nature. Responses to postmodern deconstruction*. Island Press, San Francisco.

Thrift N., 1996. *Spatial formations*. Sage, London.

Whatmore S., 1997. 'Dissecting the autonomous self: hybrid geographies for a relational ethics.' *Society and Space*, 15/1: 37–53.

Whatmore S., 1999. 'Hybrid geographies: rethinking the "human" in human geography.' In Massey D., J. Allen and P. Sarre (eds), *Human geography today*: 24–39. Polity Press, Cambridge.

Whatmore S. and L. Thorne, 1998. 'Wild(er)ness: reconfiguring the geographies of wildlife.' *Transactions of the Institute of British Geographers*, 23/4: 435–54.

Wolch J. and J. Emel (eds), 1998. *Animal geographies*. Verso, London.

# Situating knowledges, sharing values and reaching collective decisions

## The cultural turn in environmental decision making

*Jacquelin Burgess*

## Introduction

In this chapter, I want to pursue some of the contributions that cultural geography can make to the development and implementation of environmental policies. In particular, I shall explore how cultural geographers working in fields dominated by the natural sciences and economics can articulate an alternative way of shaping society–nature relations. The tide of social and institutional change is running fast, and my argument is set within a political context in which calls for more inclusionary processes of policy making are becoming insistent. I believe that cultural geography has much to offer and it can make a significant difference to the processes and outcomes of environmental decision making. One challenge is to persuade policy makers that there is an alternative theoretical and methodological approach to reductionism. It is possible to design decision making processes based on discussions between a wider range of people than would normally be the case. A second is to persuade cultural geographers that society–environment issues are worthy of attention. In many ways, this is more daunting, given cultural geographers' passion for 'high' social theory rather than 'low' cultural practice, especially when the latter is policy-relevant research.

The significance of context in studies of meanings, values and knowledges has emerged powerfully over the last two decades, especially in studies of the sociology of science (see, for example, Knorr-Cetina, 1981; Latour and Woolgar, 1986; Haraway, 1991). Engaged in many different forms of representational practices – using linguistic, mathematical, instrumental, visual and graphic communications, for example – scientists are active participants in different 'epistemic' or knowledge communities who 'compose and use particular representations in a contextually organised and contextually sensitive ways' (Lynch and Woolgar, 1990, p. 1). As work by cultural geographers has shown, such knowledge worlds exist beyond the laboratory walls and constitute a dynamic field where different kinds of authority are contested fiercely by those who may not have academic expertise but who do have considerable depths of local or 'tacit' knowledge. Take, for example, the promotion of agri-environmental schemes to benefit nature conservation in the wider countryside (Harrison et al., 1998). In offering advice to farmers and landowners, advisers to agri-environmental schemes gain authority from relying on scientific models and experiments to define appropriate management practices. Such prescriptions, however, involve judgements about the environmental benefits of particular practices and frequently disguise the absence of consensus amongst scientists about precisely what features of nature benefit from given management practices. Differing opinions on the comparative value of particular management practices are commonplace as reserve and land managers try to adapt the universal knowledge of science to the range of circumstances provided by different localities and institutional contexts.

The feminist geographers are thus right to assert that 'the field' – as in 'fieldwork' – is not just the particular setting for research (Katz, 1992; Nast, 1994). Rather, it describes a set of political relationships which connect researchers' different worlds. As Nast says, being in the field demands that feminist geographers 'link large scale political objectives to smaller scale methodological strategies which break down hierarchical objectivist ways of knowing in the field' (1994, p. 58). In other words, it demands a different valuing of relations between the situated, contextualized knowledges of informants and those of researchers; a sharing of understandings rather than an imposition of 'truths' derived from abstract social theory. This is part of the search for what Katz (1996) calls 'minor theory', and it is my contention that more deliberative and inclusionary processes are central to its construction.

Being in the field also signifies a notion of 'between-ness' (Nast, 1994), a position of not quite belonging which requires translations of the different languages of the different worlds. Between-ness is a good metaphor: certainly, it catches something of the nature of our professional, social and personal experiences in working in the space constructed by involvement in the geographical community, statutory environmental policy-communities, voluntary sector groups, and local people. The metaphor also highlights the range of relationships researchers have with the different audiences who also constitute the field. Translation is crucial when working between a subdisciplinary context where the language games demand texts of maximum opacity, and an outside world where any sign of social science 'jargon' is guaranteed to disrupt or fracture what are always conditional relationships, where trust has to be worked hard for.

In this chapter, I will discuss some of these issues through a piece of action research I, and colleagues, carried out with the UK Environment Agency in the autumn of 1997 (Clark *et al.*, 1998). Action research was a feature of planning, especially advocacy planning, in the 1970s when academics would provide their professional expertise freely to local community groups. My research unit works with statutory and voluntary organizations to support innovative ways of reaching environmental decisions, especially in terms of achieving sustainability goals. The aim of the project I will discuss here was to develop a methodology for making decisions about what environmental actions to undertake in a local plan. The project was carried out in the New Forest district in Hampshire, Southern England. The political context of the project is relevant. The establishment of the Environment Agency in 1995 brought together Her Majesty's Inspectorate of Pollution, the Waste Regulation Authority and the National Rivers Authority. Charged with providing an integrated and holistic approach to the environment, the Environment Agency has a legal duty to consider the costs and benefits of its actions; that is, not only its internal costs and benefits but also costs and benefits incurred by other organizations, individuals, society as a whole, and the environment. The Agency is not committed to any particular methods of appraising costs and benefits, although historically cost-benefit analysis had been regularly employed by its parent organizations for appraising larger scale programmes and projects. How the Agency might interpret these duties in the specific context of producing Local Environment Agency Plans (LEAPs) for all the river catchments in England and Wales is the subject of the case study to be discussed below.

## The cultural turn in environmental contexts

The 'cultural' or 'linguistic' turn is not unique to geography. None of the social sciences has remained entirely unaffected by the profound questioning of positivist epistemologies and methodologies which has occurred over the last two decades. The environmental policy field is still, however, dominated by natural scientists, ably supported by resource and environmental economists. These disciplines share a reductionist way of understanding individuals and society which has had profound implications for the ways in which policies are developed and evaluated. Fundamentally, the reductionist model posits atomized individuals processing information from the external world through which they construct their beliefs, values, attitudes, preferences and patterns of behaviour. The reductionist model aggregates individual characteristics into larger social categories through the use of questionnaires and statistical techniques. For example, environmental economists may seek to measure the costs and benefits of landscape schemes such as the remediation of derelict land to recreational land; to establish the value of the work to local people they would use a technique known as contingent valuation which uses a questionnaire survey to ask individuals how much they would be willing to pay (WTP) for new recreational green space. The WTP figures are averaged, and then put into a cost-benefit analysis to evaluate the economic efficiency of the scheme. In a stinging criticism of economic valuation methods promoted by neo-classical economists who remain wedded to positivism, Funtowitz and Ravetz (1994) coined the term 'post-normal science' to describe the fundamental reappraisal of epistemology that the new cultural perspectives demanded. Post-normal science, they argued, is characterized by uncertainty, complexity and recognition of the plurality of values, many of which are not commensurable. In other words, the value of a songbird cannot be expressed in terms of a money value (see Burgess *et al.*, 1998a).

Three key characteristics of the cultural turn are important in rethinking how environmental decisions and policies can and should be made in this time of post-normal science. First, the foundation of the cultural turn has been a recognition of the fundamental significance of language and other signifying systems in the formation of knowledges. As Barnett has put it recently: 'Both epistemologically and in the construction of new empirical research objects, the cultural turn is probably best characterised by a heightened reflexivity towards the role of language, meaning, and representations in the constitution of "reality" and knowledge of reality' (Barnett, 1998, p. 380). With this understanding, it will be readily apparent that scientific knowledge can no longer claim the status of objective truth. It becomes one discourse among many, playing to its own rules of the game. In the words of Irwin and Wynne (1996, p. 7), research in the sociology of science 'has convincingly demonstrated the *socially negotiated* nature of science (with) *varying* (i.e. heterogeneous) constructions and representations' (emphasis in original). A few cultural geographers have begun to address questions of 'social nature' and the hybrids of nature-and-culture that form the empirical objects of environmental policy (Whatmore and Boucher, 1993; Harrison and Burgess, 1994; Cloke *et al.*, 1996); while others are exploring social constructions of environmental issues through the discursive practices of environmental NGOs (Non-Governmental organisations), and the representational practices of the media (for example, Burgess and Harrison, 1993; Routledge, 1997).

Recognition of heterogeneity in social groups, of the pluralism of values, and the significance of difference mark the second dimension of the cultural turn. Whilst these debates have progressed most significantly in the literatures of feminism, postcolonialism and postmodernism, they are also penetrating environmental and planning discourses, undermining the widespread rhetorical appeal to 'the public' or to 'the community' in institutional, environmental discourses (Macnaghten *et al.*, 1995; Burgess *et al.*, 1998b). Contemporary social life is marked by individualistic lifestyles, the breakdown of communities of place, and shifting subjectivities reflecting gender, ethnicity, age and expertise, as well as class and locality. This fragmentation makes environmental and local planning extremely difficult. To what extent is there consensus about the value of nature in any particular place? Or the quality of a landscape? Or the need to reduce levels of water abstraction? Who is to decide, and on what grounds? Will the decisions be equitable between different interests? What about questions of environmental and social justice? These new uncertainties are being expressed in the need to acknowledge the existence of many different 'stakeholders' which, in some formulations, include future generations and nature. The shift in planning discourse reflects a desire to 'manage co-existence in shared spaces which seek to be efficient, effective and accountable to all those with a "stake" in a place' (Healey, 1997, p. 29); acknowledging that there is a wider range of experience, knowledge and understanding than that which resides in particular bureaucracies.

The third factor is an appreciation of the pluralism of values and know-ledges, and the ways in which values, knowledges and practices of actors and institutions are transformed in the light of new experiences (Giddens, 1991). Social learning, or reflexivity, marks a significant step forward from the for-mulations of a cultural politics defined in terms of endless struggles between differentially empowered groups. The norms of set-piece battles between environmental protesters and business interests, for example, or the top-down dissemination of scientific and technological information are being questioned as never before. With the cultural turn, power relations are being reworked to try and ensure a less confrontational process where decisions are based on mutual understanding. This shift from 'win–lose' to 'win–win' solutions is sometimes described as 'ecological modernization' (Hajer, 1995); and the growth of new approaches to dialogue such as citizens' juries, consensus conferences and mediation show that the old BT slogan 'It's good to talk' is rapidly gaining ground in environmental and planning contexts (see Aldred, 1998; Corbett, 1998). In part, this new kind of discursive practice is building on planning's long history of public participation (Rydin, 1999). But it is symptomatic now of the new public mood in which public and private institutions cannot take public support for granted, where trust is contingent on performance, and where the local arena has gained in significance.

## Stakeholder decision analysis and the shaping of the New Forest Local Environment Agency Plan

The project we completed with the Environment Agency can be seen to demon-strate all three dimensions of the cultural turn discussed above as I shall now go on to show. First, some context. Local Environment Agency Plans (LEAPs) are being developed by the Environment Agency to promote an integrated and sustainable approach to managing the natural environment. The purposes of LEAPs, which are not legally binding, are to focus attention on the environment of a specific area, particularly in terms of the Agency's remit to deal with air, water (freshwater and marine) and waste matters. The strategy for the develop-ment of LEAPs is to involve all interested parties in planning for the future well-being of that area; to agree to a vision for the area to guide Agency activities; and to establish an integrated strategy and plan of action for five year periods. This is an ambitious agenda for a new Agency. LEAPs are a new addi-tion to the arsenal of local plans and, as such, are regarded with some suspi-cion by local authorities and other statutory agencies responsible for planning

**Table 1:** Members of the New Forest Leap Group

*Public sector*
New Forest District Council (officer)
New Forest District Council (member)
English Nature
Environment Agency

*Voluntary sector*
Royal Society for the Protection of Birds, and Hampshire Wildlife Trust
Hampshire Council for the Protection of Rural England, and New Forest Association
New Forest Friends of the Earth
Brockenhurst Manor Fly Fishing Club
Calshot Sailing Club and Southampton Water Sailing Association

*Private sector*
National Farmers Union and Country Landowners Association
Commoners Defence Association
Exxon Chemical, plc
Southern Water, plc
Associated British Ports, plc

and environmental protection. The cultural politics of a new Agency inserting its own concerns (and authority) into what are already complex institutional arrangements are sensitive. The Agency needs to create a programme of consultation over the production of LEAPs which is inclusive of many local interests, and which is open and transparent. At the same time, it is charged under s.39 of the Environment Act 1995 to demonstrate that it has taken into account the likely costs and benefits of its programmes, projects and plans. In the research project we carried out with the Agency, we tried to meet these different aims through a decision making process which involved sharing knowledges and values through conversation, and undertaking a collective, rigorous assessment of policy options.

The range of actions the Agency *could* undertake in its LEAPs is very wide. Should they, for example, invest in alleviating local flows in rivers? Or tackle the loss of biodiversity along the coast? Or deal with problems of toxic leachate from landfill sites? Or undertake basic research on the impacts of different scenarios for climate change? Resources – money, time and people – are constrained. How should these different but equally worthy issues be prioritized in terms of when/whether they should be tackled? The standard approach has been to decide on the basis of scientific expertise, mediated by measures of economic efficiency.

We worked with Agency staff to develop a deliberative and inclusionary methodology for appraising the selection of environmental issues to be included in the LEAP for the New Forest. Our aim was to demonstrate we could produce a robust alternative to cost-benefit analysis to determine the order in which the environmental issues should be tackled over a five year period. To achieve these goals, we recruited a stakeholder group of 14 people. As Table 1 shows,

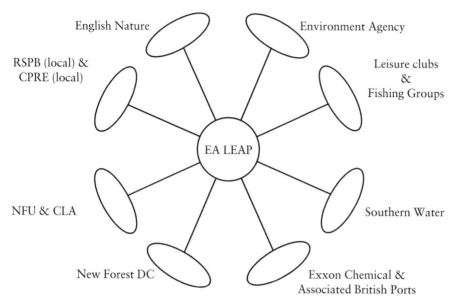

English Nature

Environment Agency

RSPB (local) &
CPRE (local)

Leisure clubs
&
Fishing Groups

EA LEAP

NFU & CLA

Southern Water

New Forest DC

Exxon Chemical &
Associated British Ports

**Figure 1** New Forest stakeholder map

the stakeholders represented a mixture of public, private and voluntary sector interests in the New Forest. The Agency itself had one stakeholder. Each stakeholder was invited to participate both as an individual and as a representative of their particular institution. These willing volunteers then undertook a series of quite intensive activities over four, three hour workshops. Working together, we developed a set of criteria through which the issues in the LEAP could be discussed, evaluated and prioritized – and then completed that task (see Figure 1). The outcome is a process of environmental decision making which combines the sharing of knowledge and values through discussion-negotiation with a more rigorous assessment known as multi-criteria analysis (MCA).

So far as I am aware, this is the first time such a project has been attempted in the environmental field. The approach combined systematic appraisal with group deliberation in a procedure where the emphasis was as much on the discursive process as on the product. The approach to the process had three distinctive elements. The first was to acknowledge explicitly, and to value, the different perspectives that individuals brought to the process. We worked with the stakeholders to build coalitions and achieve a measure of consensus between different interests through the negotiation of what count as costs and benefits, and how these should be appraised. The second was to make explicit, and to value, the different knowledges stakeholders possessed. This was done by facilitating enhanced dialogue between scientific and economic knowledge, and other kinds of knowledges that are embedded in local places and communities. The third was to encourage networking between the participants to increase local support and access to expertise and resources for local Agency staff.

The participants worked through a process which began with discussion of the policy options in the draft LEAP. Were there others that should be included? Were these the right ones to tackle? What were the risks, costs and benefits of each of the policy options included in the LEAP?

Each individual was then asked to think about the potential criteria (or factors) that they considered important in deciding the relative importance of diverse environmental policy options. They were also asked to make explicit their personal value judgement underpinning each criterion. For example, one person might frame a criterion along the lines of 'to what extent will this policy option enhance the biodiversity of the New Forest?' Their value judgement might reflect legal concerns – 'the UK government is charged under the EU Habitats Directive to protect biodiversity', or it might be based on cultural considerations – 'contact with a rich and diverse nature is important to people's quality of life'.

We could not expect automatic support from participants and nor could we expect the group simply to follow instructions. They would need persuading of the rationale and importance of the decision making method we used, even if they were not fully convinced of its utility. They would also need opportunities to reflect on the method and the process. The integration of deliberation and formal analysis required considered planning. The MCA was tackled in stages during the four workshops. First, the group derived a list of criteria acceptable to everyone. Second, the criteria were weighted according to their importance and the least important criteria were discarded. Third, the group 'scored' each issue against each of the criteria in terms of having a high, medium or low priority, or not being relevant to the issue. Finally, a ranked list was produced from this by summing weighted scores and separating the issues into priority groups on the basis of those scores. However, these tasks were not tackled mechanistically. At each stage participants had an opportunity to deliberate: in determining the criteria, in assessing the issues against the criteria and in reviewing the results. In the last session, in the presence of observers from the Environment Agency, the stakeholders reviewed and evaluated the whole process. See Figure 2.

Each stakeholder was asked to bring their own initial criteria – with underlying value judgements made explicit – to the second workshop. Members were put into pairs, matched by their interests and experiences. Thus, the business stakeholders worked together, as did the environmentalists, and local government representatives so there was common ground at this critical stage of sharing and negotiating values. Once each pair had agreed their list of criteria, they moved into a larger group of combined pairs to go through the same process of discussion, sorting and evaluation of criteria. In the final stage, the whole group was brought together to evaluate all the criteria and produce a final list. In the following workshop, each pair was given two criteria that were close to their mutual expertise, and asked to consider all the policy options against each criterion. Members of the research team worked with the pairs to help them produce a working definition of the meaning of each criterion, and to apply it systematically to the list of options.

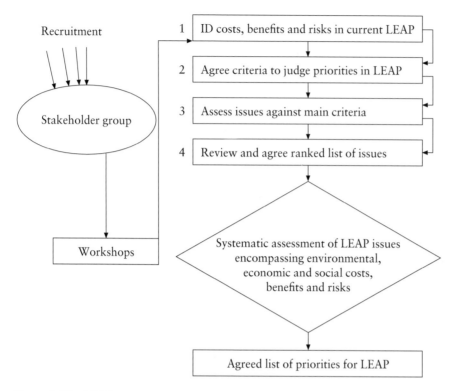

**Figure 2** The LEAPs process

In both workshops, these were rich and fruitful discussions; all the members of the group participated fully, sharing the criteria they had brought to the second workshop, and negotiating to produce an agreed set for the whole group to work with. Members were very willing to debate the value judgements that underpinned their criteria. A wide variety of issues were discussed. These included the precautionary principle and whether it could be used as a criterion; what sustainable development might mean in terms of relations between the national and local economy; problems with using the word 'traditional' in the context of local economic and social relations; the special character of the New Forest and what that meant; the relative value of private property and common goods; what constitutes irreversibility in environmental terms; relationships between economic pressures and environmental needs; the need for adequate scientific knowledge; national and international law; co-operation and partnerships between different agencies, and between different sectors of local activities; public health and individual risk; political pressures in decision making and the need to maintain public support; and how to relate costs and benefits – who gains?

The value judgements were particularly important in making meanings clear and helping people to see quickly where there was agreement and where there was difference. Conflict over the importance of different criteria was avoided because participants did not have to reach consensus on this, only on the validity of different criteria. The task of developing criteria was achievable, participants tackled it willingly, and seemed to find the process stimulating and enjoyable. The atmosphere was one of co-operation rather than confrontation, with people listening to others as well as expounding their own views. Everyone actively sought new ways of expressing their concerns, and more creative ways to frame the value judgements. This creativity and collaboration was promoted by the emphasis on inclusiveness (the reassurance that all criteria would count for something) and the emphasis on making the value judgements explicit. Participants did not have to agree to endorse others' values but only the validity of the criteria that they put forward because the point at issue was whether a criterion *could* be used to assess the issues, not whether it *should* be. This meant that participants could focus on understanding, on teasing out their own and others' views, and on working out similarities and differences. In addition, all participants felt that they could contribute as the debate was perceived to turn less on specialist knowledge. At the same time people seemed less afraid that they lacked technical expertise and more confident of debating meanings of specialist terms such as 'ecosystem' and 'biodiversity' (cf Goodwin, 1998).

The strategy of splitting the stakeholder group into compatible pairs before they moved into larger groups worked well in building consensus about the validity of criteria. We were surprised by the lack of conflict between stakeholders who, in other contexts, would be much more confrontational. It was not because people were not willing to be disagreeable; views were often expressed trenchantly, especially in the groups. Rather, conflict was defused because people were meeting in a forum where they had to work together but were not forced to defend their interests adversarially. Putting different interests together in this situation not only made for productive discussion but also meant that the underlying politics could be acknowledged without getting in the way. Thus, the process enabled people who might otherwise have been locked in fierce opposition to talk together more peaceably.

Table 2 shows the 10 criteria (with their underlying value judgements) which were negotiated through the process and then used to prioritize the 33 issues in the LEAP. The criteria cover environmental, social, economic and legal considerations. Legal requirements include legislation driven specifically by science such as the EC Habitats and Birds Directives, as well as other statutory requirements such as water quality and air quality which comprise the Agency's regulatory duties. Scientific knowledge is accorded high value, in terms of maintaining biodiversity, irreversibility of changes to natural systems, and current levels of scientific knowledge of problems which was interpreted in terms of a measure of effectiveness. Specific landscape and cultural values are recognized in a criterion that assesses local character or the local distinctiveness of the New Forest; and also signifying the importance of quality of life issues. Due regard is given to economic concerns in the list of criteria: the most highly weighted of these was in terms of due regard being given to maintaining the local economy, whilst holding those needs in balance with social and environmental needs. The

**Table 2:** Ranked criteria agreed by the group

| Code | Criterion | Underlying value judgement |
|---|---|---|
| 1 | To what extent is resolution of this issue a legal requirement? | Legal obligations should be met. |
| 2 | To what extent would tackling this issue benefit non-human species and habitats? | Biodiversity should be protected and the Environment Agency should contribute to the UK Biodiversity Action Plan in line with government policy. |
| 3 | To what extent would tackling this issue maintain the unique status/international importance of the New Forest? | The Environment Agency's actions should not affect the 'New Forestness' of the area. |
| 4 | To what extent is the problem identified in this issue likely to get worse? | Issues which are likely to get worse should be tackled sooner rather than later; in particular high priority should be given to issues where delay would lead to irreversible decline. |
| 5 | To what extent would tackling this issue require the Environment Agency to work in partnership with other agencies? | The Environment Agency should work in partnership with other organizations within a cross-organization strategic approach. |
| 6 | To what extent would tackling this issue benefit public health? | Public health should be safeguarded; danger to human life is unacceptable. |
| 7 | To what extent is the issue well understood scientifically? | Priority should be given to tackling issues which are well understood. |
| 8 | To what extent would tackling this issue benefit the quality of life for residents in the LEAP area? | Improving amenity and redressing nuisance should be given high priority. |
| 9 | To what extent would tackling this issue benefit the local economy? | Maintaining/creating employment should be given high priority. |
| 10 | To what extent are actions relating to this issue likely to be affected by potential future legislation? | Future legislation will have to be complied with so its potential impact should be considered. |

outcome of applying these negotiated criteria to the original set of priorities produced in the draft LEAP was to change, quite dramatically, the order in which specific environmental issues should be tackled; and to add a new one that had not previously been included.

The Agency agreed to abide by the decisions of the stakeholder group, and duly issued the revised plan for wider public consultation in spring 1998. This was a brave step, and is certainly not characteristic of institutions who engage in public participation. The norm is to agree to take note but not to commit to the outcome of a process. The project was written up and disseminated to all the Agency LEAP teams around the country (Clark et al., 1998). Subsequently, we have presented the work to regional and national conferences of Agency staff, and have had the pleasure of seeing it endorsed as best practice. In future, all the LEAPs will incorporate an SDA (stakeholder decision analysis) in their planning phase. Our roles now are to work closely with Agency staff to provide training and insights into the methods we used. In other words, to engage in reflexive process whereby social learning is advanced.

## Discussion and conclusion

In the years since the 1992 Earth Summit in Rio, there have been major changes in the rhetorics of governments and institutions, reflecting the success of NGOs in getting public participation on to the agenda. There are explicit commitments in Agenda 21 to greater equity and inclusion in policy making debates, with women, young people and ethnic groups targeted specifically. The LA21 process is connected to the development of new forms of planning theory and practice around the world. These innovations reflect growing anxieties about what is being called 'the democratic deficit' – the increasing alienation, mistrust and scepticism which characterize relations between elected representatives, institutions and citizens (Healey, 1997). Thus, debate is concentrating on both how, and why, it is necessary to widen the repertoire of democratic practices utilized in central and local government. Radical changes in policy evaluation include challenges to predominantly instrumental and rational evaluation of policies which ignore or suppress the ethical/value dimensions of those policies, one reason why the articulation of values is vital. As O'Brien and Guerrier have argued:

> Values are important in the debate about the environment not because some value or other in itself can or should be described as 'right' or 'wrong', but because value systems refer to the underlying principles about the 'proper conduct' of life in general and about ways of interpreting specific events in terms of more extensive commitments to particular social arrangements and political orders. They indicate the cultural plurality – and often ambiguity – within which notions of 'rightness' and 'wrongness' are formulated, maintained, contested and changed. (O'Brien and Guerrier, 1995, p. xiv)

Thus, values are fundamentally concerned with the judgements of what is right or proper conduct in relation to the living world and to each other. But more than this, as the environmental philosophers Holland and Rawles (1993) remind us, values are not things we always argue from, but what we reason towards. Often it is only through open and sustained debate that values are clarified. It is vital, therefore, that the deliberative process focuses attention on the way value judgements contribute to decisions about what to do. As the LEAP case study has shown, when stakeholders are involved in this process, policies based on agreed values will gain wider support than those 'values' elicited through a process confined to the institution alone. Of course, greater inclusion of the public in decision making is not necessarily a panacea, as both Rydin (1999) and Goodwin (1998) point out. There is always a danger of capture by special interest groups, and increasing bureaucratization; local people may find themselves being co-opted into an institutional project as 'hired hands' as Goodwin puts it, rather than 'local voice'. The LEAP process avoided these problems by having participants from across the range of interests in the New Forest District; by designing a process that was completed in just three months; and ensuring everyone played an equal role in the tasks. We did not widen the recruitment of stakeholders to include so-called 'ordinary people' in this first project – but are now addressing ways of extending membership of subsequent workshops.

## Turning environmental policy making . . .

The cultural turn is impacting most strongly on environmental policy in three ways. First, ways in which it reveals 'the social nature of environmental issues in a plural world' (Munton, 1999) raises awareness that new forms of decision making are required. Second, it is impacting in terms of challenging the rules of the game; the formal rules of decision making based on regulatory requirements underwritten by unquestioned expertise are being softened by the new discursive, collaborative and reflexive strategies for 'deciding together'. Third, it is emphasizing that social relations with the natural world encompass much more than scientific knowledge and rational, utilitarian argument; moral, aesthetic, emotional and local ways of knowing and valuing are equally significant in decision making.

So, while I would not agree with Barnett's argument that 'the cultural turn is primarily . . . a move in a game internal to the discipline of geography' (1998, p. 390), I think his interpretations of some of the consequences of the game are correct. He identifies a cult of personality (reflecting the intellectual 'star' system of literary and cultural studies), in which the 'epistemological authority of theory' is asserted through its promotion/association with the names of individual cultural geographers (see Mitchell, 1995; Jackson et al., 1996, for example). He is also correct, I think, in arguing that 'the rise of academic celebrity reinforces the well-established difficulty of acknowledging and practising more collective and collaborative forms of academic writing', going on to argue that this 'effacement' of collective work inhibits social transformation in the discipline. I have shown in this chapter that the cultural turn is about sharing knowledge, respecting different ways of seeing the world, and working in good faith to come to an agreed outcome. This kind of academic work certainly challenges the language of 'mastery' and 'control' which characterizes so much social theory in geography:

> a way of dealing with knowledge in a progressive, linear and commanding way that garners respect for those who play by its rules. The different subjectivities and material conditions of those who produce and exchange knowledge continue to be erased under the sign of mastery. Yet these different conditions have everything to do with what knowledge is produced and how it is handled. (Katz, 1996, p. 497)

In a series of collaborative projects undertaken since 1985; through the work of our graduate students; and through the establishment of ESRU (Environment and Society Research Unit) to work in partnership with public and voluntary sector organizations, we are constructing a different kind of cultural geography from that of the 'masters'. Our project is about making cultural geography work. It is about making a difference by playing an active role in stimulating change in institutions charged with environmental responsibilities. We have a political project inside and outside the academy. Inside, we are clearly aligned with feminist geographers, challenging patriarchal forms of knowing in social theory. Outside and in-between, we are enabling a greater range of voices to speak and be heard; challenging received wisdoms and routines; and supporting individuals within environmental institutions who want to change aspects of their policies and practices.

## Acknowledgements

The research in the New Forest was funded by the Environment Agency under Research and Development Contract W114, 1997–8.

## References

Aldred, J. (1998) 'Land use in the Fens: lessons from the Ely Citizens' Jury', *ECOS*, 19 (2), 31–7.

Barnett, C. (1998) 'The cultural turn: fashion or progress in human geography?' *Antipode*, 379–94.

Burgess, J. and Harrison, C.M. (1993) 'The circulation of claims in the cultural politics of environmental change', in Hansen, A. (ed.) *The mass media and environmental issues*, Leicester University Press, Leicester, 198–221.

Burgess, J., Clark, J. and Harrison, C.M. (1998a) 'Respondents' evaluations of a contingent valuation survey: a case study based on an economic valuation of the wildlife enhancement scheme, Pevensey Levels in East Sussex', *Area,* 30, 19–27.

Burgess, J., Harrison, C.M. and Filius, P. (1998b) 'Environmental communication and the cultural politics of environmental citizenship', *Environment and Planning, A*, 30, 1445–60.

Clark, J., Burgess, J., Dando, N., Bhattachary, D., Heppel, K., Jones, P., Murlis, J. and Wood, P. (1998) *Prioritising the issues in Local Environment Agency Plans through Consensus Building with Stakeholder Groups.* R & D W114, The Environment Agency, Bristol.

Cloke, P., Milbourne, P. and Thomas, C. (1996) 'The English National Forest: local reactions to plans for renegotiated nature–society relations in the countryside', *Transactions, Institute of British Geographers*, NS21, 552–71.

Corbett, J. (1998) 'Creating a new conservation vision: the stakeholders of Wychwood', *ECOS*, 19 (2), 20–30.

Funtowitz, S.O. and Ravetz, J.R. (1994) 'The worth of a songbird: ecological economics as post-normal science', *Ecological Economics*, 10, 197–207.

Giddens, A. (1991) *Modernity and self identity: self and society in the late modern age*, Polity Press, Cambridge.

Goodwin, P. (1998) ' "Hired hands" or "local voice"?: understandings and experience of local participation in conservation', *Transactions, Institute of British Geographers*, NS23, 481–500.

Hajer, M. A. (1995) *The politics of environmental discourse: ecological modernization and the policy process*, Clarendon Press, Oxford.

Haraway, D. (1991) *Simians, cyborgs and women*, Free Association Books, London.

Harrison, C.M. and Burgess, J. (1994) 'Social constructions of nature: a case study of conflicts over the development of Rainham Marshes SSSI', *Transactions, Institute of British Geographers*, NS19, 291–310.

Harrison, C.M., Burgess, J. and Clark, J. (1998) 'Discounted knowledges: farmers' and residents' understandings of nature conservation goals and policies', *Journal of Environmental Management*, 54, 305–20.

Healey, P. (1997) *Collaborative planning: shaping places in fragmented societies*, Macmillan, London.

Holland, A. and Rawles, K. (1993) 'Values in conservation', *Ecos*, 14 (1), 14–17.

Irwin, A. and Wynne, B. eds (1996) *Misunderstanding science? the public reconstruction of science and technology*, Cambridge University Press, Cambridge.

Jackson, P., Cosgrove, D., Duncan, J., Duncan, N., and Mitchell, D. (1996) 'Exchange: there's no such thing as culture?' *Transactions, Institute of British Geographers*, NS21, 572–81.

Katz, C. (1992) 'All the world is staged: intellectuals and the projects of ethnography', *Environment and Planning D: Society and Space*, 10, 459–510.

Katz, C. (1996) 'Towards minor theory', *Environment and Planning D: Society and Space*, 14, 487–99.

Knorr-Cetina, K. (1981) *The manufacture of knowledge: an essay on the constructivist and contextual nature of science*, Pergamon Press, Oxford.

Latour, B. and Woolgar, S. (1986) *Laboratory life: the construction of scientific facts*, Princeton University Press, Princeton, NJ.

Lynch, S. and Woolgar, S. (1990) *Representation in scientific practice*, MIT Press, Cambridge, Mass.

Macnaghten, P., Grove-White, R., Jacobs, M. and Wynne, B. (1995) *Public perceptions and sustainability in Lancashire: indicators, institutions, participation*, Lancashire County Council, Preston.

Mitchell, D. (1995) 'There's no such thing as culture: towards a reconceptualisation of the idea of culture in geography', *Transactions, Institute of British Geographers*, NS20, 102–16.

Munton, R.J. (1999) 'Environmental governance and the politics of difference', address to the RGS/IBG Conference, Leicester, January.

Nast, H.J. (1994) 'The politics of knowing and political/empirical emphases in feminist geography', *Professional Geographer*, 46, 54–66.

O'Brien, M. and Guerrier, Y. (1995) 'Values and the environment: an introduction', in Guerrier, Y., Alexander, N., Chase, J. and O'Brien, M. (eds) *Values and the environment: a social science perspective*, Wiley, Chichester.

Routledge, P. (1997) 'The imagineering of resistance: Pollock Free State and the practice of post-modern politics', *Transactions, Institute of British Geographers*, NS22, 359–76.

Rydin, Y. (1999) 'Public participation in planning', in Cullingworth, J.B. (ed.) *British planning policy: 50 years of regional and urban change*, Athlone Press, London.

# Socialized nature
## England's royal and plantation forests

*Judith Tsouvalis*

This chapter aims to contribute to the ongoing reconceptualization of society–nature relations in geography (for a review see Braun and Castree, 1998) by considering how particular, socially defined reality formations come into being and change over time and space in terms of their material and imaginary manifestations. It focuses on English royal forests and plantation forests and describes some of the multiple material, socio-cultural and symbolic relations formed in the course of their production. Particular attention will be given to the many threads of meaning weaving them together in a fabric of reality where questions of identity and origins seem ever present but never answered.

Both royal forests and plantation forests are seen here as complex, durable, spatially extensive and socio-culturally significant phenomena, and I suggest the notions of *meaningful, composite formation* and related to it that of *formative context* as heuristic devices for their exploration (see Tsouvalis, forthcoming). Meaningful, composite formations do not denote end-products of society–nature relations. Akin to Latour's (1993) 'quasi-objects' or Haraway's (1991) 'cyborgs', they refer to socio-natures that are constantly in the making. Their form is driven by the bio-physical transformation processes they embody and the dialectical relations of these processes with the production processes set in motion following the power struggles over their functional role and cultural significance.

Meaningful, composite formations depend for their achievement on a complex web of social and bio-physical processes and the establishment of certain relations between living and non-living entities. Their meanings and boundaries (physical and imaginary) are negotiated and renegotiated in the course of their existence, and they are deeply suffused in time-space and subject to time and space constructions. These two interrelated notions of time and space need to be considered in the analysis of meaningful, composite formations as they constitute an important moment in the physico-socio-cultural interaction. On the one hand, there is the expression of time in space: time-space (for example, growth, maturation, decay and death) (see Lefebvre, 1991, p. 95). On the other hand, there are abstract notions of time and space: time and space as conceptual constructs, manipulated and constructed imaginatively and applied in practice (for example, timetables, taxation periods, regions, boundaries, and so forth) (see Harvey, 1990). As this chapter will demonstrate, both interact with and shape each other in complex ways (see also Adam, 1997; Gerber, 1997).

For a composite formation to become *meaningful* – and hence, differential – for example, for it to become a royal forest *or* a plantation forest, a gravel pit *or* an Area of Outstanding Natural Beauty, London *or* Manchester, requires the successful mediation of ideological and symbolic representational practices which will render it unique and justify its existence as *that* and no other formation. I would argue here with Swyngedouw (1996, p. 72) that what are often considered as 'real' material processes are 'encapsulated with and engulfed by the equally real discursive/linguistic/cultural constructions of reality'. Meaning making and knowledge construction and the power relations they embody and express intermingle with material processes in complex ways and are part and parcel of the production of particular, meaningful reality formations. They also impose form, order and discontinuity on an otherwise continuous stream of being, becoming and passing away. The same holds true for the formative context within which they can occur.

The notion of formative context stresses the influential, political nature of the surrounding conditions on which the achievement of meaningful, composite formations depend and is used here in preference to that of structured system or bounded field. For example, it is not at any time or in any place that royal forests or plantation forests could have been achieved. Hence, such formations are never disinterested but suffused with power relations, and always geographically specific. Both meaningful, composite formations and formative contexts are deeply rooted in a world that Swyngedouw (1996, p. 70) has described as a

> historical-geographical process of perpetual metabolism in which 'social' and 'natural' processes combine in an historical-geographical 'production process of socio-nature', whose outcome (historical nature) embodies chemical, physical, social, economic, political and cultural processes in highly contradictory but inseparable manners.

As long as the processes and relations that constitute formative contexts and meaningful, composite formations remain relatively stable, change is almost imperceptible. This is the moment when human beings start to take such formations for granted, when for them they begin to take on a thing-like appearance (Harvey, 1996). However, because of the multiple and contradictory forces that constitute them, change is inevitable.

In what follows, various aspects of the changing nature of royal forests and plantation forests as meaningful, composite formations will be explored. As this brief and by no means exhaustive account will show, the rules governing their production and the definitions of and relations between the entities activated for their achievement changed considerably over time, as did their ever contested meanings. It should be stressed here that every meaningful, composite formation holds different meanings for different people or groups of people at different times in different places. For example, both royal forests and plantation forests have served as vehicles in the articulation of diverse and often opposing local, regional and national identities and have held considerable symbolic significance in representations of Englishness. The importance of the place specificity of such articulations and representations has received considerable attention by geographers concerned with constructions of national and regional identities (see, for example, Matless, 1990; Daniels, 1993; Cosgrove, Roscoe and Rycroft, 1996; Withers, 1996; Edensor, 1997; and Brace, 1999). In this chapter, however, rather than focusing on the spatiality of particular meaning constructions, I want to focus on their temporality by investigating how meanings associated with royal forests have influenced the production of plantation forests. The fluidity of the identities of the beings activated for their achievement and their social and bio-physical implications will also be problematized.

## Royals, ruffians, deer and oaks – Britain's earliest forests

During the early middle ages, the word *nemus* was the standard word for woods and woodlands and comes closest to the present day meaning of the word 'forest' in Britain (although the term woodland is primarily used nowadays to refer to deciduous woods and that of forest to coniferous woods). Harrison (1992) has suggested that the most likely origin of the word forest is the Latin word *foris*, meaning 'outside'. Forests during the early medieval period were considered as places lying outside civilization, in other words, outside the feudal and religious social order of the time. They were places haunted by spirits; places where pagan tree worshippers met, outlaws hid and chaos reigned. A trouble spot for those wrestling for power and social control – including the church – forests were where culture was not.

The imaginative opposition of forests with culture is thought to date back to a time when human beings were still forest dwellers but gradually began to become aware of clearings in the forest and the contrasts they provided: light as opposed to darkness, openness as opposed to enclosure, hospitality as opposed to hostility (Steiner, 1993). Hence the significance of the term *agriculture*. In legends of the middle ages, 'clearings get bigger, forests become fields, agriculture replaces nature' (Bechmann, 1990, p. 289). This process of deforestation, of culturing the wilderness and socializing nature, has continued to this day. Although the early myths, legends and fairy tales of forests are not a direct concern of this chapter, they are of immense importance for an understanding of the dreams and nightmares forests manage to inspire to this day (for an introduction, see Porteous, 1928; Harrison, 1992; and Schama, 1995).

Royal forests first appeared in France during the ninth century, when the term *forestis* was used as the juridical term to refer to land delimited by royal decree (Glacken, 1967; Harrison, 1992). According to Rackham (1976), in England the word forest was first mentioned in the Domesday Book compiled in 1086, and Manwood's *Forest Laws*, published in 1598, offers a good description of what the term implied at the time:

> A forest is a certain territory of wooddy grounds and fruitful pastures, privileged for wild beasts and foules of forrest, chase and warren, to rest and abide in, in the safe protection of the King for his princely delight and pleasure. (Quoted in Brown, 1883, p. 12)

Royal forests, to which this definition refers, were introduced to Britain by the Norman King William I, who invaded England in 1066. Often, they were designated in areas recognized as hunting grounds in late Anglo-Saxon times (Hooke, 1998). Most royal forests contained a multitude of land uses as well as villages and towns. They formed an exclusive component of the medieval landscape, but were in complex ways linked to the pre-existing manorial system of administration, of which the manor was the principal unit. This formative context came to be referred to as feudalism, a term describing a system of vassalage and serfdom. Vassalage denotes an intra-elite relationship where a subordinate vassal (retainer) holds landed property from a lord, ultimately the Crown, in return for military or other service for the Crown. Serfdom implies the legal subjection of peasant tenants to lords through the manorial jurisdictions of lords, of which unfree tenants were legally considered part (Harrison, 1984). Everyday practice, apart from being governed by the laws of the king, was also subject to a multitude of customary rights and obligations, passed on from generation to generation by word of mouth. Many of these rights were celebrated with local festivals aimed at their preservation well into the twentieth century (Hole, 1944). Apart from property relations, technological developments (particularly in the sphere of legal, fiscal and administrative technologies to control time and space) gradually led to a transformation in this formative context where status was no longer vested in interpersonal relations but came to be embodied in property.

The magnitude of the destruction wrought to achieve this meaningful, composite formation is hard to estimate, especially as it became part of an oppositional discourse of the English to challenge the legitimacy of this newly, and initially undoubtedly violently, achieved foreign reality. For example, concerning the establishment of the New Forest, tales of the destruction of the established way of life were passed on from generation to generation until in the eighteenth century they reached such proportions that it was thought that 'the Conqueror and his heirs had been so determined to swathe Old England in woods populated only by boar and by buck that they had gone to the length of planting good arable fields with trees' (Shama, 1995, p. 140). Although this claim was rejected as absurd by New Forest historians, assertions that William I 'cleared agricultural land and destroyed some houses' to create the New Forest can still be found in contemporary history books (Tomkeieff, 1966, p. 27).

Whatever the facts might have been, royal forests were politically contested not only by the dispossessed English landowners and common people living in afforested areas, but also by the nobility and the church who had their privileges and powers curtailed by the king. Interestingly, the Forestry Commission took up the example of the New Forest nearly nine centuries later when it appropriated it as a model for its proposed National Forest Parks, arguing that it was 'the nearest approach to a National Park in Great Britain' (quoted in Revill and Watkins, 1996). According to Revill and Watkins (1996), one of the Forestry Commission's reasons for making such a move was to justify their forestry practice and access and amenity policy at the time. In the care of 'professional foresters working for the state, the forest is restored as an economic resource and a public domain after a thousand years of expropriation and neglect' (ibid., p. 108). Apart from that, the reworking of the New Forest as a model for forest parks is significant because it constructs an identity for forests located in highland Britain 'founded on a history which is exclusively English, substantially lowland and southern' (ibid., p. 108). This is one example in which royal forests were appropriated temporarily and linked to the production of forest plantations symbolically.

A similar incident seems to have occurred after the introduction of royal forests by William I, for it is now thought that Canute's *Constitutiones de Foresta* – a written law of the forest – was a Norman forgery 'designed to show that forests were not a foreign introduction' (Hinde, 1985, p. 19). Considering that Canute himself was a Danish Viking, the reason given for this forgery by Hinde is itself an interesting proposition. Questions of identity and origin are a recurrent theme in forest history, and I shall come back to this thread weaving royal forests and plantation forests together over the centuries in the discussion of the issues raised by the introduction of 'foreign' or 'alien' tree species and continental methods of forestry later in this chapter.

Although forest law did not so much supersede as supplement common law, it nevertheless constituted a violation of the traditional rights of the common people (Shama, 1995). The forest laws introduced by William I, which were said to be very severe, continued in force until the reign of Henry III (1216–1272), who issued the 'Charte Forestie' to revise them. They gave power to the king to designate any tract of the country as forest, a process known as 'afforestation', and governed the relations between human and non-human beings. As much as a quarter of the country was designated royal forest during the first century of Norman rule, and individual forests could be spatially extensive. For example, the whole of Surrey and parts of Bedfordshire were afforested by Henry II, and the New Forest in Hampshire, the Forest of Dean, and the forests of Yorkshire extended to about 95,000 acres, 100,000 acres and 90,000 acres respectively (Serovayskaya, 1998).

To realize the objective of a forest as a pleasure ground for the king, particular relations had to be constructed between the living and non-living entities it contained. As Manwood pointed out, territorie of ground so privileged [areas designated as forest] is *meered and bounded* with irremoveable markes, meeres, and boundaries . . . ; for *the preservacion and continuance of which said place*, together with the vert and venison, there are *certen particular laws, priviledges, and officers* belonging to the same, meete for that purpose that are *onely proper unto a forrest*, and not to any other place. (Manwood, quoted in Brown, 1883, p. 12 – emphasis added.)

Achieving these relations and enforcing the rules required many different officials, and the Norman kings established special courts of justice (forest swainmotes) to deal with forest affairs and offences. By far the most powerful administrator of a royal forest was the chief forester, who travelled all over the country and visited each forest periodically to preside over the eyres, or forest courts, where forest offenders were tried. In charge of particular forests were wardens, also called keepers, stewards, or master foresters, who had to attend forest courts and assist the *verderers* or judges. They constituted the group of forest officials who, because of their privileges and everyday presence, suffered most directly as targets of social opposition to royal forests. For example, in a study of Whichwood forest in Oxfordshire, Freeman (1996) found that between 1788 and 1836 four keepers were assaulted and an assistant keeper murdered, all in connection with incidents involving deer. Occasionally, keepers had their dogs or venison stolen, and at times there were regular feuds between keepers and local men (ibid.).

Other forest officials were *regarders* who had to inspect the whole forest every three years, enquire into all offences and survey all assets, wastes and the like. The duty of the *forester* was to preserve the venison (the flesh of the deer or wild boar) and the vert (the pasturage for them). They also had to tackle offenders and present offences at the forest court. The purpose of *agistors* was to receive and account for the profits arising from the herbage (the right to pasture, generally cattle) or pannage (the right to allow pigs to feed on acorns and beech mast during autumn) of the king's woods and the lands in the forests. *Woodwards* had to look after the woods as well as present offences concerning these to the forest courts. Other offices were either appointed by the king or selected according to local custom (see Brown, 1883; James, 1981; and Rackham, 1976).

From the time of their inception, royal forests were potent symbols of power. Their main objective was to provide a means whereby the social order of the day could be displayed and kept in place, and this was primarily achieved through the ritual of the royal hunt. As Shama (1995, p. 145) has observed, the royal hunt was a 'blood ritual through which the hierarchy of status and honor around the King was ordered'. For example, it entailed the exact knowledge on the part of the skilled hunter – often the symbolically rich figure of the mounted knight – of whom to formally present specific parts of the kill to, and served for the participants as a means of competition to gain proximity to the king. On a subtler level, royal forests served as an instrument of fiscal exploitation of peasants and artisans, but also feudal proprietors of all rank. Between 1179 and 1188, for example, the largest category of income provided by royal forests was fines (Serovayskaya, 1998, p. 37).

Royal forests were also complex socio-legal institutions. Living beings were allocated their titles and duties according to their status in the social cosmology of these meaningful, composite formations. A forest's most essential features were not the trees it contained but the animals. These were perceived as 'wild beasts' and their role was to act as 'game'. 'A forest', as Manwood (quoted in Brown, 1883, p. 13) asserted, 'must always have beasts of venery abiding in it, otherwise it is no forest.' Such 'beasts' included the red, roe and fallow deer and the wild boar. The fallow deer was imported to England by the Normans, and the deer normally belonged exclusively to the king (Rackham, 1976). Dogs owned by forest dwellers were routinely 'declawed', that is, their claws were cut off using a mallet and a chisel, so that they could not harm the deer (Shama, 1995). The status and role of animals was determined by the ritual of the hunt. For example, the smallest of the deer, the native roe deer, ceased to be a forest beast by a judicial decree of 1339 and remained no longer under forest protection. Like wolves, badgers, foxes, martens, wild cats, otters, hares, rabbits and squirrels it could be hunted by anyone outside royal forests except in areas where that right (or warren) had exclusively been granted to an individual or institution (Hinde, 1985). Foxes were considered as pests and could be killed by anyone except in royal forest, where a licence was needed to do so. Hares were usually included in the fox hunting licences. The killing of rabbits (another Norman introduction) and occasionally that of squirrels also required a licence.

Animals were thus significantly involved in the achievement of the meaningful, composite formation called royal forest. Without providing a vehicle for the roles they were assigned to play in the ritual of the hunt, royal forests would, quite frankly, have been meaningless. Similar to the Roman gladiatorial games, royal forests depended on what Whatmore and Thorne (1998, p. 437) have come to refer to as 'topologies of wildlife', namely, 'a relational achievement spun between people and animals, plants and soils, documents and devices, in heterogeneous social networks that are performed in and through multiple places and fluid ecologies.'

The initial role of trees in achieving royal forests was of secondary importance. They provided wood, food and shelter for animals and people, and were a useful means for the king to pacify the church with timber gifts for cathedral building (Simpson, 1998). In terms of their role in the realization of the meaning of royal forests as hunting grounds, they were most important for the purpose of game preservation, and forest law forbade encroachment (*purpestre*), the damaging of trees (waste), and the rooting up of trees (assarting). A freeholder living under forest law jurisdiction needed permission to fell his trees (Thomas, 1984). The main function of foresters was to look after the animals, who had to give pleasure to the king and his huntsmen at the cost of their lives. The achievement of royal forests required that forest officials and commoners had a thorough knowledge of forest law and common law. Apart from an understanding of socially created rules and norms (even if such an understanding was habitual rather than academic), all participants had to have some knowledge of forest ecology, which was vital for their survival. In short, they all had to possess particular qualities, skills and knowledges to produce and consume these forests. Many of these roles were to be reversed in the formation of plantation forests, where the deer became a pest, the trees became the pets, and the role of the forester became to look after the trees and protect them from the deer, rabbits and squirrels.

Not everyone, however, respected these imposed boundaries, roles and relations. Some human beings, for example, adopted the unlawful role of 'poacher' in spite of the beatings, torture, death, and, more commonly, prison and heavy fines this implied (Rackham, 1976; Shama, 1995). Deer, too, broke the rules by straying across the physical boundaries of royal forests, where they were 'killed without hesitation' by the owners of what were known as purlieu woods (Freeman, 1996, p. 13). Although death was their inevitable destiny within and without royal forests, *within* their social and symbolic role was very complex. For royalty, they provided a means for displaying and asserting their power and status in medieval society; for the commoners they were an aid to alleviating their poverty. Occasionally they were used for social and political protests, not only directed at keepers. For example, deer stealing by gangs in Whichwood forest in 1714 was motivated by Stuart sympathizers protesting against the Hanoverian succession (Freeman, 1996). For younger commoners, they provided an opportunity for reinforcing male gender identity, where poaching was treated as a game (ibid.). Outside society's webs of meaning, deer played their biological role in the forest, where they competed with human beings and other animals for herbage within the wood–pasture regime of the forest.

In the mid-thirteenth century, the original purpose of these forests was already in decline. Many were disafforested by successive monarchs who wanted to relieve their financial difficulties. Throughout the middle ages, the Crown's interest in deer hunting decreased, while its interest in timber – particularly from the oak – rose. Although royal forests slowly lost their permanence as real world occurrences in the form of hunting grounds, the same was by no means true of their hold on the imagination as a symbol of the power of the monarchy. This symbol, however, was no longer so much epitomized by the highly esteemed deer of the king, as by the mighty oak. So dear had that tree become to royalists and the aristocracy because of its role in shipbuilding and hence imperial expansion, that by the seventeenth century it had begun to be referred to as the heart of England – England's Heart of Oak (Shama, 1995, p. 155).

## Changing times, changing things: the rise of the trees or the return of the native

Towards the end of the fifteenth century the office of Purveyor of the King's Timber was created, and in 1543 the first important timber preservation act was passed. This act demanded that wherever a wood was cut down, a minimum of 12 young trees must be left on every acre. It also determined the minimum size a mature tree had to have attained before it could be felled (Simpson, 1998). Because large amounts of timber were required to build up naval power, and political intrigue made importing (especially from the Baltic) an ever less reliable business, its shortage begun to be felt. Grazing, wood cutting, and the conversion of woodland into arable land had for centuries been the main factors in the reduction of tree cover in Britain, and now the needs of the commoners (still rooted in feudalism) conflicted with the wishes for the exploitation of woods for timber production (already rooted in capitalism). Imperial politics created an industrial demand that led to a rise in timber prices during a period when the transition from feudalism to capitalism turned '[w]hole populations ... from habitual users and gatherers of the woods into dispossessed consumers, required to purchase firewood at market prices' (Shama, 1995, p. 154).

A turning point in royal forest history was the Civil War (1642–1648), when all royal forests were abolished. As a result, the woods, so long fenced off by physical and imaginary boundaries, were now invaded by vast numbers of commoners, who took anything they could find, from brushwood to standing timber, fallen branches and boughs (Shama, 1995; Simpson, 1998). This highlights that the formative context for tree planting in the form of large scale plantation forests was not yet achieved. As Arthur Standish, campaigner for woodland conservation, observed in 1613, if a tree is planted one day, 'the poor plucketh or cutteth them up the next day, if not the same night' (quoted in Thomas, 1984, p. 200). This process accelerated during the years of the Civil War.

The invasion of what had previously been royal forests, in other words, the breakdown of this meaningful, composite formation in the light of a changing formative context was far subtler in meaning than a basic need for wood. Thus while revolutionaries and the poor attacked the 'thing' itself, its symbolic counterpart was depicted by royalist sympathizers to be suffering a similar fate. A satirical print representing Oliver Cromwell's rise to power serves as a good example (Cromwell was the ruler of the newly constituted Commonwealth after Charles I's execution in 1649). The print (see Figure 1), entitled 'The Royall Oake of Brittayne' and on display in the British Museum, shows

> Cromwell standing on 'a slippery place,' above the mouth of Hell, and beneath the avenging fires, 'late but determined,' of Heaven, directs the cutting down of the Royal Oak, which represents the English constitution. Monarchy ('Eikon Basilike'), Religion (the Bible), Liberty ('Magna Charta'), Law and Order ('Statutes' and 'Reportes'), hang on its branches and fall with it. A group of men gather up the fallen boughs; some swine, 'fatted for slaughter,' represent the common people in whose interest this destruction is nominally wrought, and who are destined to be its real victims. (Green, 1893, p. lxxi)

Nowhere was this intermingling of symbolic, political, economic and physical reality more explicit than in John Evelyn's *Sylva, or A Discourse of Forest Trees,*

**Figure 1** The Royall Oake of Brittayne

*and the Propagation of Timber in his Majesty's Dominions*, first published in 1664. Formerly an editor publishing royalist propaganda in stout support of Charles I, Evelyn's objective in *Sylva* went much further than what the Royal Society and the Crown Commissioners of the navy had in mind when requesting a book on how to counteract the impending shortage of timber for the building and repair of warships (Edlin, 1972). It cleverly managed to establish a link between the wanton felling of trees with republican politics. Combining scientific precision with poetry, his political message was clothed in words addressing Charles II as Cyrus, the restorer of the Temple, and Hiram, the king of the cedars of Lebanon, the prince who 'by cultivating our decaying woods will contribute to your Power as to our Wealth and Safety' (quoted in Shama, 1995, p. 160). He firmly believed that the restoration of the king would bring with it the restoration of the royal forests, for 'who better, after all, to effect this than the monarch whose life and reign was owed to the oak in which he sheltered after the defeat of the Battle of Worcester?' (ibid., p. 160). Charles II had hidden in an oak tree near White-Lady's after the battle of Worcester in 1651, seeking cover from the rebel soldiers sent after him. Another oak associated with Charles II was the 'Royal Oak' at Boscobel in Shropshire, under whose boughs the king is said to have held council with Colonel Carlos. Evelyn thus devoted the fourth edition of his *Sylva* to

> you then, Royal Sir, . . . since you are our *Nemorensis Rex* [Sylvan King]: as
> having once had your temple, and court too, under that sacred Oak which you
> consecrated with your presence, and we celebrate, with just acknowledgement
> to God, for your preservation. (Johns, 1894, p. 34)

The oak had literally become the saviour of the king, and with it the monarchy
and the nation itself. Not surprisingly, Charles II did reinstate the forest laws,
and he was an ardent supporter of tree planting. Increasingly, forests where the
Crown still owned the soil were seen as places to grow timber for naval
purposes, and subsequent Acts of Parliament were passed to ensure this was
done. English oak or *quercus robur* turned into ships became in Evelyn's work
England's wooden wall, and soon came to define what it meant to be English:
'heart of oak were our men' (Garrick, 1770s, quoted in Wilkinson, 1981,
pp. 34–5). Never mind that there was no proof that it was indeed a native, as
Wilkinson nicely put it, 'we will give it the benefit of the doubt in the circumstances'! (ibid., p. 35).

Between 1660 and 1700, about 11,000 acres of the Forest of Dean, and
1,400 acres of the New Forest were enclosed and planted, this being probably
the first instance of modern forestry to occur in Britain (Rackham, 1976,
p. 156). After centuries of ruthless exploitation, debates about afforestation in
the modern sense of the term had begun, although Evelyn's *Sylva* had to wait
for nearly a century for the plantation movement it envisaged and inspired.

Fuelled by patriotism, fashion and war, tree planting became the nationwide
passion in the eighteenth century not so much of the monarchy but of the
aristocracy, and it owed much to the favourable economic climate provided by
the agricultural revolution taking place at the same time, which released the
capital, land and labour resources it necessitated. Much has been written about
the socio-economic, political and symbolic significance of these early plantations (see, for example, Daniels, 1993; Daniels and Watkins, 1994a; Miles,
1967; Seymour, 1998; Shama, 1995; and Thomas, 1984), and more shall be
said about them in the next section as they form part of the dawn of the
modern age when the sun was already setting on the medieval one.

It is important at this stage to point out that in that transition period an
aesthetic appreciation of nature rooted in the Picturesque and Romanticism
grew up that still lingers in England today. In relation to trees and forests, the
former favoured a form of landscape design that 'celebrated old features like
steep-banked lanes and ancient trees' (Daniels and Watkins, 1994b, p. 12). The
latter, particularly as expressed in the writings of the poet William Wordsworth,
celebrated unsocialized nature and was to be of vital importance, for example,
in the fight against conifer afforestation in the Lake District in the 1930s. This
was one instance in which a particular regional version of Englishness was
employed to defend a landscape against the 'alien' invasion of forest plantations.

Amidst the changes that took place up to that time, the meaningful, composite formation of royal forests changed considerably. The deer lost their central although by no means enviable status of 'game' to obtain, as the next section
will reveal, the even less desirable one of 'pest'. The importance and status of

NEAR THIS SPOT STOOD
THE FAMOUS ELM
PLANTED BY THE REV JOSIAH PULLEN
ABOUT 1680 AND KNOWN AS
JOE PULLEN'S TREE
DESTROYED BY FIRE
ON 15 OCTOBER 1909

**Figure 2** Joe Pullen's Tree

trees, particularly the English oak, grew in economic and cultural terms. Their metaphysical importance expanded as they were increasingly employed as metaphors for power, social status, and family lineage (see Daniels, 1988). Even trees that had been planted in memory of people occasionally became more famous than those they were supposed to commemorate (see Figure 2, Joe Pullen's Tree).

Throughout this period, the formative context within which royal forests had been formed was in transition. The power of the monarchy declined and that of the aristocracy rose. By the later eighteenth century forest courts had already been abandoned. Then, during the mid-nineteenth century, the aristocracy's power began to wane and the rights of the common people gained weight. Politics were democratized and the economy globalized – the formative context of feudalism was replaced with capitalism. This erosion of the political power of the aristocracy coincided with the agricultural depression, whose most intense period lasted from the late 1870s to the late 1890s. Worldwide, agricultural prices collapsed, leading to a dramatic fall in estate rentals and correspondingly, land values. Landed wealth could no longer compete with industrial wealth, and increasing estate taxation reduced what wealth remained. Inevitably landlords in England and elsewhere began to sell off their land. The responsibility for all royal forests was finally passed to the Forestry Commission in 1923 (James, 1981), and it is the production of the early forests this Commission became famous for – forest plantations – that constitute the focus of the next section.

# On alien invasions: the rise of forest plantations

It was the new formative context of capitalism that enabled the large scale production of plantation forests. Capitalism refers to a form of economic and social organization in which the direct producer is separated from the ownership of the means of production and the product itself, a separation effected by turning labour power into a commodity bought and sold on a labour market.

The first use of the term forestry as 'the science and art of farming and cultivating forests' can be found in the Oxford English Dictionary of 1859 (Stewart, 1989, p. 48), indicating how much the meaning of the term 'forest' had changed by then. However, large scale, high forest, plantations forestry was introduced in Britain with the Forestry Commission (hereafter FC). The FC was inaugurated as a government department in 1919 with the prime objective to create a strategic reserve of timber to counteract the wood shortages experienced as a result of the First World War.

With the FC, forests began to taken on a form quite unlike the one of royal forests, and quite unlike anything England had yet seen. Forest plantation forestry had originated in Germany and was introduced to Britain through its Empire (the Indian Forest Service) (see Tsouvalis-Gerber, 1998; and Tsouvalis and Watkins, forthcoming). With the introduction of these methods and the establishment of even-aged, monocultural, coniferous plantations on bare land with straight boundaries and clear-felled once the trees have reached maturity, forests became geographically extensive spatial entities of tracts of land covered by trees.

Although the Commission's main aim of producing a strategic reserve of timber was discarded with the Zuckerman review in 1957, forest policy to this day has been characterized by the setting of a particular planting objective expressed in spatio-temporal terms: so and so many acres of land to be planted with trees per year. Nowadays, however, the FC has to achieve many additional aims, such as balancing forestry and environmental interests, encouraging broadleaved planting, and creating multi-purpose forests.

The ability to produce the meaningful, composite formation of forest plantations was due to unprecedented advances in science and technology, many rooted in military efforts, that proceeded at a hitherto unimaginable speed. Time and space are of vital importance in the formative context of capitalism, and their manipulation characterized the forest plantation production process from the start. To understand plantation forests and their symbolic significance, one has to understand these manipulations, and they will be considered below in reference to the forest workers, the land on which forests were produced, and the trees that were planted.

## THE FOREST WORKERS

The number of people employed by the FC (excluding divisional and district officers and office staff) amounted to a minimum of 210 during the summer of 1920 and a maximum of 935 during the winter of the same year. This number

**Table 1:** Time-study findings 1920

**Finding of co-ordination officer:**

Labourers were . . . doing five hours' useful work per day, the remaining three hours being *lost time*.

**Reasons for lost time:**

Upon further examination it was found that the three hours were occupied as follows:

*Lost time* per man per day:

| | | |
|---|---|---|
| (i) | walking to and from meals | 40 minutes |
| (ii) | walking time (between finish of one line and commencement of next) | 30 " |
| (iii) | Time (additional to (ii)) spent in setting pickets to assist in keeping lines straight | 30 " |
| (iv) | Delays at beginning and end of lines | 30 " |
| (v) | Various delays (waiting for plants, resting, smoking, talking, etc.) | 50 " |
| | | 3 hours |

**Conclusions:**

The above items were attacked singly and *waste* eliminated so far as possible; e.g. a *portable* shelter allowed the men to take their meals *near* their work, they were trained to work without pickets, and so on.

Having thus secured that the *maximum* time is spent in *useful* work, it remains to secure that operations are carried out at the *highest possible speed* compatible with good work.

Source: FC, 1921, pp. 31ff.; emphasis added.

rose to a minimum of 2,640 during the summer of 1929 and a maximum of 3,595 during the winter of that year (FC, 1929, p. 61). As soon as the FC was set up in 1919, a co-ordination officer was employed to study methods and procedures in planting operations with special reference to securing economy in the employment of labour (FC, 1921, p. 31). For this investigation, the officer used the method of time-study and found that an unsatisfactory gang of planters planting steadily dealt with a rate of three plants in two minutes, or 90 per hour per man, and that each man planted approximately 450 trees in a normal (eight hour) working day. Table 1 shows the final results of this study.

Various strategies were tried to overcome these losses and various rationalization measures were introduced. Although there were no assembly lines as depicted in Charles Chaplin's *Modern Times*, the production process of forest plantations was from the start based on principles of Taylorism, implying the fragmentation of production activities into their simplest constituent components and linking them together into precisely co-ordinated and closely supervised sequences. By using a stop watch to record the time spent on each operation carried out, it was possible to alter that particular operation in order to save time that could then be 'invested' elsewhere. This drives home Harvey's (1990, p. 419) point that, 'The spread of capitalist social relations has often entailed a fierce battle to socialize different peoples into the common net of time discipline implicit in industrial organization.'

As will be shown below, these relations also socialized nature into that net. Some of the solutions proposed by the FC to tackle these time losses were as follows:

- Planters were to be divided into small, self-contained units of five to ten men, under a leader or 'ganger'.
- The ganger was to keep to the rear flank of the gang so as to control the work of each worker.
- Two to three boys per gang were to make sure men were kept supplied with plants at all times, so that they could give their 'full and undivided attention' to the actual planting process.
- Workers were to be employed on piece-work contracts which would improve the quality of the work carried out. This, so the report stated, 'may be attributed to the workers' additional care, lest they should be made to rectify unsatisfactory work without remuneration' (FC, 1921, pp. 33ff.).

Time-study was not just a passing fashion of the time. In 1956 formal work study was introduced by the FC, and in 1966 the Work Study Branch responsible for machinery research and development was set up (FC, 1973). The success of the FC in rationalizing forestry operations has been summed up in the FC's Diamond Jubilee report, which found that the planting rate was more than three times that achieved during the interwar years and timber production had more than doubled. This, it was stressed, was because: 'Since 1949 technological advances have revolutionized every aspect of work in the forest, and transformed a largely rural craft into a highly organised and efficient industrial operation' (FC, 1980, p. 8).

Woodmanship in the traditional sense had disappeared (see below; also Tsouvalis and Watkins, forthcoming). The structure of forestry work had changed considerably, and workers were particularly affected by the mechanical revolution. The axe and the crosscut saw as logging tools had been exchanged with the lightweight chainsaw, and the horse in extraction work had been replaced with custom-designed vehicles with articulated frame steering and four wheel hydrostatic transmissions, radio controlled winches and hydraulic grapples (FC, 1980). It comes hardly as a surprise that the peak number of forest workers employed by the FC of 13,337 in 1950 decreased to 2,179 in 1980 (Ryle, 1969), and has continued to decrease ever since.

Other measures to overcome the loss of time in forestry operations in relation to forest workers was through the elimination of space, namely the space between the worker and the forest. Prior to the establishment of a good road network, forest workers often had to walk long distances to reach their working site; a walk of two hours in the morning to get there and two hours in the evening to return home was not uncommon (Ryle, 1969). It was in 1924, under Lord Lovat's chairmanship, that the idea of smallholdings for forest workers first took hold. Forest workers were to be moved to the forests.

After 1936, when industrial depression led to the designation of Special Areas, the pressure on the FC by the government to create more smallholdings was increased, but after the war smallholdings were abandoned in favour of a policy of 'forest villages'. Although several such villages were created, this policy was on the whole unsuccessful due to the multitude of problems they encountered because of their remoteness. Attempts to settle forest workers in forests have long since been abandoned. What has remained to this day is the belief that: 'The faster a man works . . . the more economical will be his use of machinery, the smaller will be the unit cost of overheads and the greater will be his contribution to his own and the country's economy' (Johnston *et al.*, 1967, p. 486).

Although many advances were made concerning the personal comfort of forest workers, the social implications of these 'improvements' in terms of feelings of isolation and the loss of a community spirit were far-reaching (see Karpowicz, 1987). Human beings, however, were not the only hurdles in the achievement of these plantations.

## THE LAND

One social definition of land is that of it as a limited resource, a finite entity. Historically, modern forestry had to compete for land with agriculture, and generally had to be content with marginal land of too poor a quality for farming. Often, such land was located uphill, and occasionally included sand dunes. It is interesting to note how such land was described. For example, in relation to the early land acquisition of the FC, Ryle observed that land transactions were by no means:

> limited to the poor-grade grazings: there were heathlands and other really *unused wastelands* . . . There were sand-dunes on the coasts . . . There were the bracken and heath *wastes* . . . There were, too, the large expanses of *derelict* or nearly derelict woodlands. (Ryle, 1969, p. 178; emphasis added)

This continued reference to wastelands in relation to the acquisition of land is indicative of how thoroughly forests had been socialized by that time, considering that the origin of the word forest – the Latin word *foris* – had implied an expanse of land lying outside the bounds of civilization. Now unproductive land was to be turned into forests, an inversion of what would once have been considered as waste and as civilized terrain. To bring such land within the bounds of the valuable and purposive, many inventions had to be made and complex structures of knowledge and practice put to work.

For example, the socialization of the land encompassed the following activities. First, the land had to be *cleared*, which took place by hand, machine, or a combination of both. Secondly, an area might have to be *drained* to prevent excess water from flowing on to the area or to remove excess water from the area. Concerning the former, intercepting channels had to be cut across a slope above the area to be planted, and concerning the latter, a system of open drains had to be constructed. Both these operations depended on soil conditions and topography. Furthermore, forest drains had to be kept open and free from tree roots. Fourthly, fences had to be erected around the boundaries of the plantation to prevent rabbits and deer – the main enemies of the saplings – from getting in. 'Today', the FC observed in 1980,

> Developments in fertilisation and herbicidal protection, coupled with improved techniques allowing wider spacings in planting, have markedly changed standards of land plantability, permitting poorer and higher land, particularly in northern Scotland, to be planted successfully. (FC, 1980)

Ploughs enabled the manipulation of water, fertilizers the manipulation of soil, herbicides the manipulation of weeds, pesticides the manipulation of insects and fungi, fences the manipulation of rabbit and deer, and machines the manipulation of human beings. Never, however, was success complete; never was there complete domination. Just as the deer in royal forests escaped across its boundaries, so chemicals infiltrated the groundwater, wind blew down the trees, deer and rabbits broke through the fences and ate the saplings, and the soil deteriorated. All this demanded further research, further development, further attempts to socialize nature as completely as possible. These attempts extended to the trees.

## THE TREES

In the post-war years, non-native spruces and pines accounted for over 80% of plantation forests. Lodgepole pine rose in significance to second behind Sitka spruce while the planting of hardwood, in other words broadleaves, nearly ceased. The reason for this impending rise in the popularity of Sitka spruce was because it produced 'timber much more rapidly than Norway spruce and under a wider range of conditions' (FC, 1933). Speed and space is how research into tree growth could be summarized, and in both respects, Sitka spruce proved something of a miracle tree; so much so that in 1992 it begun to be referred to as 'super Sitka' (FC, 1992). Forgotten was England's oaken heart.

The conquering of the time-space of the trees began with the selection of their seedlings. Pre-war research on seed was confined to the pre-treatment of seed to 'accelerate germination' (FC, 1974, p. 45). As a nursery practice, this did not alter much in post-war years, except that it now became possible to assess the success of pre-treatment. The role played by physical factors in germination became of particular interest during the 1950s, and other investigations were undertaken into sowing density, soil reactions (the response of seeds in certain soils and under the addition of certain fertilizers), the transplant stage, and so forth. The grading of seeds was one important step in the identification of superior stock.

A major step towards the conquering of time-space in trees came in the 1960s, when forestry stood 'on the threshold of a genetical revolution' (Johnston *et al.*, 1967, p. 224). To know where seedlings came from became vital, and classification systems of seeds and seed certification were introduced. It became possible not only to produce quicker growing but stronger plants, which, when science was coupled with technology, led to the mechanization of planting in the form of tubed seedlings. This, in effect, led to an alteration in the planting season, or in other words, to time-space:

> With the introduction of tubed seedlings and container plants the planting season can be extended over a much longer period. The Forestry Commission have found that when tubed seedlings are used in peat, planting can be continued from April to August. (James, 1981, p. 32)

Once seedlings had grown to saplings in the nursery and were planted at their final destination, growth depended to a great extent on the factors already described above in relation to land. The topography of the land, the exposure to wind, the properties of the soil, the presence of 'pests', all determined whether the formation of a plantation forest could be achieved.

Considering that the form and function of these early forest plantations was relatively simple – even-aged, square shaped blocks of trees with the objective of providing as much timber as quickly as possible – their realization required an incredibly complex structure. In the 1960s, Johnston *et al.* observed that although there had been a worldwide trend towards more regular silvicultural techniques, this 'apparent simplification of silviculture has been made possible by advances in our fundamental knowledge of genetics, nursery work, ground preparation, drainage, fertilizing, thinning and protection' (Johnston *et al.*, 1967, p. 284).

Forest plantations, in the form of spatial entities, were vastly different mean-ingful, composite formations than royal forests had been, and yet opposition to the former was linked to the latter symbolically. The power of the oak and what it stood for lingered on far beyond the time it served as a pillar of the economy and thus the power of the nation. Although visually inspired, debates against conifer plantations as a symbol of a new age were deeply rooted in cultural per-ceptions of social change.

## Continuity and change: the role of symbolism in the production of reality

Forest plantations were not something people in England were used to seeing. According to their cultural sensibilities, bare hills, moorlands, and heaths – which officially came to be defined as wastes – were what constituted their habit-ual landscape. Furthermore, the advances in science and technology, discussed above, led to a situation that when 'sweeping changes were presented abruptly and with a Teutonic precision after a long spell of inactivity in forestry, the innate conservatism of taste was profoundly offended' (Miles, 1967, p. 67).

Forest plantations were much resented in terms of their visual impact. As far back as 1934 the FC observed that there seemed to be a 'general objection' to 'large areas of dark-coloured evergreens' (FC, 1934, p. 54) and in 1937 it noted that 'on grounds of amenity the Commission's operations continue, in England and Wales, to attract a fair share of criticism.' This, they thought, was because 'changes in familiar surroundings are rarely welcome' (FC, 1937, p. 12). In 1961, critics of the Commission continued to 'accuse [it] of blanketing whole hillsides with one species. This may sometimes be inevitable but normally both scientific forestry and the eye to landscape dictate a much more varied tree cover' (FC, 1961, p. 15). As a result of these criticisms, a landscape consultant was appointed by the Commission in 1963, and apart from addressing these issues in practice, a set of forest guides was produced to rectify such 'misconceptions' (Revill and Watkins, 1996).

From earliest times, however, resistance to forests as tracts of land covered with trees was more subtle than these comments imply. From 1827 onwards, the introduction of conifers from North-West America had begun with the Douglas fir, followed by the Sitka spruce in 1831, the Western hemlock in 1851, the Western red cedar in 1853, the Lodgepole pine in 1853, and the Lawson cypress in 1854 (James, 1981, p. 184). Conifers had long served as signifiers of social change, and by assembling them into meaningful, composite formations called forest plantations they came to signify Modernity *per se*. However, even before the FC had begun to plant 'foreign' or 'alien' tree species on a large scale, their introduction to Britain had already led to contestations.

Daniels' study of the political iconography of woodlands in later Georgian England, for example, found that woodland imagery was deployed to signify, and furthermore naturalize, varying and conflicting views of social order. The landscape theorist Uvedale Price, writing in the Picturesque tradition, could thus compare native deciduous trees and coniferous trees (all non-native with the exception of the Scots pine) in the words 'the slow progress of beauty with the upstart growth of deformity' (Daniels, 1988, p. 61). Far from just comparing trees, such observations were part of a moralizing discourse about social change taking place in Britain at the time. Conservative observers were alarmed by the intrusion of foreign species and compared them to the 'disruptive influence of the new, often industrially, rich' (ibid., p. 52).

In the late nineteenth century, for those who believed in progress and desired the formation of plantation forests, such sentiments were problematic. Hence proponents of continental methods of forestry – based on the manipulations of time and space described above – often portrayed traditional methods of woodland management as backward and outdated. Propaganda literature was full of references against what were believed to be misplaced aesthetic appreciations of trees and forests in England. For example, John Simpson, an advocate of continental forest methods, observed in 1900:

> those under the delusion that the introduction of the German forestry system into this country will banish beauty and sentiment from the land, must be woefully ignorant of German forests and that German forest lore and romance

> . . . There is one direction, however, in which German sentiment does not run, and that is in extravagant veneration for very old and useless dead trees such as encumber so many woods and parks in Britain. They believe in live trees and plenty of them. They do not excel in collections of relics like the Old Caledonian Forest for example, or like Sherwood . . . Their rotation periods have long since put an end to all that, and they point with pride instead to the grand tracts of forest that clothe their mountains and waste lands almost everywhere. (J. Simpson, 1900, p. 36)

With the formative context in transition, conditions were soon to be right for Simpson's wish to be fulfilled. Indeed, the early zeal of the FC to cover waste lands and clothe mountains in conifers was such that forester St Barbe Baker in 1944 complained that while in Germany much emphasis was being placed on planting mixed woods and remedying past mistakes of a short-sighted policy of planting pure coniferous woods, 'the Forestry Commission as well as many private landowners have fallen into the same mistake . . . following their out-worn policy which they [the Germans] themselves revised nearly fifty years ago' (ibid., p. 87). It was not until the 1980s that these issues were addressed in earnest (see Tsouvalis, forthcoming).

Protests against plantation forests, however, often took the form of lamenting the passing of a (not seldom idealized) pre-modern age and could take liberal as well as conservative overtones. For example, H.J. Massingham, a critic of modernism and the Victorian cult of progress, in a book published shortly before his death in 1952, regretted what had become of a former woodland (consisting of deciduous trees) on the Welsh border. He observed that all had been savagely cut over and that 'hardly a tree is to be seen except the conifers of the Forestry Commission.' (Massingham, in Ableson, 1988, p. xvii).

In another place in the same area he found that the original moorland landscape had been 'effaced and what has taken its place were walls of light green and dark green in vast uniform blocks' of larch and spruce (ibid., p. 177). The hillside was strewn with fallen logs of former oakwoods, and he felt that all had been transformed from a place of freedom to a prison. Proponents of modern forestry would, of course, have called this a 'waste'. He was dismayed by the displacement of woodland and grass with forest plantations, whose 'monstrous uniformity' he saw as an insult to the eye. Not primarily concerned with aesthetics, Massingham lamented the loss of rural traditions in the pursuit of industrial expansion and what he called the phantasm of export markets:

> Correlate the dispossession of sheep-farmers and the felling of oakwoods . . . to plant conifers with the artificial deserts of opencast mining and ironstone mining among the wheatfields of Northamptonshire . . . and it becomes clear that we are waging a merciless war upon our own country, . . . against our own reason and nature. (ibid., p. 178)

This reference to *our own* reason and nature drives home the point made by Matless (1990, p. 179) that the landscape preservation movement prominent in the early part of the twentieth century, 'in seeking to protect, and to project, the countryside as symbolic of national identity, itself enacts and generates definitions of Englishness'. Just as the oak had become a symbol of Englishness in the late sixteenth century, so forestry plantations became a symbol of what Englishness was not.

Massingham was particularly opposed to universal state control, because he feared that it would lead to an exploitation of the earth far beyond what was safe and reasonable. The state's role in the production of forest plantations and the 'deeper concerns over the invasive and modernist impacts of conifers on the British landscape/nation' (Cloke *et al.*, 1996, p. 556) was also critically questioned in an article published in *The Guardian* more recently. Again raising the issue of identity, of what it means to be English or, in this case, British, it was observed that

> We British (and we do appear to have at least this much in common) resent the way the Forestry Commission's firs march across the landscape in artificial rows. High-rise council flats may carry the same message in the city, but in rural areas it is the Forestry Commission's plantations that signify the totalitarian presence of the modern state: . . . those wretched fir trees are as deprived of individuality as people under communism.
> (Wright, 1992, pp. 6–9 and 52)

The xenophobia non-native trees often inspire was then tackled, and the author observed that conifers were seen as alien imports, 'plainly lacking the cultural credentials of the native broadleave . . . [L]ike other immigrants, these fir trees "all look the same" to the affronted native eye' (ibid.). Although, it was pointed out, conifers had been introduced to Britain in 1600 and many broadleave trees were 'foreign', nothing 'has prevented the Forestry Commission's growing mass of trees being declared identically "Teutonic" . . . as if each plantation was a Nazi platoon advancing over British acres' (ibid.).

Considering that questions of ethnicity have become a major source of political conflict in numerous geographical locations in recent years, it is important to understand how in England such questions have often found expression in discourses associated with forests and trees (the National Forest, initiated in 1991, being one of the latest examples). From earliest times the meaningful, composite formations of royal forests and plantation forests were embroiled in negotiations of identity. The strong, durable and native – such as the oak – became symbols of the power of the monarchy, the aristocracy, and the upper class English generally, while the short-lived, alien conifer came to symbolize the under class, the intruder in that exclusive social universe, the ruthless opportunist. For representatives of the new formative context of capitalist social relations, on the other hand, ancient oaks and their associated landscapes became derelict trees and bare wastelands, while conifers – associated with modern civilization – came to be seen as pillars of national security and prosperity.

Apart from questions of identity and origin, the transformations that took place in these meaningful, composite formations also expressed changing attitudes to nature and how they altered as the relations between human- and non-human beings co-evolved. For example deer, loved in royal forests by William I 'as though he were their father' gradually became a 'pest' (from the *Anglo-Saxon Chronicle*, quoted by Shama, 1995, p. 145). Although their status in royal forests had not saved them from their fate of being killed, they had nevertheless been of considerable symbolic importance in that social universe. The fluidity and constructedness of relations and identities is also made explicit by the changes that have taken place in attitudes to dogs. Nowadays, it is unimaginable that dogs would be declawed using a mallet and a chisel to prevent them from attacking deer. Yet although the leash has become a more acceptable form of restraint, it is a restraint nevertheless: a physical and symbolic boundary imposed by human beings between non-humans to socialize them into the complex web of society–nature relations characterizing the present.

Looking at socio-natures such as royal forests and plantation forests as meaningful, composite formations that come into being in complex, ever changing formative contexts provides a means of addressing subtle changes that take place in the production of reality over time and space. Their analysis highlights the intricate ways in which human beings relate with and reflect on their relationships with each other and other beings. In recent years, the meanings of forests in England have again been shifting: ancient woodlands and the status of deciduous trees have been revived (Tsouvalis-Gerber, 1998), and many forests are now more profitably run as hunting grounds (especially for pheasant shoots) than as timber factories. The recognition of such changes raises urgent political questions in terms of the roles human and non-human beings are made to play in a social universe constantly defined and redefined, configured and reconfigured over time and space. It invites us to consider whether as individuals involved in particular activities and groups we do indeed partake in the construction of the best of all possible worlds, and how we could discard reworkings of history that have no other purpose than to fuel nationalism and possibly war. Searching for alternative futures necessitates a better understanding of how changes in reality occur, and the analysis of how socio-natures come about provides one way of gaining such an understanding.

## Acknowledgements

I would like to thank Simon Naylor for his helpful comments on an earlier version of this chapter.

## References

Ableson, E. (ed.) (1988): *A mirror of England. An anthology of the writings of H.J. Massingham (1888–1952)*. Bideford, Devon: Green Books.

Adam, B. (1997): 'Time and the environment.' In: Redclift, M. and Woodgate, G. (eds): *The international handbook of environmental sociology*. Cheltenham: Edward Elgar Publishing.

Bechmann, R. (1990): *Trees and man – the forest in the Middle Ages*. New York: Paragon House.

Brace, C. (1999): 'Finding England everywhere: regional identity and the construction of national identity, 1890–1940.' In: *Ecumene* 6 (1), 90–109.

Braun, B. and Castree, N. (1998): *Remaking reality. Nature at the millennium*. London: Routledge.

Brown, J.C. (1883): *The forests of England and the management of them in byegone times*. Edinburgh: Oliver and Boyd, Tweeddale Court.

Cloke, P., Milbourne, P. and Thomas, C. (1996): 'The English National Forest: local reactions to plans for renegotiated nature – society relations in the countryside.' In: *Transactions of the Institute of British Geographers*, 21, 3, 552–71.

Cosgrove, D., Roscoe, B. and Rycroft, S. (1996): 'Landscape and identity and Ladybower Reservoir and Rutland Water.' In: *Transactions of the Institute of British Geographers*, 21, 3, 534–51.

Daniels, S. (1988): 'Woodlands in later Georgian England.' In: Cosgrove, D. and Daniels, S.: *The iconography of landscape*. Cambridge: Cambridge University Press.

Daniels, S. (1993): *Fields of vision. Landscape imagery and national identity in England and the United States*. Cambridge: Polity Press.

Daniels, S. and Watkins, C. (1994a): 'A well-connected landscape: Uvedale Price at Foxley.' In: Daniels, S. and Watkins, C. (eds): *The Picturesque landscape. Visions of Georgian Herefordshire*. Nottingham: Department of Geography, University of Nottingham in association with Hereford City Art Gallery and University Art Gallery, Nottingham.

Daniels, S. and Watkins, C. (1994b): 'The Picturesque landscape.' In: Daniels, S. and Watkins, C. (eds): *The Picturesque landscape. Visions of Georgian Herefordshire*. Nottingham: Department of Geography, University of Nottingham in association with Hereford City Art Gallery and University Art Gallery, Nottingham.

Edensor, T. (1997): 'National identity and the politics of memory: remembering Bruce and Wallace in symbolic space.' In: *Environment and Planning D*, 29, 175–94.

Edlin, H.L. (1972): *Trees, woods and man*. London: Collins.

Forestry Commission (1921): *Second Annual Report of the Forestry Commissioners*. London: Her Majesty's Stationery Office.

Forestry Commission (1929): *Tenth Annual Report of the Forestry Commissioners*. London: Her Majesty's Stationery Office.

Forestry Commission (1933): *Forestry practice – a summary of methods of establishing forest nurseries and plantations with advice on other forestry questions for owners and agents*. Forestry Commission Bulletin No. 14. London: Her Majesty's Stationery Office.

Forestry Commission (1934): *Fifteenth Annual Report of the Forestry Commissioners*. London: Her Majesty's Stationery Office.

Forestry Commission (1937): *Eighteenth Annual Report of the Forestry Commissioners*. London: Her Majesty's Stationery Office.

Forestry Commission (1961): *Forty-second Annual Report of the Forestry Commissioners*. London: Her Majesty's Stationery Office.

Forestry Commission (1973): *Work Study in Forestry*. Edited by W.O. Wittering, Forestry Commission Bulletin No. 47. London: Her Majesty's Stationery Office.

Forestry Commission (1974): *Fifty years of forestry research – a review of work conducted and supported by the Forestry Commission, 1920–1970*. In: Wood, R.F.: Forestry Commission Bulletin No. 50. London: Her Majesty's Stationery Office.

Forestry Commission (1980): *Sixtieth Annual Report of the Forestry Commissioners.* London: Her Majesty's Stationery Office.

Forestry Commission (1992): *Super Sitka for the 90s.* In: Rock, D.A. (ed.). London: Her Majesty's Stationery Office.

Freeman, M. (1996): 'Plebs or predators? Deer-stealing in Whichwood Forest, Oxfordshire in the eighteenth and nineteenth centuries.' In: *Social History*, 21, 1, January, 1–21.

Gerber, J. (1997): 'Beyond dualism – the social construction of nature and the natural *and* social construction of human beings.' In: *Progress in Human Geography*, 21, 1, March, 1–17.

Glacken, C. (1967): *Traces on the Rhodian shore. Nature and culture in Western thought from ancient times to the end of the eighteenth century.* Los Angeles: University of California Press.

Green, J.R. (1893): *A short history of the English people.* London: Macmillan.

Harrison, J.F.C. (1984): *The common people. A history from the Norman conquest to the present.* London & Sidney: Croom Helm.

Harrison, R.P. (1992): *Forests: the shadow of civilization.* Chicago: The University of Chicago Press.

Haraway, D. (1991): *Simians, cyborgs and women – the reinvention of nature.* London: Free Association Books.

Harvey, D. (1990): 'Between space and time: reflections on the geographical imagination.' In: *Annals of the Association of American Geographers*, 80, 418–34.

Harvey, D. (1996): *Justice, nature and the geography of difference.* Oxford: Blackwell.

Hinde, T. (1985): *Forests of Britain.* London: Victor Gollancz.

Hole, C. (1944): *English custom and usage.* London: B.T. Batsford.

Hooke, D. (1998): 'Medieval forests and parks in southern and central England.' In: Watkins, C. (ed.): *European woods and forests. Studies in cultural history.* Wallingford: CAB International.

James, N.D.G. (1981): *A history of English forestry.* Oxford: Blackwell.

Jenkins, D. (1975): 'The old mansions and their woodlands.' In Edlin, H.L. (ed.), *Cambrian Forests.* Forestry Commission Guide. London: Her Majesty's Stationery Office.

Johns, C.A. (1894): *The forest trees of Britain.* London: Society for Promoting Christian Knowledge.

Johnston, D.R., Bradley, R. and Grayson, A. (1967): *Forest planning.* London: Faber & Faber.

Karpowicz, Z.J. (1987): *Forestry: the sociology of an occupation.* Oxford University: Unpublished Doctoral Thesis.

Latham, B. (1957): *Timber. Its development and distribution. A historical survey.* London: George G. Harrap.

Latour, B. (1993): *We have never been modern.* New York: HarvesterWheatsheaf.

Lefebvre, H. (1991): *The production of space.* Oxford: Blackwell.

Matless, D. (1990): 'Definitions of England, 1928–89. Preservation, modernism and the nature of the nation.' In: *Built Environment*, 16, 3, 179–91.

Miles, R. (1967): *Forestry in the English landscape.* London: Faber & Faber.

Porteous, A. (1928): *The lore of the forest. Myths and legends.* London: George Allen and Unwin.

Rackham, O. (1976): *Trees and woodland in the British landscape.* London: J.M. Dent.

Revill, G. and Watkins, C. (1996): 'Educated access: interpreting Forestry Commission Forest Park Guides.' In: Watkins, C. (ed.): *Rights of way: policy, culture, and management.* London: Pinter.

Ryle, G. (1969): *Forest Service – the first forty-five years of the Forestry Commission of Great Britain*. Newton Abbot: David & Charles.

Serovayskaya, Y.J. (1998): 'Royal forests in England and their income in the budget of the feudal monarchy from the mid twelfth to the early thirteenth centuries.' In: Watkins, C. (ed.): *European woods and forests. Studies in cultural history*. Wallingford: CAB International.

Seymour, S. (1998): 'Landscape estates, the "spirit of planting" and woodland management in later Georgian Britain: a case study from the Dukeries, Nottinghamshire.' In: Watkins, C. (ed.): *European woods and forests. Studies in cultural history*. Wallingford: CAB International.

Shama, S. (1995): *Landscape and memory*. London: HarperCollins.

Simpson, G. (1998): 'English cathedrals as sources of forest and woodland history.' In: Watkins, C. (ed.): *European woods and forests. Studies in cultural history*. Wallingford: CAB International.

Simpson, J. (1900): *The New Forestry, or the continental system adapted to British woodlands and game preservation*. Sheffield: Pawson & Brailsford.

St Barbe Baker, R. (1944): *I planted trees*. London: Lutterworth Press.

Steiner, D. (1993): 'Wald, Seele, Kultur: Einleitung zum Themenheft.' In: *Geographica Helvetica*, Nr. 2, 59–60.

Stewart, P.J. 1989: *Growing against the grain. United Kingdom forestry policy*. Second edition. A report commissioned by the Council for the Protection of Rural England. ISBN 0 964044 05 8, 19.

Swyngedouw, E. (1996): 'The city as hybrid: on nature, society and cyborg urbanization.' In: *Capitalism, Nature, Socialism*, 7 (2), 26, June, 65–80.

Thomas, K. (1984): *Man and the natural world. Changing attitudes in England 1500–1800*. London: Penguin Books.

Tomkeieff, O.G. (1966): *Life in Norman England*. London: B.T. Batsford.

Tsouvalis, J. (forthcoming): *Socializing nature/naturalizing society: forests and their meaning in Britain since 1900*. Oxford: Oxford University Press.

Tsouvalis, J. and Watkins, C. (forthcoming): 'Imagining and creating forests – an evaluation of work practices and their implications for trees, landscapes and people in early twentieth century England.' In: Angelotti, M. (ed.): *History and forest resources*. Proceedings of an international conference on forest history, Florence, Italy, 20–23 May 1998.

Tsouvalis-Gerber, J. (1998): 'Making the invisible visible: ancient woodlands, British forest policy and the social construction of reality.' In: Watkins, C. (ed.): *European woods and forests. Studies in cultural history*. Wallingford: CAB International.

Whatmore, S. and Thorne, L. (1998): 'Wild(er)ness: reconfiguring the geographies of wildlife.' In: *Transactions of the Institute of British Geographers*, 23, 4, 435–54.

Wilkinson, G. (1981): *A history of Britain's trees*. London: Hutchinson

Withers, C.W.J. (1996): 'Place, memory, monument: memorializing the past in contemporary Highland Scotland.' In: *Ecumene*, 3 (3), 325–44.

Wright, P. (1992): 'The disenchanted forest.' In *The Guardian Weekend*, 7 November, 6–9 and 52.

# Natural communities

## 'The pauper question' in the *Atlantic Monthly* 1880–1884

### *Sheila Hones*

## *Common ground*

Speaking in Seattle in November of 1998, 'at the edge of the so-called "American" continent' to an international and multidisciplinary group of people attending the 1998 meeting of the (US) American Studies Association, Janice Radway focused her Presidential Address on issues of culture, geography, nation, identity and disciplinarity. Building her argument around the implications of the organization's name, the *American* Studies Association, Radway sparked enormous debate with her consideration of three alternatives invented to explore her concern that the organization's formal title might serve to limit its activities by enforcing 'premature closure through an implicit, tacit search for the distinctively American "common ground" ' (p. 3). Intense discussion of the Address and its themes swiftly spread out from the conference site across the internet, as Americanists worldwide engaged in a collective reconsideration of their traditional interest in the question of exceptionalism and definition-by-exclusion as it related both to subject matter (what is 'American?') and methodology or academic identity (what is American Studies?) (Campbell and Kean, 1997). Radway's remarks and the response they provoked suggest strongly the ways in which work associated with American Studies now inter-sects or coincides with work in the traditions of cultural and critical geography. Not only the theme but the very language of Radway's emphasis, for example, on work in American Studies that engages with 'the question of how American nationalism was actively constructed at specific moments, at specific sites, and through specific practices' will sound familiar to many cultural geographers today (p. 10).

Radway's call for an American Studies Association that was 'not hemmed in by the need to peg cultural analysis of community and identity-formation to geography', was not an attempt to distance the field from academic geography but rather an attempt to discourage simplistic assumptions regarding the significance of location, and as such it serves to suggest how closely connected (or at least how tantalizingly adjacent) work in American Studies and in cultural geography has become. Insisting that culture is a 'meaning effect produced by hierarchical relationships established between different spaces and the communities that give them significance', Radway argues that work being produced and discussed within American Studies today does not 'diminish the importance of place or geography', but rather 'demands a reconceptualization of both as socially produced' (pp. 14–15). The coincidence of interest here is obvious, and without going so far as to call for the disciplinary border erasures of consciously postdisciplinary cultural studies (Bartolovitch, 1995), it is still possible to acknowledge the potential value of an increased mutual awareness between cultural geography and American Studies as currently practised.

The celebration of a new border crossing, talk of the creation of a 'mutual awareness', and the associated idea of any new kind of 'we', should remind us, nonetheless, that the definition of communities is not a simple and happily creative activity for all involved. Speaking in her Presidential Address not only about the problematic idea of a 'common ground' shared by 'Americans' but also about the problems of finding a collective identity for practitioners of American Studies, Janice Radway for this reason 'sought to avoid using the pronoun "we" throughout . . . as a way of refusing . . . presumptive and coercive enclosure' (p. 3). Within disciplines, within nations, and within communities of all kinds, as Radway reminds us, the corporate voice of the speaking 'we' depends on the silent voice of the excluded 'they'. 'The very notion of the "American" ', as she argues, 'is intricately entwined with those "others" produced internally as different and externally as alien through practices of imperial domination and incorporation' (p. 14).

This essay also avoids the first person plural, in part from similar principles but also on the practical grounds that having been written for a collection of essays in cultural geography by a literature specialist who is the British member of an American Studies team at a Japanese university it belongs to an almost laughably complex reader/writer community. Not surprisingly, it makes no claim to be normal. But the focus of the essay is, nonetheless, on the textual production and maintenance of a normative communal identity, and its subject is the confidently spoken 'we' that excludes and incorporates in its production of a normative version of collective American identity at one specific (five year) moment, in one specific (virtual) site, and through one set of specific (textual) practices.

## Communities and texts

David Sibley (1995) has written of the ways in which academic communities establish self-definition through exclusion. Drawing a parallel between the exclusion of forms of knowledge with the exclusion of discrepant others, Sibley describes the way in which the 'ideas which gain currency through books and

periodicals, [are] conditioned by power relations which determine . . . boundaries . . . and exclude dangerous or threatening ideas and authors' (p. vxi). This
similarity between spatial boundaries and the guarded gateways of the world
of texts is suggestive of the role of power structures in the creation of common
sense and general knowledge not only in academic life but in the world of
mass-market publishing in general, and strongly editor-controlled periodical or
magazine publishing in particular. A well edited periodical inevitably reflects
its sense of a reader/writer community and its time-specific social context, self-
consciously marketing itself as the meeting place of a particular imagined community, and participating knowingly in the production of and struggle over
social power relations and accepted values. As the literary historian Louis James
(1982) has explained, the 'total effect of [the] contents, tone and style' of a periodical can give it its own distinct identity so that it can be read 'as a microcosm
. . . of a cultural outlook' (p. 351). Editorial input and a clear 'house style'
strengthen this aspect of the periodical 'microcosm'; a piece of text that feels
'out of place' in a periodical can be highly disturbing, even leading the surprised
reader to lose confidence in their own sense of community.

Tim Cresswell (1996), writing on the way in which particular people and
behaviours are made to seem 'out of place' and how expectations of the
appropriate depend on particular power hierarchies and ideologies, has focused
on the analysis of transgressions as a methodologically efficient approach to
the study of common sense. Any attempt to get people 'to state what usually
remains unformulated', he explains, is not only difficult but can result too
easily in misinterpretation (p. 21). But as Cresswell himself demonstrates in
his later work on metaphors of exclusion (1997), there is a valid alternative
approach to the identification of a common sense of 'what's natural' that
comes out of the investigation of shared narratives – in this case, approaching
the 'normal' not by looking at what the editors throw out, but at the underlying assumptions that structure the pieces they put in. For this reason, this essay
looks at a five year run of the firmly edited American monthly magazine, the
*Atlantic Monthly* 1880–84 (Sedgwick, 1994) and focuses on the ways in which
the text functioned for a socially powerful group of people as a way of defining
society as a community which included them but excluded 'others', a form of
group definition visible in the text's boundary-drawing use of the first person
plural, and its confident references to what 'all readers must feel'.[1]

American Studies, of course, like cultural geography, has its traditions in the
study of social forms and social history through the analysis of language and
literary text. Carl Smith, for example, places his own work in the tradition of
the eminent Americanist Alan Trachtenberg, who insisted that forms of language are 'of prime historical interest' and are 'forces in their own right' (1982,
p. 8). In the introduction to his own subsequent use of language in the study
*Urban Disorder and the Shape of Belief* (1995), Smith argues that figurative
language and literary convention are 'at once descriptive and interpretative, conveying the details of life and an understanding of them' (p. 10). The analysis
of narrative and figures of speech has also been convincingly supported within
cultural geography (Cresswell, 1997; Kearns, 1997; Rycroft, 1996), Rycroft

arguing particularly for a sensitivity towards the original context of 'contemporary production and reception' and Cresswell stressing that the 'significance of metaphor to geography extends . . . well beyond the use of metaphor in geography' (p. 343). Cresswell's point that the 'creation and maintenance of metaphor is . . . an inherently political project with material effects and consequences' is particularly relevant here, as is his insistence that metaphors 'arise out of specific hierarchical structures of power and serve to reproduce such asymmetrical power relations' (pp. 334, 341).

This essay, then, focuses on the way in which social power relations in late nineteenth century urban America were in part produced and maintained through the currency of an underlying structure of assumptions worked out textually through the narrative manipulation of story-telling conventions and figurative language. It points out ways in which textual strategies of structure and language connect a set of assumptions about 'nature' with expectations about social order and city life. As Neil Smith (1996) has argued, any particular interpretation of 'nature' successfully projected as authoritative and universal has a powerful social effect; the apparent 'givenness and unalterability' of nature make it difficult to challenge social structures and patterns of behaviour identified as 'natural'. These social structures become even harder to recognize, let alone subvert, when they are textually represented as natural through the conventions of 'realistic' narratives and the assumption of a shared frame of reference for images drawn from the natural world. This essay considers how the *Atlantic's* reliance on apparently natural narrative conventions and patterns of figurative language functions in constructing a particular view of appropriate urban social order as moral and reasonable. In this way, the essay relies on what Brian Jarvis (1998) has called 'the geographical preoccupation of American Letters' at the stylistic level, in a consideration of ways in which the text's uses of narrative convention and figures of speech rest upon its assumptions about the natural world and appropriate social place.

## The pauper problem

In the *Atlantic* of the years 1880–84, the smooth functioning of modern city life is understood to depend upon a general recognition of clear boundaries of social differentiation and frequent use of cross-class channels of social connection and interaction. The divisions mark individuals as members of particular groups and classes, and the lines of connection hold individuals and groups together within the social whole. This structure of boundaries and channels is taken to be both the hallmark of civilization and the reflection of natural order, and it depends upon individuals recognizing clearly marked social and spatial locations and territories: 'knowing their place'. Civil virtue, respectability, and healthfulness are all connected with being 'what one was made to be' and occupying 'the right place' – and because it seems to be generally assumed that such a 'real', 'wholesome', and 'comfortable' way of life is easier to achieve in rural areas, the *Atlantic* is particularly concerned in these years with the problem of creating and maintaining a 'natural' form of urban society for the American city.[2]

The biggest challenge to the natural, healthy functioning of urban society is understood by the *Atlantic* to come from 'the floating classes', from the ranks of which 'uncertain but ever-increasing army . . . the hosts of tramps, paupers, "repeaters," and vagrants are chiefly recruited.'[3] In a text where even ants are more respectable when they lead settled lives – an essay on ant life distinguishes the bad (cannibal) ants of the 'thieving, vagabond race', from the more reliable ants who have 'regular, settled homes' – the 'floating' classes represent unnatural social dislocation.[4] The indigent poor, both travelling 'tramps' and the institutionalized 'indoor paupers', are placed beyond the boundaries of normal society and marked as problematic. The *Atlantic's* response to the problem is to assume that it is the responsibility of the socially powerful to reconnect the 'excluded' group with mainstream, natural social order. The marking, excluding and reclaiming process is rehearsed and validated metaphorically in the text through narrative form and figures of speech, resting on what Felix Driver (1988) has identified in an earlier British text on a comparable topic as the 'symmetry between the language in which problems were identified and that in which solutions were articulated' (p. 281). By organizing its discussion according to the terms of a particular version of 'nature', and by using value-laden metaphorical language, the *Atlantic* (and, by implication, the reader/writer community) presents its analysis of 'the pauper problem' and its solutions as both moral and natural.

The *Atlantic's* natural world is taken to be the physical expression of a divine benevolent organization, an absolute universal order: 'we all believe (do we not?)' asks one writer, 'that the world is a universe, governed throughout by one Mind, so that whatever holds in one part is good everywhere.'[5] This concept of 'nature' emphasizes its reliability and meaningfulness, and assumes the universal and 'natural' tendency of all things towards improvement, achievement and appropriate destination. Like a conventional *Atlantic* short story, it relies upon a central, controlling intelligence (Hones, 1998). It is one 'of the truths of science' for the *Atlantic* that 'everything hastens where it belongs' both physically and socially.[6] Social disorder arises when people become misplaced, resist their natural placement, or refuse to hasten where they belong. In a society in which property-owning is virtuous, 'wandering' thus becomes subversive, and to be unplaceable (geographically or socially) is to be unnatural. To 'know their place' people need either to be sensitive to natural law themselves or to follow the social conventions created by those who are. The results of these different forms of obedience will be the same, as the power to define appropriate social behaviour and the right to interpret nature and the universal mind belong in the *Atlantic* generally to the socially powerful – 'the stable, property-owning classes'.[7]

Typically, the 'author' persona in the *Atlantic* is presented as a member of these socially powerful classes, even when the first person narrator is not. The characteristic author thus writes as a representative of the implied reader/writer group, articulates social issues in particular value-laden metaphors, and controls plot construction and resolution in such a way as to make sure that everybody ends up in the right place. The 'author' stands in the same relation to the text as the universal Mind stands in relation to physical and social nature. The final working out of plot and character with which this kind of author rounds up a conventional short story in the *Atlantic* achieves its satisfying sense of conclusion through a confirmation of conventional social place and social order: the unhappiness and disruption that have characterized the crisis of the narrative are resolved with a careful rearrangement that mirrors in social terms the conviction that nature's moments of disruption are only stages on a journey towards re-established order.

Appearing in the *Atlantic* of May 1880, 'McIntyre's False Face' is a short story that illustrates this characteristic of the text's narrative conventions as it works through the difficulties of social misplacement and the pauper problem in a local colour drama of romance at cross-purposes.[8] At the centre of the story is a young New Yorker, Francis Fosdick, who has arrived on an island off the coast of rural Maine. There he meets the local girl Idella Bowker, and Alrick Cooley, a young farm worker who has been rescued from his previous precarious existence as a homeless pauper boy by Idella's father. Alrick's homelessness had come about as the result of a practical joke played on his mother by the McIntyre of the 'false face'. Suddenly appearing before her while wearing a grotesque mask, McIntyre had frightened the woman so badly that she had eventually died, although she had first produced two children, Alrick's 'confessedly imbecile' brother Albon, and Alrick himself. The theme of the 'masquerader' making a 'false imposition', which is set up at the beginning of the story by McIntyre's trick, is then worked out in an unfolding drama of social stratification some twenty years later involving Idella and three young men: Fosdick; the farm boy Alrick; and the local mailman. The narrative structure of this story moves from an initial moment of drama through social confusion and misunderstanding to tragedy, as Alrick (in love with Idella, and jealous of the city visitor) attempts to murder Fosdick. Social order is put on hold while Alrick is captured and sent to an asylum, and the story finally comes to an end when the unplaceable pauper boy is released and allowed to head off 'for parts unknown', while Fosdick returns to the city, and Idella marries the lighthouse keeper. Each class representative has, in the end, hastened where they belong – the pauper boy Alrick, most notably, hastening straight off stage and beyond the bounds of all known society.

## Miserable confusion

'McIntyre's False Face' provides a reassuring story of crisis and reordering amid the social flux of the American 1880s, towards which the *Atlantic* sees tidal waves of disorder sweeping from Europe. 'Fast-succeeding waves of immigration' seem to be bringing with them 'an excessive urban population, the increase of disease and poverty, and the necessity that the benevolence and brain of the community should solve for the ignorant the problem which they could not solve for themselves.'[9] The *Atlantic* believes that American society is faced with the task of absorbing a large alien group into a harmonious system, a system which it takes, like the natural world, to be the physical manifestation of a controlling intelligence: God in nature, 'the benevolence and brain of the community' in New England. One of the brains, writing on 'The People of a New England Factory Village' reports that 'until within a quarter of a century ... the mill operatives were almost exclusively Americans. ... Old customs prevailed, and the community was homogeneous.' But by the 1880s, it seems, the 'native Americans' were a small minority, and the 'new Americans' were a heterogeneous, uneducated, undifferentiated mass.[10]

This lack of differentiation is highly disturbing, undercutting as it does the assumed natural social system of places, boundaries and connections. The extent of the uneasiness caused by the image of an 'undifferentiated mass' of new urban residents is suggested by a short story published in the *Atlantic* in 1883 under the title 'A Landless Farmer' – a paradoxical concept which, taken in relation to the text's emphasis on land-owning and stability, is clearly ominous. In this story, the social disruption and confusion brought on by the deeding away of rural property is represented figuratively by the jumbling together of the farmer's papers in his secretary desk when it is sold. 'They were huddled together in miserable confusion, though he had always known where to put his hand on each, when they were in their places.'[11] 'They were huddled together in miserable confusion': in the *Atlantic* at this time this could quite as easily be a description of immigrant factory workers or urban paupers as, here, of misplaced paperwork. The key note to the social disorder is a lack of differentiation, a lack of proper relations, a 'huddling together' and a 'confusion'.

The apparent result of the new urban immigration, an absence of coherence and unity in the American social and national system was taken by the *Atlantic* to be sharply threatening – actually, unnatural. And the absorption, amalgamation and consolidation of 'others' within the ideal, most natural American system was thus a prime concern. The threat of the 'new Americans' becomes personified in the figure of the pauper, the most visible result of society's failure to amalgamate or 'raise' its newest Americans to an appreciation of conventional ideology. The *Atlantic* therefore marks the pauper class off as an extreme out-group, distinguishing them sharply from the unfortunately displaced, both the 'unhappy and guiltless poor' heartlessly evicted from 'vine-covered cottages, or tidy rooms', and the middle-class victims of economic disasters. In the view of the *Atlantic*, the great majority of paupers 'belong to what are called the lowest classes, and seek the almshouse not because of unmerciful disaster, but because of very common vices'. It believes that 'between half and two thirds of them are of foreign birth' and cites the insistence of a New York investigating committee that 'by far the greater number of paupers have reached that condition by idleness, improvidence, drunkenness or some form of vicious indulgence.'[12]

The pauper class was thus perceived to be dangerously detached from both the natural and the national systems of order, unhealthily 'separated' from the motives and assumptions of the dominant ideology and thus cut off from the views of those who had 'brought [the nation] into existence' as well as the 'Mind' that provided the world with order. It is on these grounds – extremity of separation – that the pauper is distinguished from the criminal; in the *Atlantic*, the criminal represents a predictable aspect of society while the pauper represents the disorder beyond its boundaries. Criminals are subject to fixed laws and behave predictably. They increase their activities in times of social upheaval, but they are also capable of 'graduating' into employment and social conformity in times of stability: 'there are inevitable laws regulating crime', writes a commentator on the 'Origin of Crime in Society', and 'civilization is in the ascendant'.[13] The criminal classes are occupying a reliably predictable place in the social order, and everything, in their case, has at least the potential to hasten where it belongs.

The paupers, however, are neither placed nor hastening. They are the personification of 'parts unknown'. They represent the breakdown or absence of social and also moral ties, and so for the *Atlantic* pauperism is a 'spiritual debasement', the result of 'a deeper incapacity to sustain social relations than crime'. 'It is a lower abyss of physical and mental inaptitude.'[14] Paupers are thus an urgent problem, and the question of a society's responsibilities and appropriate attitude towards its indigent poor recurs throughout the *Atlantic* of this period, even providing the central subject matter for three substantial essays: 'The Indoor Pauper', published in 1881, 'The Pauper Question', in 1883, and 'Penury not Pauperism', in 1884.[15] These essays, as well as the various short stories that touch on the same social issues, consider the question of the most appropriate (most beneficial, most natural) forms of poor relief. Should charity be private and spontaneous – or public and professionally regulated? The clear answer to this question, in the understanding of the *Atlantic,* is that private charity is real, natural and healthy, while municipal charity is artificial, unnatural and unhealthy. The reasoning behind this distinction comes from two related sets of ideas, the first and more literal having to do with contemporary attitudes towards different forms of water supply, and the second having to do with the critical issue of social separations and relations. The solution to the pauper problem is to reconnect them directly with mainstream social values in the person of the middle-class philanthropist, through the application of direct personal charity, and this direct reconnection is presented metaphorically as the reconnection with natural spring waters.

## Wholesome springs of relief

The force of the private charity/natural springs metaphor that underlies the *Atlantic's* rhetorical construction of an appropriate social response to the pauper problem comes in large part from its general understanding of nature as a benevolent system that represents in physical form the goodness of a divine order. The 'natural springs' of private charity are therefore understood to be the source of wholesome social connections, and a vital part of healthy social order. At a more literal level, the contrasting public charity/dangerous pipes connection that also animates the magazine's discussion of charity and social relations relies on popular attitudes towards real plumbing systems. Sewer systems are not 'picturesque' and they lack 'the noble charm' that belongs to great natural disasters, but as a writer on the topic of 'Hurricanes' remarks, 'a single bad sewer may, in its time, kill more people than all the shocks of earth and air within the bounds of an empire.'[16] Indoor plumbing was, at the time, thought to be the source of a lethal 'sewer gas', seeping dangerously out of water pipes citywide. Water pipes were thus a useful image for the representation of the dangers of an intervening urban system separating source and recipient in a dangerous way. As Suellen Hoy (1995) has pointed out, city water sources at this time had strongly negative associations; in 1885, for example, Harriett M. Plunkett's book *Women, Plumbers, and Doctors*, detailed how sewer gas and germs could enter a home through 'overlooked channels of infection', such as 'leaky sewer pipes, contaminated wells, broken drains [and] impure ice' (p. 72).

American cities in the 1880s 'were bewildering and perilous places' (Hoy, 1995, p. 73). Little wonder, then, that the image of a countryside spring seemed so much more obviously positive than the image of municipal waterworks. Urban water systems were regarded as literal health hazards and at the same time as images of impersonal city life. Personal letters, for example, are contrasted with newspapers in the same way that private charity is contrasted with large scale charitable organizations: where letters were once the channels into which 'love and friendship pou[red] out their overflowing waters', in the urban 1880s newspapers seem to be taking their place. 'Once we used to draw much of water from private wells and pumps', a member of the Contributor's Club remarks, wondering 'Why We Don't Write Letters', but now the newspaper 'is a great reservoir with pipes, and we have only to turn a faucet, and there is abundantly more than we can use'.[17] This abundance may be convenient, but it has its dangers – primarily, contamination and the replacement of direct relationships with impersonal systems.

In the first of a pair of articles on the pauper question, appearing in May 1883 and June 1884, personal charity is compared consistently to a natural spring, while professional organizations are figured as mechanical systems of water pipes. The problem with municipal charity, the essay argues, is that 'the fountain of beneficence is concealed. Paupers do not drink at the clear spring.' Here, the official almoner (the faucet?) seems merely to distribute what the recipients see as morally owed them. 'They can recognize but little more ground of gratitude to the mechanism of distribution' in this exchange than they can 'to the hydrant which brings the water that it taints into their dwelling.'[18] The natural human sympathy that provides the charity (from its natural springs of wealth) is invisible: the professional almoner is merely a contaminating conduit. And it is the human connection that the *Atlantic* takes to be the 'higher and harder charity', the more socially necessary, and without this personal exchange 'organization is not method but mechanism.'[19] Impersonal charity may alleviate physical suffering, but it cannot transform the recipient, and thus it fails in its social obligation to address urban 'moral confusion' by 'reknitting the ties which hold all parts of the community in healthful social order'.[20] As the writer on 'The Pauper Question' insists, 'the mechanisms of charity can never shape the hard rock of pauperism into the features and forms of beauty', because 'the hand without the mind is but a tool', and 'society must become an inspired artist' in order to solve the pauper problem by absorbing it.[21]

The image of personal charity as a benevolent spring is so well established in this particular text that even when it is not being used explicitly its overtones still colour the argument. Key words and phrases, fragments of images rather than complete metaphors, are enough to make the point and remind the reader of the base image. Thus when the separation of donor and recipient marks 'the closing of the doors to high human fellowships', those 'fellowships' are subliminally linked to the image of a spring, 'with their moral basis of order and concord, with their bright conventions of courtesy and refinement, with their rich play of responsive sympathies, with their hope exciting vistas of still ampler and purer prospects.'[22] These hints of height and brightness and playfulness and purity, in the context of the *Atlantic's* patterns of figurative language, are clearly reminiscent of the water source of a spring.

The second of this pair of essays, 'Penury not Pauperism', develops the idea that society most naturally takes care of its poorest members through personal relationships by contrasting the 'artificiality' of charitable organizations with the 'naturalness' of local self-help. The essay describes an experiment in poor relief which took place in Glasgow in the early 1800s, when a Dr Chalmers abolished the established official system of poor relief to return to a more individual form of charity, believing that the 'natural offices of neighborly kindness' would provide 'more copious as well as more wholesome springs of relief' than the mechanisms of public organization. The essay notes that he 'called his experiment a "retracing process," or an effort to return from a highly artificial scheme to "the natural sufficiency" of society', and explains that the doctor 'believed that there were innumerable fountains of affection and good-will, ready to burst into action as soon as they were released from the ice of professional or legal charity'. The source of charity is here taken to include not only the 'wealthy patrons' but also the doctor's ' "plebeian" parishioners', and the intention is partly to demonstrate 'the natural sufficiency of lowly society to provide for its own'.[23] Here, in a variation on the spring/hydrant image, the contrast is drawn between water that is flowing and water that is frozen.

According to the author, the doctor's expectations were fulfilled, and the 'bounty' of the private sources turned out to be far greater than anything that even 'the most extortionate taxes' and 'the most opulent societies' had been able to command. The ice melted. The focus of this second essay is not so much on the fear that an artificial system for moving water was also contaminating it, but on the apparent fact that the most natural water had ceased to move at all – somehow frozen in its channels by the competition created by a mechanical system. But in concluding his argument that the two forms of charity (private and public) could not work together the author returns to the plumbing image: 'both sources of supply did not flow together; the mechanical stifled the spontaneous movement, and hence the overthrow of the former liberated ampler aid for the unfortunate.'[24] This explanation differentiates the two 'sources' by making one frozen and the other moving, or one 'spontaneous' and the other 'mechanical', and then makes the impossibility of coexistence seem reasonable by having the one 'stifle' the other. Only after the mechanical system has been 'overthrown' can the natural source reassert itself.

The rhetorical force of the *Atlantic's* construction of the case for private over municipal charity comes from its identification of benevolence with fresh moving water, springing up 'naturally' and spontaneously from fountains and well-springs, and its identification of publicly managed charity with mechanic- ally propelled or channelled water, pumps, pipes and water faucets. While this distinction comes partly from the distrust with which well-to-do philanthropists regarded officials working in the area of poor relief at this time, it is rooted most firmly in a particular understanding of the parallels between a divinely ordered nature and conservative social order. In her work on charity and class in the nineteenth century USA, *Women and the Work of Benevolence* (1990), Lori Ginzberg gives details of the disagreements separating philanthropists from officials, but she also forcefully makes the more general point, that 'the rhetoric of postwar society focused on class . . . as the source of social change or, to the middle class, social control' (p. 212).

## Concerning separateness

In its engagement with 'the pauper question', the *Atlantic* is not primarily con- cerned with excluding the 'floating classes' from mainstream society or poli- cing its own borders. Rather, it associates exclusion with confusion and takes the prime duty of its assumed reader/writer community to be the 'reknitting' of a social fabric which will connect 'all parts of the community in healthful social order'. For this reason, the most important result of private charity for the *Atlantic* seems to be the development of a personal relationship between donor and recipient. On the one hand, of course, this relationship is designed to fos- ter cross-class ties of obligation and gratitude that will sustain the particular social arrangement most suited to the values of the reader/writer community. The historian Paul Boyer quotes one of the leaders of the charity organization movement in the 1880s, Josephine Shaw Lowell, as arguing that society should 'refuse to support any except those whom it can control' (1978, p. 148). Carl Smith writes of the ways in which charitable organizations at this time in the US 'aimed at staving off social disintegration associated with the rising number of immigrants . . . by reconstructing the character of the lower classes', explain- ing how the ultimate goal was 'to bring [a] "they" back into the circle of the useful and productive "we"' (pp. 66–7). But as the *Atlantic* explains it, the importance of direct contact between the pauper and the philanthropist comes from the fact that the more powerful party represents not only the 'benevolence and brain of the community' but also provides the disconnected pauper with a direct, redeeming link to the universal Mind in all its truth and beauty.

Boyer points out that the charity organization movement in the 1880s rested on 'the assumption that the urban poor had degenerated morally because the circumstances of city life had cut them off from the elevating influence of their betters.' (p. 149). The *Atlantic* takes its reading of the harmful effects of separating those who do good from those to whom good must be done one step further: the separation is damaging to both parties, affecting both donor and recipient. A small anonymous article published in the *Atlantic* in 1884, on the dangers of letting any metaphorically pipe-like mechanism intervene between source and recipient in public charity, emphasizes the harmful effects of social as well as physical separation. The writer of the piece, 'Concerning Separateness', disputes the idea that there is no 'interaction' between donor and recipient, and that charity has 'no effect upon society at large other than its effect upon such persons as are afflicted with poverty'. Rather, the author argues, in 'estimating the value of a charitable relation . . . we should consider its effect on everybody concerned'. Charity, in this view, affects both parties, and 'we are all needy together'.[25]

This way of looking at charity is dramatized in a short story published in the *Atlantic* of January 1882. The story, 'And Joe', centres on the experiences of the significantly named Theodora Justice as she first encounters workers from her father's factory. Among them, she meets the difficult and probably epileptic boy Joe, who has been abandoned by his family and left to beg on the street. Her friend Margaret encourages her to get to know the factory workers and 'seek to be their friend'. Theodora resists: ' "A friend", she said, – "that is what *they* need; but for *me*! Was *I* made for Joe?" ' Margaret, whose 'soul saw farther than Theodora's dimmed eyes', speaks for the narrator when she replies: 'Not more than Joe was made for you. You need some one to work for. It may be God made him to keep you from aimless idleness.'[26] In its discussion of Joe's brother-in-law Andrew, who comes from a family of 'Irish Protestants of a low class', the text moves easily from narrative to confident social criticism. 'It is a notorious fact', the narrator remarks, 'that the children of Irish parents are a turbulent, disturbing growth in our civilization.' But the narrative voice in this story also moves to criticize and control the behavior of the 'we' who make up the text's implied readership and who possess 'our civilization'. Employers ought to 'elevate their work people', the narrator insists, even if it takes the impetus of class fear to bring them to action: the sense that the nation itself is under threat 'may supplement the tardy moral senses of the rich and rouse them to the necessary action to secure the enlightenment of the poor'.[27] Within the text, at least, the narrative voice succeeds in controlling all classes of character and by the end of the story Joe's family has been broken up and, in narrative terms, eliminated and Theodora has come round to the narrative point of view and accepted her moral duty to add 'and Joe' to her list of responsibilities.[28]

The story 'And Joe' thus works through the question of the importance of establishing direct relations between the powerful classes and the indigent poor, creating a normative view of the most natural and most moral way of sustaining social relations. It dramatizes the concern raised by the contributor writing on 'separateness', that the insertion of a professional organization between philanthropist and pauper would cause the donor, in his or her 'secure superiority', to lose sight of the fact that even the most 'wretched part of humanity' is connected to the 'unknown, unknowable God'. This would mean that the pauper would lose 'the halo of kinship with divinity', and the 'rich' (Theodora, 'we' the readers) would 'withdraw . . . our own spirits' and 'learn to despite our fellow-men'.[29] Insofar as the personal connection satisfies the needs of both giver and receiver, the best form of charity is therefore the most personal and the least mechanical, for the mechanical, the professional, would separate the water from its source and the goodness from its giver.

## Conclusion

Central to the *Atlantic's* view of the natural world in the 1880s is the concept of 'appropriate place'. At their most natural, it is assumed, all things tend to move towards recognizable destinations. This emphasis on location within an ordered and integrated system is not only central to the *Atlantic's* understand-ing of nature but also underpins its views on urban social relations. Being in the right place, behaving like a fountain, owning land – these things are all nat-ural. Being in the wrong place, floating about, having no fixed social relations – these things run counter to natural order. To be natural, to be in the right place, is to be real and to be genuine: to be 'what one was meant to be'. 'Well for us', writes one author, at the end of a long piece on bird-watching in the White Mountains, 'if we are . . . able to stand in our place and do faithfully our allotted task, like the mountain-spruces'.[30] And because it is easier to find our place, to know our task, to be natural – and to be poor – in the country than in the city, the 'city/country' division at times becomes conflated with the difference between the 'artificial' and the 'real'. 'The country is kind to all', one writer insists, because 'everything that is real is wholesome. . . . How genuine all things seem. . . ! I . . . feel that I am not expected to be anything but what is natural to me. . . . It is so comfortable to be what one was made to be and every-thing becomes so easy if one is only so fortunate as to slip into the right place.'[31] If the country is kind to all in the *Atlantic's* 1880s, the city is much harder, more confusing, less ordered; but city dwellers in this text nonetheless rely on their rurally oriented sense of the natural in their expectations of urban community.

This reading of the *Atlantic Monthly* 1880–84 has been able to reveal some of the connections the text assumes between natural placement and urban social order by disengaging itself from an implied reader/writer community and focus-ing, from some distance, on its conventions of narrative closure and figurative explanation. It has by this strategy been able to suggest how short stories like 'And Joe' and 'McIntyre's False Face' dramatize the apparently 'natural' impulse of the 'narrating class' towards class stratification and the reassuring pre-dictability of people 'knowing their place', while patterns of imagery in the non-fiction essays – particularly the association of private charity with natural springs and public charity with unhealthy plumbing – rely upon and suggest a particular sense of the natural as an integrated and purposeful whole.

This way of reading the text, engaging with it from a reader position consciously outside its implied textual community, is in some ways itself a transgressive activity, and as such shares the limitations of other transgressive acts: as Tim Cresswell (1996) has remarked, transgressions not only subvert but also at the same time rely on established boundaries. This approach to the text of the *Atlantic* and its role in the construction of 'natural communities' has therefore, like any other act of transgression, been able to do no more than reveal, deconstruct and destabilize. But it nonetheless insists on the value of questioning and destabilizing 'realistic' fiction and the rhetorical construction of reasonable argument, and presents its limited case study as an exercise in transgressive awareness, in this case focused on the social specificity of assumptions about nature and the textual articulation in particular reader/writer communities of 'the presumption that what is natural is really real and even normal' (Bennett and Chaloupka, 1993, p. ix).

## *Notes*

1. Quotations from the *Atlantic Monthly* 1880–84 will be cited in the following notes, which will list title, year, volume number, relevant page(s), and the author's name when it was included in the original Table of Contents. 'Contributor's Club' pieces come from a collection of unsigned comments and short essays printed at the end of each monthly issue.
2. 'Sylvan Station' 1883, 52: 116 (Caroline E. Leighton).
3. 'Municipal Extravagance' 1883, 52: 89 (Arthur Blake Ellis).
4. 'Under the Maples' 1884, 54: 326–33 (Mary Treat).
5. 'Bird-Gazing in the White Mountains' 1884, 54: 53 (Bradford Torrey).
6. 'Is God Good?' 1882, 49: 852–5 (Contributor's Club).
7. 'The Political Field' 1884, 53: 127 (E.V. Smalley).
8. 'McIntyre's False Face' 1880, 45: 600–16 (W.H. Bishop).
9. 'A Study in Sociology' 1882, 50: 215 (M.A. Hardaker).
10. 'People of a New England Factory Village' 1880, 46: 460–4 (Unsigned).
11. 'A Landless Farmer' 1883, 51: 637 (Sarah Orne Jewett).
12. 'The Indoor Pauper: A Study' 1881, 47: 756 (Octave Thanet).
13. 'The Origin of Crime in Society' 1881, 48: 462 (Richard L. Dugdale).
14. 'The Pauper Question' 1883, 51: 651 (D.O. Kellogg).
15. 'The Indoor Pauper: A Study' 1881, 47: 749–64 (Octave Thanet); 'The Pauper Question' 1883, 51: 638–52 (D.O. Kellogg); 'Penury not Pauperism' 1884, 53: 771–9 (D.O. Kellogg).
16. 'Hurricanes' 1882, 49: 330 (N.S. Shaler).
17. 'Why We Don't Write Letters' 1880, 45: 140–1 (Contributor's Club).
18. 'The Pauper Question', 651.
19. 'The Pauper Question', 652.
20. 'The Pauper Question', 638.
21. 'The Pauper Question', 652.
22. 'The Pauper Question', 652.
23. 'Penury not Pauperism', 75.
24. 'Penury not Pauperism', 778.
25. 'Concerning Separateness' 1884, 53: 291 (Contributor's Club).
26. 'And Joe' 1882, 49: 42 (S.A.L.E.M.).

27. 'And Joe', 43.
28. 'And Joe', 56.
29. 'Concerning Separateness', 292.
30. 'Bird-Gazing in the White Mountains', 59.
31. 'Sylvan Station', 116.

## References

Bartolovitch, Crystal (1995) 'Have theory; will travel: constructions of "cultural geography" ', *Postmodern Culture* 6,1 pmc@jefferson.village.virginia.edu.

Bennett, Jane and William Chaloupka (1993) 'Introduction: TV Dinners and the Organic Brunch', in Jane Bennett and William Chaloupka (eds) *In The Nature of Things*, Minneapolis, University of Minnesota Press.

Boyer, Paul (1978) *Urban Masses and Moral Order in America, 1820–1920*, Cambridge, MA, Harvard University Press.

Campbell, Neil and Alasdair Kean (1997) *American Cultural Studies: An Introduction to American Culture*, London, Routledge.

Cresswell, Tim (1996) *In Place/Out of Place: Geography, Ideology, and Transgression*, Minneapolis, University of Minnesota Press.

Cresswell, Tim (1997) 'Weeds, plagues, and bodily secretions: a geographical interpretation of metaphors of displacement', *Annals of the Association of American Geographers*, 87, 330–45.

Driver, Felix (1988) 'Moral geographies: social science and the urban environment in mid-nineteenth century England', *Transactions of the Institute of British Geographers*, NS13, 275–87.

Ginzberg, Lori (1990) *Women and the Work of Benevolence: Morality, Politics, and Class in the 19th-Century United States*, New Haven, Yale University Press.

Hones, Sheila (1998) ' "Everything hastens where it belongs:" nature and narrative structure in *The Atlantic Monthly* 1880–84', *The Japanese Journal of American Studies*, 8, 35–61.

Hoy, Suellen (1995) *Chasing Dirt: the American Pursuit of Cleanliness*, Oxford, Oxford University Press.

James, Louis (1982) 'The trouble with Betsy: periodicals and the common reader in mid-nineteenth-century England', in Joanne Shattock and Michael Wolff (eds) *The Victorian Periodical Press: Samplings and Soundings*, Leicester, Leicester University Press, 349–66.

Jarvis, Brian (1998) *Postmodern Cartographies: The Geographical Imagination in Contemporary American Culture*, New York, St Martin's Press.

Kearns, Robin A. (1997) 'Narrative and metaphor in health geographies', *Progress in Human Geography*, 21, 2, 269–77.

Radway, Janice (1999) 'What's in a name? Presidential Address to the American Studies Association, 20 November 1998', *American Quarterly*, 51, 1, 1–32.

Rycroft, Simon (1996) 'Changing lanes: textuality off and on the road', *Transactions of the Institute of British Geographers*, NS21, 325–428.

Sedgwick, Ellery (1994) *The Atlantic Monthly 1857–1909: Yankee Humanism at High Tide and Ebb*, Amherst, University of Massachusetts Press.

Sibley, David (1995) *Geographies of Exclusion: Society and Difference in the West*, London, Routledge.

Smith, Carl (1995) *Urban Disorder and the Shape of Belief: The Great Chicago Fire, The Haymarket Bomb, and the Model Town of Pullman*, Chicago, University of Chicago Press.

Smith, Neil (1996) 'The production of nature', in George Robertson, Melinda Mash, Lisa Tickner, John Bird, Barry Curtis and Tim Putnam (eds) *FutureNatural: Nature, Science, Culture*, London, Routledge.

Trachtenberg, Alan (1982) *The Incorporation of America: Culture and Society in the Gilded Age*, New York, Hill and Wang.

# Spaces and subjectivities

United Kingdom. 1989. In Bradford, young British muslims of Bangladeshi origin are in their best for a wedding. © Abbas/Magnum Photos

# Introduction

*James R. Ryan*

Understanding the relationship between human society and space is one of the most central and long-standing concerns of human geographers. In the course of the cultural turn such concerns have flourished along a range of new and significant trajectories. This final section of this book is concerned with mapping some of these new, culturally inspired ways of thinking about space and social identity. The three essays in Part V emerge and take their cues from a relatively recent body of enquiry, situated within geography but closely connected with allied social science disciplines, concerned broadly with both the conduct and constitution of the self in space. A brief outline of the emergence of this field of study might be helpful at this point.

In its early and conventional form, much social and cultural work in human geography saw its task as mapping patterns of social identity, such as ethnicity, in space correlating social interaction with spatial distance (see, for example, Peach, 1975). Whilst such work has been very influential it has been largely eclipsed by work which seeks to show how social categories – such as those of 'race', class and gender – are constructed and reproduced spatially (see, for example, Jackson and Penrose, 1993; Rose, 1993). Moreover, drawing on postmodernist and postcolonial currents, geographers have sought new ways of conceptualizing space and place in order to consider how new kinds of social identity are formed in relation to different kinds of spaces. This general shift in emphasis has necessitated and been stimulated by a greater concern with new forms of social and cultural identification which cut across conventional categories of difference (Duncan, 1993; WGSG, 1997; Smith, 1999). This is evidenced in much exciting new work on geographies of 'the self' and 'the body' (see, for example, Pile and Thrift, 1995; Pile, 1996; N. Duncan, 1996; Bell and Valentine, 1995; Matless, 1997; Nast and Pile, 1998), which is part of a broader re-mapping and 'decentring' of the autonomous, modern western self, with its normative contours of maleness, whiteness and able-bodiedness.

Like many aspects of the cultural turn such work has its own particular contexts and limitations. Geographers such as Harvey (1998) explain the focus on the body in the light of diminishing confidence in conventional categories of knowledge. Historians also point to how many contemporary debates on the self, despite claims of novelty, have significant pre-modern historical antecedents (Porter, 1997). Furthermore, despite the importance of such closer engagement with the body and its performances of new subjectivities and identity in geography (Thrift, 1996) the production of genuinely 'embodied' geographies is far from straightforward (Longhurst, 1997). Indeed, whilst concern with 'the body' has become fashionable across all kinds of contemporary cultural practice, from academia to art, scrutiny of the spaces of bodies does not, in and of itself, breathe life into their form (see Hauser, 1998).

How to capture the performative nature of bodies in space, to make them alive rather than dead, something raised in Nigel Thrift's introductory essay to this volume, is a core concern of the first essay in this section by David Matless. Matless moves us towards what we might term, borrowing a term discussed in Marcus (this volume, see pp. 13–25), a 'multi-site ethnography' of cultural objects and subjects. Taking five 'objects' (three books, a journal and a song sheet) in turn as his 'subjects', Matless performs what he terms the 'geographical enfolding of subject and object'. Again, as with Marcus' model of 'non-obvious' strategies of multi-site ethnography, Matless does not place his subjects/objects within some pre-existing historical structure but in a distinctly imaginative move creates new spaces of objects and hence subjectivities through new conjunctions and configurations. By working through such non-representational strategies, and by considering his subjects/objects not simply as 'texts', Matless captures the specificity of distinctive cultural forms, from music to pageants, and their multiple meanings as performed in space and in conjunction with one another.

The second essay in this section, by David Sibley, is different in style from that by David Matless but shares its concern with the conduct and construction of geographical subjectivities. The celebration of cultural difference is often held up as the heterotopia to which enthusiasts of the cultural turn aspire. Yet why, Sibley asks, is such a condition so hard to achieve? In order to explore this and other questions, Sibley considers the origins of personal and collective anxieties which shape the desire of people in modern societies to feel safe and insulated from the differences 'out there'. He does this by locating a psychoanalytic perspective in relation to the idea of place and of anxiety, effectively spatializing Freud's conception of the self as a socially shaped and enframed being. Sibley thus shows how anxiety is 'placed', less as definite object than a vague location, and how collective strategies are produced to keep the abject at a comfortable distance. However, Sibley also points out that these boundaries of self are not only employed as a means of keeping 'otherness' at bay, but also produce new realms of desire and possibilities for transgression. In this way spaces can thus take on connotations of impurity, compounding the abject status of those that occupy and utilize such space. These socio-spatial processes of boundary making that Sibley discusses are crucial in the formation of the self and its borders. Whilst the latter are certainly dynamic and porous, their significance in securing senses of selfhood and belonging are often taken for granted unless they are threatened or disrupted (see, for example, Valentine, 1998).

Explorations of the geography and culture of subjectivity, selfhood and the body frequently employ metaphors of the map and mapping. Indeed, the cultural and spatial turn in the social sciences and humanities has given mapping metaphors and practices a revitalized currency beyond their traditional domain of cartography and geography. As a number of commentators have observed (Pickles, 1999), the place of maps and mapping in the making of social and spatial identities, at a range of historical and geographical scales, has become a rich vein of enquiry within mapping studies. In the third essay in Part V, and the final essay of the book, Keith D. Lilley discusses the practices and subjective qualities of map making. Whilst there has been much work in geography and the history of cartography exposing cartographic practices as forms of institutional and authorial power, and treating maps as 'texts' to be deconstructed, far less attention has been given to mapping as an interpretative technique and creative process. Similarly, although much contemporary human geography makes use of maps and mapping as metaphors and analogies, particularly in broadly 'cultural' work, the value of mapping in a literal sense is rarely noted. However, as Lilley argues, and as other geographers are beginning to show, for example in relation to social statistics (Dorling, 1998), mapping can and should have an important place, as a practical, creative technique as much as a metaphorical device, within contemporary geography.

Lilley proposes, in accord with David Matless, to study maps not merely as texts, or representations of a more 'real' terrain, but as an embodied way of experiencing, imagining and creating cultural landscapes. By marrying concepts and methods of morphology with those of iconography, Lilley devises a critical approach for landscape interpretation. In his map making of the morphology of one medieval town, Lilley thus shows something of the opportunities which exist currently for dialogue among cultural geographers, cartographers and historians.

These three essays are not presented here as any kind of conclusion to the volume. However, they do pick up on a number of threads introduced at the start of the book and which run throughout the text. In particular, these essays suggest and promote greater concern for both the methodologies and theories that are used to record and analyse cultural forms and processes. They show that it is important to revisit concepts and methods that are either taken for granted or dismissed as unfashionable. In addition, it is possible to detect here a shift away from treating cultural forms merely as 'texts' to be deconstructed. Rather these essays argue for, and themselves perform, a more imaginative engagement with objects and subjects. Through such engagement, these essays also make a powerful argument for the need to appreciate a variety of forms of sensory experience besides, or along side, the visual in the performance of subjectivity and space. Finally, these essays show that, at least in accounts of the relations between subjectivity and space, the cultural turn, far from being dead, is very much alive.

# References

Bell, D. and Valentine, G., eds, (1995). *Mapping Desire: Geographies of Sexualities*, London: Routledge.

Dorling, D. and Simpson, S., eds, (1999). *Statistics in Society: The Arithmetic of Politics*, London: Edward Arnold.

Duncan, J.S. (1993). 'Landscapes of the self/landscapes of the other(s): cultural geography, 1991–1992', *Progress in Human Geography*, 17, 367–77.

Duncan, N., ed., (1996). *Bodyspace: Destabilising Geographies of Gender and Sexuality*, London: Routledge.

Harvey, D. (1998). 'The body as an accumulation strategy', *Environment and Planning D: Society and Space*, 16, 401–21.

Hauser, K. (1998). 'Sensation: young British artists from the Saatchi Collection', *New Left Review*, 227, 154–60.

Jackson, P. and Penrose, J., eds, (1993). *Constructions of Race, Place and Nation*, London: UCL Press.

Longhurst, R. (1997). '(Dis)embodied geographies', *Progress in Human Geography*, 21, 487–501.

Matless, D. (1997). 'The geographical self, the nature of the social and geoaesthetics: work in social and cultural geography, 1996', *Progress in Human Geography*, 21, 395–405.

Nast, H. and Pile, S., eds, (1998). *Places through the Body*, London: Routledge.

Peach, C., ed., (1975). *Urban Social Segregation*, London: Longman.

Pickles, J. (1999). 'Social and cultural cartographies and the spatial turn in social theory', *Journal of Historical Geography*, 25, 93–8.

Pile, S. (1996). *The Body and the City: Psychoanalysis, Space and Subjectivity*, London: Routledge.

Pile, S. and Thrift, N., eds, (1995). *Mapping the Subject: Geographies of Cultural Transformation*, London: Routledge.

Porter, R., ed., (1997). *Rewriting the Self: Histories from the Renaissance to the Present*, London: Routledge.

Rose, G. (1993). *Feminism and Geography: The Limits of Geographical Knowledge*, Cambridge: Polity.

Smith, S. (1999). 'Society–space', in Cloke, P., Crang, P. and Goodwin, M., eds. *Introducing Human Geographies*, London: Arnold, 12–23.

Thrift, N. (1996). 'The still point: resistance, expressive embodiment and dance', in Pile, S. and Keith, M., eds, *Geographies of Resistance*, London: Routledge, 124–51.

Valentine, G. (1998). ' "Sticks and stones may break my bones": A personal geography of harassment', *Antipode*, 30, 305–32.

WGSG (Women and Geography Study Group), (1997). *Feminist Geographies: Explorations in Diversity and Difference*, Harlow: Longman.

# Five objects, geographical subjects

*David Matless*

## Introduction: subjects/objects

This essay considers five objects. Three are books, one is a journal, one a songsheet; all are very different kinds of cultural product, and all carry various senses of geography. The aim of the essay is to show what might proceed from understanding such items in terms of the enfolding of subject and object, rather than through more established cultural geographic concerns for 'text' and 'representation'.

The latter terms might seem the most obvious ways to think about words, pictures, songs, etc., but such terms will be used here only when appropriate to specific genres, rather than as general categorizations of the nature of cultural material. The essay treats books as books, songs as songs. My sense is that much work in cultural geography increasingly resists being categorized as a general matter of text and/or representation, in part because of two possible consequences of aligning culture with those terms. If all culture, and all the world, becomes a matter of representation, then we may lose purchase on differences of material substance, whether that material is concrete, earth, paper, celluloid, and similarly the power of the 'textual' metaphor may be lost through over-extension. Conversely if culture is only in part representation, and if the concern of cultural geography is for that particular part, then we are drawn into unresolvable issues of delimitation and comparison between real, substantive materiality and unreal, insubstantial representation, not to mention debates concerning what kind of cultural product – high, low, popular, unpopular, etc. – we should be studying, debates which often end up reproducing rather than deconstructing hierarchies of value which place the high, or the popular, at an apex of cultural life.

This essay then takes a different approach, considering words, pictures, voices in terms of the geographical enfolding of subject and object. The objects considered here are 'geographical subjects' in three senses: they are our subject for enquiry; they present geographical practices which go to make up senses of self; they are produced through and seek to produce geographical subjectivities. Each object then can take us into various dimensions of the geographical self: the production of identity through the internalization of wider spatio-temporal relations, the moral geographies of conduct in place, the constitution of the human through relations with the animal, vegetable and mineral non-human, and the historicity and spatiality of experience.[1] I use the term 'geographical self' (Matless, 1997) here in a very different way from that proposed by Sack in his recent *Homo Geographicus*, where the geographical self finds itself caught at the heart of 'a framework for action, awareness and moral concern' (Sack, 1997). While I sympathize with the relational approach of Sack to the 'inescapable' question of 'being geographical', I do not share his approach of building an *a priori* moral, aesthetic and empirical framework from which we may then derive principles for action, and would not follow his normative agenda that to highlight the geographical nature of the subject 'helps us see more clearly our world and our place in it' (p. 2). The aim in this essay is less to use geography as a basis for clear sight than to suggest the problematizing nature of the geographical, and we might go some way towards this by simply acknowledging and respecting the literacy of our subjects of enquiry regarding those issues of space and subjectivity which we tend to mark out and classify as our own intellectual property. Each of the subjects/objects considered here posits complex geographies of the self which often uncannily anticipate recent concerns.

If this essay connects to a range of work on geographies of the subject, it also draws on an increasing concern for the geographical object, a concern which seeks a path other than that of fetishization or disaggregation. Just as in the geography of the subject proposed by Thrift and others (Thrift, 1996; Pile and Thrift, 1995), where the intent is neither to reproduce a philosophy of interiority nor to dissolve the subject as a 'mere' effect of things beyond itself, a geography of the object-as-subject emerges whereby an understanding of the object in terms of things beyond itself, effects neither to reinforce an object's assumed integrity, to fetishize its finish, nor to disaggregate, to leave the object in pieces. The life and power of objects may be understood less through an assumption that hidden relations are concealed in a finished form which thereby requires dismantling, than by considering that finished form as one significantly congealed state within a wider field of relations of which it is an effect (Latour, 1993; Bingham, 1996). This latter point of, as Law and Mol put it, 'relational materialism' (1995, p. 277) should not downplay an object's significance; after all, if all things are to be understood as effects of fields of relations, then the relational, the effectual, the contingent becomes the inevitable substance of all our enquiries, and the field of objects is in no sense diminished. The concern of this essay is then in effect for the social geography of the object/subject; or in Latour's terms the 'collective' geography of objects/subjects, where the 'collective' encompasses not only relations between people but also 'the association of humans and nonhumans' (Latour, 1993, p. 4). The geography of the things figuring in this essay is never solely human.

The objects/subjects considered here have common ground in the cultural geography of Britain over the past 120 years, but there is no general historical thesis running through the essay. While the historical contexts of each object are addressed, I am not seeking here to suggest any general model of, say, the transformation of the geographical self in twentieth century Britain. I have considered elements of such a story elsewhere (Matless, 1998), but while the subject matter of this essay forms part of that story, no claim is made here for special influence of these particular objects/subjects. That is not to say though that each is not significant in its own way. We will take in turn a mystical autobiography, a dietary journal, a communal song, a book on the circus, and a guide to prehistory. To begin we move back over 100 years, through a book.

## Soul topography

Our first object is a self-consciously geographical rendering of self. Richard Jefferies (1848–1887) published *The Story of My Heart* in 1883. Jefferies made his living as a writer, and was well known for his nature essays, accounts of rural society and children's adventure stories. *The Story* told of Jefferies' 'soul-life' (Jefferies, 1979, p. 32), rendering his experience through solitary, located meditations, and extended the nature-mysticism present in his earlier work, making him a touchstone for those later constructing an English rural tradition combining observation, imagination and spirituality (Keith, 1965, 1975). The stress in Jefferies' writing on contact with place led his first biographer, poet Edward Thomas, to present him as the genius of his country (Thomas, 1909); Thomas's book, and *The Story*, become a form of what painter Paul Nash would term a 'geogbiography' (Abbott and Bertram, 1955, p. 214).

Jefferies' nature-mysticism was not bound to any particular religion, indeed he declares in *The Story* that: 'There is no god in nature, nor in any matter anywhere, either in the clods of the earth or the composition of the stars' (1979, p. 65; Matless, 1991). Jefferies instead posited a transcendent reality reachable only through the experience of an immanent life-force, and *The Story* thus becomes a document of his most precious memory, with the placing of that memory crucial. Each chapter begins through a place: a hill, a valley, a hollow by the sea, the Roman sea wall at Pevensey, barrows on the downs, some elms, an aspen by a brook near London. Jefferies recalls 'pilgrimages' to these spots, movements which are in part an escape from being stuck at home: 'It is injurious to the mind as well as to the body to be always in one place and always surrounded by the same circumstances. A species of thick clothing slowly grows about the mind' (Jefferies, 1979, p. 29). Jefferies searches out points of departure for the soul through a sensuous geography, with tactile encounters with earth and light and air constituting a form of 'prayer' (p. 31; on Jefferies and touch see Keith, 1965, p. 87).

Jefferies' stress on the senses is no rejection of thought. Much of *The Story* engages with contemporary scientific concerns for energy, time and matter, Jefferies both drawing on and departing from late nineteenth century mystical and evolutionary thinking to produce a kind of philosophical anti-philosophy of experience: 'The sun was stronger than science; the hills more than philosophy'

(Jefferies, 1979, p. 40). Jefferies can on the one hand progressively argue for rational organization as the means to achieve progress, yet on the other defiantly reject evolution as being inadequate to explain the world, holding out for some as yet unknown future principle of understanding which might be intuitively explored through a direct contact with nature. Edward Thomas writes that Jefferies' 'favourite lines' were Goethe's: 'All theory, my friend, is grey, / But green is life's bright golden tree' (Thomas, 1909, p. 204). Praying by the elements, Jefferies enfolds rare and common knowledge: 'Life-force is not a secret, inviolable and sacred thing: it is as ordinary and common as water; and water is sacred, holy, and beautiful – a marvel and a miracle' (Jefferies, 1948, p. 55).

Jefferies opens his book by recalling a late teenage walk from his home at Coate to the hill fort of Liddington Hill on the crest of the Wiltshire downs: 'the labour of walking three miles to it, all the while gradually ascending, seemed to clear my blood of the heaviness accumulated at home' (Jefferies, 1979, pp. 29–30). Achieving the summit, Jefferies gained a different experience of body and soul – 'I felt myself, myself' – feeling himself in letting his self go: 'I restrained psyche, my soul, till I reached and put my foot on the grass at the beginning of the green hill itself' (p. 30). A particular geography, with specific senses of space and time, is at work here. The walk up acts as preparation, and the topography of ramparts on grassed downland is a prompt; Jefferies lies where a slump in the outer bank of the fort allows a view over the plain, the eye being led to a further groove through distant horizon hills. Time beside topography comes into play, with distant history – the fort, the tumuli of the downs – jogging Jefferies' heart: 'Like a shuttle the mind shot to and fro the past and the present, in an instant' (p. 39). The absence of others allows a tactile elemental prayer: 'I was utterly alone with the sun and the earth' (p. 30). Over two pages the author lies back and prays by earth, grass, sun, sky and the 'thought of ocean', before rising and walking along the summit:

> Had any shepherd accidentally seen me lying on the turf, he would only have thought that I was resting a few minutes; I made no outward show. Who could have imagined the whirlwind of passion that was going on within me as I reclined there! I was greatly exhausted when I reached home. (pp. 32–3)

For Jefferies this kind of conduct should ideally remain unseen, and the accidental spotter is significantly an acceptably pastoral figure whose occupation is deemed to denote social reticence. Liddington can act as a runway for the solo soul.

And all this happens not far from Swindon. If Jefferies finds solitude in downland, he walks not far from a modern town. Late nineteenth century North Wiltshire, centred on a new railway and engineering centre, is hardly a backwater. And love for lonely Liddington does not prevent Jefferies giving two chapters of his *Story* to London. A philosophical attention to spiritual energy did not imply a flight from the modern. Jefferies had lived in and around London, and wrote on its ecology, and as naturalist Richard Fitter would later comment, 'liked London . . . much more than most of his latter-day fans' (Fitter, 1949, p. 21; Matless, 1998). While Jefferies could sometimes present the city as dystopic, as in his fiction *After London* where London is drowned to leave a hallucinogenically toxic swamp (Jefferies, 1980), he could also declare that 'London is the only *real* place in the world' (quoted by Fitter, 1949, frontispiece). Jefferies at once recorded London's ecology and immersed his self in its energy. *The Story* moves into London to further engage with the elemental, through both the tidal Thames and the streaming crowd. Jefferies sits before the Royal Exchange, calling up water metaphors for life: 'This is the vortex and whirlpool, the centre of human life today on the earth . . . Here it rushes and pushes, the atoms triturate and grind . . . it is more sternly real than the very stones' (Jefferies, 1979, pp. 79–80). Jefferies' seeing self works here as in Wiltshire, and ultimately through the same force: 'Burning in the sky, the sun shone on me as when I rested in the narrow valley carved in prehistoric time. Burning in the sky, I can never forget the sun' (p. 80).

This story of a heart is also a story of physique. In London, having walked the streets and looked from the bridges, Jefferies would rest before pictures of the human form in the National Gallery, and sculptures in the British Museum: 'The smallest fragment of marble carved in the shape of the human arm will wake the desire I felt in my hill-prayer' (Jefferies, 1979, p. 42, also pp. 74–7). Jefferies was a sick man by the time he wrote *The Story*, and died of tuberculosis four years after its publication; the book is a story of both bodily ideals and inadequacies. If beside statues Jefferies could only comment 'how despicable in comparison I am', and present his own body as not matching his soul – 'My strength is not enough to fulfil my desire' (Jefferies, 1979, p. 90) – the body as an ideal remains: 'Let me be physically perfect, in shape, vigour, and movement' (p. 92). Lying on Liddington, not strong, though never losing self-comportment whatever the whirling inside, Jefferies dreams of other bodies. Ascetism is 'the vilest blasphemy – blasphemy towards the whole of the human race. I believe in the flesh and the body, which is worthy of worship – to see a perfect human body unveiled causes a sense of worship' (p. 93). Later, pondering the possibilities of human racial improvement over 'geological time' (p. 111), the cultural model that comes to Jefferies' mind is Sparta. Downland meditation triggers thoughts not of ancient Britons but classical bodies. In his attention to the body, as in his attention to topography, Jefferies is drawn to clear and clean form: the downland and fort of Liddington, the body of the statue and the Spartan. Vision for Jefferies always comes in a clear light; there is no hint of shadow in his prose, his outdoor soul-life being conducted in and via sunlight.

Sunlight also figures strongly in our next geographical subject–object, a journal from 1916. If Jefferies is concerned for bodily contact with sunlight through exposure, here self and sunshine come together through a particular geography of consumption, with the self remade less through outdoor meditation than via a food culture mixing health, nature and nation.

## Patriotic fruit

*The Herald of the Golden Age* was the journal of the Order of the Golden Age, 'A Philanthropic Society founded to proclaim Hygienic Truth, to advocate the Humane Life, and to promote Social Amelioration' (*Herald*, 1916). *The Herald*, and the Order, were dedicated to fruitarianism, centring identity on diet, and offering a peculiar ethical and aesthetic relationship to the world. The Order's 'Aims and Objects' included: 'To advocate the Fruitarian System of living, and to teach its advantages. To proclaim and hasten the coming of a Golden Age, when Health, Humaneness, Peace, and Spirituality shall prevail upon Earth.' For the Order fruit was a vector of goodness, health, progress and patriotism.

The journal's cover (see Figure 1) heralds the future through a landscaping of health. Sun rises through the picture; things are moving onward, ripening. A girl, a presumed future mother who, thanks to her health, would not participate in what a journal article termed 'The Breeding of Degenerates', treads lightly barefoot amid sources of the recently discovered vitamin C – oranges for eating, sunlight for absorption by exposed skin. The fruits of the earth are at hand, some for eating, some for a decorative beauty. The girl has perhaps graduated from being, as another article puts it, an 'Open Air Baby', and the sun rises over her future and the national future which her healthy growth is to sustain. For this is also the *British Health Review*, where nature makes sense in terms of nation, and person in terms of future citizen. A language of 'racial health' is adopted which shifts between human, white and British referents, moving this British message into 'Circulation in Fifty-Four Countries and Colonies'.

Essays in *The Herald* argue against vivisection, dismiss the claims of carnivores that flesh foods are essential, offer Christian homilies on good living and self-improvement. Published in October 1916, this issue also aligns fruit and war. Editor Sidney Beard distances the Order from 'The Peace-Crank Complaint', registering this natural health regime as patriotic and by no means lily-livered, its harmonious sense of body and nature at odds with the militarism of 'the Hun' to the extent of supporting war. Beard states that a society dedicated to the humane and peaceful need not be pacific: 'Humanitarianism . . . is altogether unconnected with foolishness, morbid sentimentality, or weak-kneed pacifism' (*Herald*, p. 85). This is fruitarianism with backbone, with masculine fibre alongside the graceful cover femininity. An essay by Arthur Mee, patriot of boy-publication fame, is brought in from *Lloyd's Weekly News* to add moral weight, hailing the war as a chivalrous act in the cause of justice.

This is evidently a culturally complex vision of geographical health, of what it means to choose a non-meat diet. The values here may be surprising to some late twentieth century readers, but *The Herald* is by no means exceptional within the cultural milieu of early twentieth century 'progressive' thought. Modern life and utilitarian science, the latter exemplified in this case by vivisection, are taken to task, yet the Order employs its own dissident scientific language to claim a different cultural authority over the human/racial body (Matless, forthcoming). *The Herald* also shows what was at the time a common

**Figure 1** Front cover of *The Herald of the Golden Age and Britain Health Review* (1916) vol. XIX, no 4.

conjunction of the rational and the mystic under the banner of experimental life. The predominant spiritual message here is a Christian one, though a classic progressive taste for spiritualism, upheld as in tune with modern science, is registered in advertisements. F. Heslop's *Speaking Across the Border Line*,

'Being Letters from a Husband in Spirit Life to his Wife on Earth', is on sale, while Mr Alfred J. Bennett offers his services as 'Psychic Consultant and Lecturer' from an address in Knebworth. *The Herald*, in its mix of moral stricture and dietary adventure, exemplifies what Raphael Samuel has characterized as a Puritanism running through self-consciously daring thought at this time (Samuel, 1998; also Spencer, 1993). Pleasure is to be taken in self-restraint in the name of purity, wholeness and enlightenment, with all of these values open to political mutation to the extent of redirection to a great war.

Other geographies of good living emerge from *The Herald*, especially in its advertising section. Here, alongside emerging mainstream products such as Marmite and Horlicks, we finds props for the good life sourced to obscure addresses and curious companies. Beyond London a seaside and suburban pattern emerges, with health aids obtainable by post or by visit; foods such as nut butter and wholemeal bread, health drinks such as 'Instant Postum' and 'Swastika Cocoa', cures for cancer, gardening aids, vegetarian restaurants, 'food reform' guest houses. The Lady Margaret Hospital in Bromley offers fruitarian care, Gibson's Hygienic Private Hotel welcomes you to Felixstowe, Malted Nuts are available from Watford. The golden age is to be nurtured through mundane spaces in Enfield, Southport, Blackheath, Felixstowe, everyday sites for health and wholeness. The English seaside, the English suburb, give you the chance to eat yourself fitter.

The *Herald of the Golden Age* effectively offers a production of self through subscription to a proudly minority lifestyle grounded in diet. If Jefferies' soul-life entailed self-knowledge through outdoor reflection, here the self is to be improved and secured through eating and associated healthy moral action. This geographical subject, formed through discourses of nature, nation and the body, entails an identity which is at once individual and collective; an individual practice is an allegiance to a cause. We now turn to a rather different collective document which sought to produce a collective geographical subjectivity through something coming out of the mouth rather than going into it: a songsheet.

## The geography of a song

*Land of the Ridge and Furrow* (see Figure 2) was first sung in public in Leicester on 16 June 1932. From the title some might anticipate a forgotten anthem of historical geography, perhaps written by Leicester resident and researcher of ridge and furrow W.G. Hoskins, but they would be disappointed.[2] These words and music have a different cultural geography, which we explore here through a consideration of their emergence and performance.

*Land of the Ridge and Furrow* was the winning entry in a song contest organized by the *Leicester Evening Mail* to find a song for the June 1932 Pageant of Leicester. Hugh Goodacre's words won the prize, and Walter Groocock later set them to music. Goodacre had claimed to speak for the place before as a local

# Land of the Ridge and Furrow

## The Pageant Song

Words by Hugh Goodacre
Music by Walter Groocock

1. O come to the land of the ridge and the furrow!
The brook watered meadow and far spreading wold!
For here is the home of the fox in his glory,
And music of hound that is richer than gold.

*Chorus*
Come along, Come along, over the ages
Ridges and furrows that carry us back,
Leicestershire legends and Leicestershire stories,
Follow us, Follow us, Follow the pack.

2. But think not the fame of our county is bounded
By prowess in sport, though of that we are proud;
We've marched with the first in the vanguard of progress,
And Leicestershire's courage has never been cowed.
      *Chorus*: Come Along, Come Along, etc.

3. 'Twas here the Plantagenet rule was supplanted,
And Richard, the last of the line, carried dead;
'Twas here the proud Cardinal, fallen, forsaken,
Craved leave of our Abbot to pillow his head.
      *Chorus*: Come Along, Come Along, etc.

4. Here Jane, the fair queen of a moment, was nurtured,
And Bradgate's gnarled pollards still witness her end;
Here Handel composed and poured forth his world-music
And Tennyson wove his lament for his friend.
      *Chorus*: Come Along, Come Along, etc.

5. Here Liberty has ever countered oppression,
And freedom of thought has been held in esteem,
It breeds in our vales, is the growth of our uplands,
The vaunt of our people, the goal of their dream.
      *Chorus*: Come Along, Come Along, etc.

6. So come to the land of the ridge and the furrow,
The land of the spinney, the thicket, the thorn;
There's welcome for you in the tramp of our horses,
The cry of our hound and the call of our horn.
      *Chorus*: Come Along, Come Along, etc.

**Figure 2** Land of the Ridge and Furrow: The Song of the Pageant of Leicester, Hugh Goodacre.

poet and historian, and *Land of the Ridge and Furrow* would reappear as the title poem in a 1943 volume of his pastoral and personal verse (Goodacre, 1943).[3] The Pageant of Leicester exemplified a specific genre of place celebration, developed from the 1900s by Louis Napoleon Parker and taken on by figures such as Frank Lascelles, who directed the Leicester event (Woods, 1999; Ryan, 1999a, 1999b; Matless, 1990). A place's people, wearing home made costumes, would enact a place's history through scenes depicting notable events and past everyday life. Parker saw the pageant as an act of conservative communal affirmation, with place value cutting across class division: 'the great incentive to the right kind of patriotism: love of hearth; love of town; love of county; love of England' (Parker, 1928, p. 279). The pageant was not to bring up any events which might still be contentious, and should stop at 'a date not too near the present' (Parker, 1928, p. 279). All took place under a 'pageant-master', an almost dictatorial conjuror of the spirit of place. For Parker the pageant was 'absolutely democratic' in its performance, but needed a firm hand: 'If I were asked to indicate the ideal Master of the Pageant I should unhesitatingly point to Signor Benito Mussolini' (Parker, 1928, p. 284).[4] Lascelles shared Parker's sense of conservative theatricality; he had staged the Wembley Pageant of Empire in 1924, was known for mastering pageants in Canada, South Africa and India as well as the UK, and styled himself as a lover and liver of traditional olde English life at his medieval-style new manor house at Sibford Gower in the Cotswolds (Ryan, 1999b). Pageantry at this time was not necessarily a conservative dramatic form (Wallis, 1994, 1995, 1999a, 1999b), but for these men the form was anything but disruptive, and *Land of the Ridge and Furrow* works within these dramatic conventions.[5]

The Pageant of Leicester was performed in the city's Abbey Park every day for 10 days in June 1932, with *Land of the Ridge and Furrow* sung after each performance, its words reprinted at the end of the souvenir 'text of the episodes' (Pageant of Leicester, 1932a, p. 55, 1932b). The song gathers meaning from the preceding drama: seven episodes, each mixing common people with significant heroes and heroines, beginning in 'a village of mud huts on the bank of the Soar' with the Roman legion arriving to build a town, and ending in Abbey Park in 1882 at its opening by the Prince and Princess of Wales. Intervening scenes included a hunting pack following an aniseed trail around the arena, a flock of sheep moving across the ground to commemorate scientific sheep breeder Robert Bakewell, and performers carrying a cardboard cut-out train for Thomas Cook's first excursion from Leicester to Loughborough by rail. The pageant, the song, and concurrent events, mixed narratives of tradition and progress, giving the city and county a long proud history as a build-up to a progressive present of civic improvement. An industrial exhibition ran with the pageant, while the souvenir brochure text was framed by photographs of the cherished old and the fine new: the Elizabethan Ragdale Old Hall and the newly opened Charles Street running north–south through the city, officially opened by the Lord Mayor of London processing along it to the pageant arena, with the Mayor of Leicester and pageant cast in costume (Pageant of Leicester, 1932a; Seaton, 1951).

If a song makes sense in terms of the events around it, it also takes meaning from the way it is sung. *Land of the Ridge and Furrow* was sung by both audience and performers; sung by or in the presence of costumed citizens in outfits mixing the frivolous and serious. Photographs of the event and rehearsals, held in the Leicestershire Record Office and seemingly taken by the local College of Art, record a democracy of dressed-up pleasure; lords and ladies lording it up, knights posing chivalric. Girls in impish outfits perch in the mouth of a monster: 'a pageant car, representing Hell's Mouth and filled with devils' (Pageant of Leicester, 1932a, p. 29). A knight in woolly chain mail and his boy standard-bearer pose on a motorbike. Home made costumes are proudly displayed, and only occasionally is a man evidently wrapped in an old curtain. This would be the dress for *Land of the Ridge and Furrow*, whether worn or observed. The song came in at the pageant's end, as a collection and recapitulation of meaning, a confirmation of costumed things gone before, joining audience and performers into the narrative of place just gone. At this time it would not have been strange to sing at the end of a performance; just as a national anthem would follow cinema and theatre performances, so a local anthem ends this civic event.

The specific setting and musical form also provide keys to this musical geography. People sing in the civic space of Abbey Park on the jubilee of its opening as a public amenity; the specific public values of the space – open yet regulated, free yet ordered – lend support to the supposed public values of the song. *Land of the Ridge and Furrow* would mean something very different sung in a cathedral, a football ground, a pub. The pageant arena was flanked by plywood replicas of the old town gates, with a semi-circle of stands between, song and setting mocking up the city. Filling this space was a song whose musical style fell somewhere between a hymn and Flanders and Swann's later 'Hippopotamus Song'. Experiments in performance show the lilt of 'Follow us, Follow us, follow the pack' anticipating the latter's 'Follow me follow, down to the hollow', lilting on to the end of the chorus as if on hunting horseback. An accessible style of song mixes the solemn and the jolly, carrying unsurprisingly proud local sentiment; in this form of communal song we can justifiably consider a tune as a carriage for a message. Verse One begins inside the county, with landscape and a hunting theme carried further by the chorus. Further verses take the county as a place whose fame spreads by being 'in the vanguard of progress'. Leicestershire becomes the home of famous individuals and universal ideals, with liberty bred in uplands and vales. The final verse returns to the firm substance of ridge and furrow, wood and hunt. Like the pageant scenes of old everyday life, the bumps in pasture left by older agrarian lives are to sustain modern local pride.

Whatever its wide points of reference, *Land of the Ridge and Furrow* did not aspire to travel. Its composers were not out for a hit, its publication being local and for a purpose. It is difficult without further oral historical research to ascertain whether the song's sentiments were subscribed to by its singers, but anecdotal investigation suggests the song did not last. While a colleague's grandmother remembered the pageant, and recalled organ lessons with Walter

Groocock, she did not recall the song, and it does not seem to have featured at later local events. The potentially contentious nature of this officially sanctioned statement of place is shown, however, by an incident prior to performance. When it won the *Evening Mail* song contest *Land of the Ridge and Furrow* had an extra verse, which was excised before it was sung. In 1943 Goodacre republished the complete lyric, and recalled the events of 1932:

> The third verse provoked a storm of controversy at the time, and was eventually withdrawn in deference to Roman Catholic opinion. As, however, the deletion of the verse not only robs the poem of its symmetry, but of two of its most salient recitals, I have reinstated it here.

The offending lines told that: 'here, in our county, the great Reformation/ Took birth from the dark, as the morning from night' (Goodacre, 1943, p. 1). History could still jar after 400 years, the county voice needing editing to an appropriate tone.

*Land of the Ridge and Furrow* was to be a song of belonging, of settlement. We now turn to a different evocation of belonging, a book which considers a culture defined not by settlement but by movement, and which offers both an argument for and an analysis of a mobile geographical self.

## Circus living

Lady Eleanor Smith's 1948 *British Circus Life* presented a year in the life of Reco's travelling circus. Smith was billed as lead author, writing the summer section of the book, while 'additional material' by John Hinde covered winter and spring, making up half of the text. Smith had died in 1945 (though her death is not mentioned in the text), but her continued lead billing reflected her status as a popular novelist on circus life. Born in 1902, the daughter of prominent Conservative politician F.E. Smith, first Earl of Birkenhead (Campbell, 1983), Smith was known as one of the 'Bright Young People' in London society in the 1920s, producing a gossip column for the *Sunday Dispatch*. Her first novel, *Red Wagon*, appeared in 1929 with great commercial success; in his diaries Evelyn Waugh records encountering Smith at a cocktail party: 'Eleanor Smith was there with a little horse in her hat made of platinum and rubies in the shape of a wagon given her by her father in honour of the book' (Waugh, 1976, p. 312). Smith entitled her autobiography, published in 1939 at the age of 37, *Life's a Circus*, and a series of novels set in gypsy communities, circuses, the ballet and Spain followed the success of *Red Wagon*. Smith's tone is classically of the romance of an Other life, with plots setting up tensions between characters signifying settled Englishness and those who choose to adventure with the circus, the Spaniard, the Romany. In her novels Smith presents a picture of roving life which claims realism in its romance, her characters using circus or gypsy language and slang, and leading characters are used to take the reader inside another culture, yet that culture is portrayed as an excessively emotional, adventurous and often dangerous other to the English home. Readers approaching *British Circus Life* with Smith's novels in their expectant minds would find a book to some degree following yet also departing from Smith's established romance of roving living.

That departure is in part due to the role played by John Hinde, then becoming well known as a photographic illustrator, and later to work with travelling circuses in Ireland for 10 years before founding the Hinde postcard company (Hinde, 1993). Hinde provided not only half of the text but the striking colour and black and white photography in the book, which was produced by the innovative colour reproduction firm Adprint. Investments in colour in book production in this period of paper rationing and strict resource control often served to highlight sites of cultural difference close to home, places easy to access yet possessing some sense of the remote or exotic: the circus, folk art, canal life, remote village life. *British Circus Life*'s companion volume was W.J. Turner's Mass-Observation based *Exmoor Village*, also with Hinde photographs, and Turner also took an editorial role in *British Circus Life*, possibly for the purpose of posthumously editing Smith's writing (Matless, 1994; Sarsby, 1998). As we shall see below, Hinde's photographic presentation of the circus goes some way towards making *British Circus Life* a different kind of romantic object from that presented in Smith's novels.

*British Circus Life* begins with Smith remembering her first encounter with the circus:

> Once, as a child, I was awakened by the sound of wheels and the clatter of horseshoes. It was not yet dawn. I ran to my window, and down in the dark street there passed a procession of creaking wagons. At the end of the cavalcade I perceived, faintly silhouetted against the grey summer night, the vast, dusky shape of an elephant. An elephant walking down Charlton village street! It seemed, and still seems, incredible. That was the first time I became aware of the circus. (Smith, 1948, p. 13)

The circus possesses the young Eleanor through an otherness which nevertheless somehow fits on the British road. Culturally distant yet appropriate proximity is the first wonder of the circus, and a similar immediacy of otherness is carried by Hinde's bright almost technicolour images. *British Circus Life* becomes a wonderland, yet one which while vivid is not flash. Smith's creaking cavalcade denotes a vernacular cultural difference which does not threaten to glamorously seduce or overtake the sedentary reader, rather the circus is itself homely; in its gentle creaking movement, in its awkward technology. Reco's circus denotes a particular way of being, a safe nomadic geography (cf. Sibley, 1995; Cresswell, 1996), its 'nomad creatures of the English road' (Smith, 1948, p. 13) happily inhabiting a series on 'British Ways of Life'. Britain and England may be classically conflated here – much of the book concerns travels in Scotland – and the multinational troupe might also be labelled 'this fascinating band of internationalists' (p. 119), yet their life is held to carry British/English national meaning for performers, audiences and readers. Here is mobile life not as transgression but as another form of national belonging, running along its established and well worn tracks and itineraries, living alongside settled town and country people with little friction. Mutual suspicion is noted, but the aim of the book is to dispel this as groundless. In this sense *British Circus Life* follows the tone of Smith's late novel *Caravan*, where gypsy life is presented as a form of national–international nomadism fitting but never settling into a national landscape: 'As Sylvester talked it was as though the music of gypsy fiddles sang in green lanes, while all over England and Scotland, across the mountains into Wales, the painted caravans crept forever along winding roads' (Smith, 1943, p. 29). The presentation of the circus in *British Circus Life* as both national and safe, however, depends in part upon its definition as not-gypsy. If both circus and gypsy life belong in their unsettled way, the former is held to fit more happily into the moral geography of a national landscape. Smith and Hinde describe gypsy culture in deliberately neutral tone, and note circus affinities with it in both language and custom, but Smith nevertheless fore-grounds perjorative gypsy associations in a manner which serves to ratify the circus's British Way: 'the darkness of the gipsy shadow cast ahead upon the tober – "Take in your washing – the circus is coming".' Such associations, says Smith, are: 'naturally infuriating to circus people whose honesty and respectability is a credit to their profession' (p. 27).

If romantic, *British Circus Life* is not idyllic. Smith and Hinde's wonderland is not all rosy. All weathers come, helping to bring pleasures but also hardships 'outside the experience, if not the imagination, of ordinary people' (p. 205). All this only adds though to a sense of circus community and solidarity. Part of *British Circus Life*'s romance of authenticity is that this life sees no division of work and leisure, with people leading 'creative lives' where their 'daily work' is 'the practice of their art':

> In the circus life is to be lived freely and without restraint . . . it just flows on like a river, sometimes turbulent, sometimes peaceful, sometimes fast, sometimes slow, but with its continuity unbroken and its freedom unfettered. Were circus a religion it would not be a bad religion, because it has to be lived seven days and seven nights in the week.

Hinde finds that circus life produces 'an elusive combination of personal experiences which lead to a state of mental poise and well-being' (p. 206).

Circus nomadism becomes appropriately British in part through a conservatism of social organization and cultural value. Just as canal life was being hailed at this time as a form of organic nomadism appropriate to English landscape (Rolt, 1944; Matless, 1998), so the circus is here an ordered vernacular space of its own, apart from and resistant to a modernity characterized as state-driven and overly rational, uniform, anonymous. In such commentary circus life comes to make cultural sense against a dystopic vision of the modern and progressive: 'All I know is that, in these enlightened times, a person can be arrested, if you please, for "wandering abroad and lodging in the open air". Shades of Borrow! Where is the Wind on the Heath?' (Smith, 1948, p. 22; also p. 97).

Smith here ties the circus into an English literary culture of wandering, citing George Borrow's gypsy tale *Lavengro* and its appeal to 'the wind on the heath' as the essence of the travelling life (Borrow, 1851; Segrott, 1999). A tradition of nomadism is asserted, and set against the restrictive state of modern life, which criminalizes wanderers and, Smith notes, prohibits child performance: 'that miserable, maiden-auntish law decreeing that no English child under fourteen may perform in the ring' (Smith, 1948, pp. 21–2). The state is presented as interfering from a cultural distance to undermine national culture. Labour regulation and itinerancy law are condemned as part of a general 'progressive' mentality which Smith also finds evident in animal rights campaigners, the 'few cranks' who picket circuses. Claiming the voice of one close to the animal – human life of the circus, Smith mocks 'a declaration on the part of Professor Joad to the effect that no circus animal can be trained without cruelty' (p. 95). Joad, famous for his radio appearances on the BBC Brains Trust as well as for his country and philosophical writings, is picked out as a prominent progressive voice. Smith argues that circus animals are trained only to do 'movements *which are natural* to them in their wild state' (p. 58 – emphasis in original). This is not to say though that *British Circus Life* denies animal pain, as when Hinde photographs a mare dying after an injury in the ring. Through such images Hinde and Smith claim a realism in their romance, claim an unsentimental love for animal and circus.

*British Circus Life* appeared at a time when the vernacular, in both culture and architecture, was emerging as central to a counter-modern culture of landscape (Matless, 1998). In this context the evocation of the everyday fabric of circus life in itself takes on a conservative value. Smith and Hinde detail the small incidents of road life, Hinde arguing that such events are special because: 'it is just small happenings which make up everyday life' (Smith, 1948, p. 155). The very language of circus also serves to register this as a distinct and valuable vernacular space: 'The circus in England has its own language, which is on the whole jealously guarded . . . a curious mixture of Romany, Italian, rhyming-slang, back-slang, and other expressions peculiar to the Big Top alone' (p. 25). Hinde's section of *British Circus Life* is told like a diary, moving from Manchester in October to springtime preparations for the shows of the new season, and his colour images show carefully composed tableaux of the everyday, with performers' bodies turned to glamour through colourful make-up and costume. In contrast his black and white photography, often reprinted at smaller scale, catches faces unaware of the camera, moments of living which pass and move on. Hinde's images, especially the colour pictures of performers, vividly standing out from text and black and white, lend *British Circus Life* a different sense of romantic order from that found in Smith's novels. Photography renders circus characters more statically romantic, still living in a distinct and romantically wonderful world, but living in the manner of a still life rather than through picaresque action. A clown's jaunty smile is caught, a balancing act stays balanced, horses remain on podia. Hinde and Smith together provide an order of romance, with otherness contained in a manner more stable than that of Smith's novel dramas.

British circus life also becomes a site of conservative value by virtue of being not a raggedy roving space but a site of strong self-regulation. The moves of the circus are, says Smith, 'planned with an almost military precision' (p. 69), and both the circus and the book are socially ordered by family. Smith presents performing families as the core of the circus, giving each a chapter. Most artistes own their own wagons, and their interiors are described in detail, the travelling community being revealed as a world of private spaces, family values, moral order (cf. Sibley, 1981): 'Circus morality is on the whole exceedingly strict' (Smith, 1948, p. 27). And these social relations are ultimately conducted through the authority of a leader, Reco. If not the ringmaster of the shows, Reco is ringmaster of the life, moving his show around the country though always keeping close ties to his home base in Yorkshire, working hard to earn the respect he carries, and supported by Mrs Reco in his family work-life. *British Circus Life* is well ordered: by Reco, by John Hinde, by Lady Eleanor Smith. British circus life, the book implies, is a way of being worth conserving, and which conserves.

Our final geographical object–subject is a contemporary document of movement around the country, which finds another scheme of value in walking through landscape. We have considered how geographical subjectivities have been made through sites as diverse as a circus, a pageant, a dietary journal, a hill fort. Our last example returns us to the prehistoric to a different end.

## A modern antiquarian

> On top of the earthwork above the white horse to the right about thirty yards, I lay amongst the sheep and discerned an incredible buzzing below me, as though I had huddled up to the Great Mother's heartbeat and was riding through space atop her great breasts. I realised that cattle and sheep are plugged into this great grid and follow lines, as do the deer. (Julian Cope, *The Modern Antiquarian*, 1998, p. 196)

In 1998 there emerged a strange piece of pop scholarship. *The Modern Antiquarian*, written by a pop star, Julian Cope, comes with a big picture budget and a hard plastic case 'to be practical for the outdoors', and plays off prehistory in a rather different fashion from Richard Jefferies. The book holds a strange mirror up to fieldwork, with author and prospective user making their identity through walking, map reading, field-noting and mystic insight. *The Modern Antiquarian* takes its place within the complex contemporary cultures of prehistory (Bender, 1998; Chippindale *et al.*, 1990; Michell, 1982), where tourists, travellers, antiquarians and archaeologists generate various meanings concerning nature and nation, amateur and expert knowledges, and the virtues and possibilities of being outdoors.

In his music Julian Cope has increasingly called up prehistory as part of a counter-culture of landscape aligned with political protest against car culture and criminal justice legislation (McKay, 1996), with songs touching on a New Age archaeoculture of alignments, Egyptian signs in the British landscape, UFOs, etc.[6] *The Modern Antiquarian*, billed as 'A Pre-Millennial Odyssey Through Megalithic Britain' and with a compass rose and megalithic quoit in the centre of its dayglo orange cover, extends Cope's growing reputation as a writer on musical cultures into other fields (Cope, 1994, 1995). Cope and family moved to Wiltshire to be near to the stone circles and avenues of Avebury, and *The Modern Antiquarian* took early shape through walks over the Marlborough downs. Cope signs himself the 'Arch-Drude', a modern version of eighteenth century antiquarian and Avebury enthusiast William Stukeley's self-presentation as 'Arch-Druid', and he builds his antiquarianism through a dissident heritage of Stukeley's late work on Avebury as a serpent-and-circle monument (Piggott, 1985, 1989), Alexander Thom's mathematical researches on the 'megalithic yard', the writings of alternative archaeologists such as John Michell, and the revived interest in dowsing as a technique tapping intuitive and folk knowledge.[7] Cope also hails the amateur vision of independent local researchers such as Margaret Curtis at Callanish (Cope, 1998, pp. 56–72), who appear in possession of non-specialist, unprejudiced eyes. While he happily draws on work from professional archaeology, Cope chooses the label 'antiquarian' rather than archaeologist to suggest that the latter denotes a narrow institutional practice, intolerant of different knowledges, looking down on the ancient people it digs up, and literally narrow in a visual focus on sites in isolation from their surrounding landscape. There is an echo here of Raphael Samuel's attempts to open up historical discourse to 'unofficial' voices (Samuel, 1994), as dissident antiquarians past and present come into alliance with Cope as gnostic artist. *The Modern Antiquarian* opens with a quote from Dadaist Hugo Ball: 'Artists are gnostics, and practice what the priests think is long forgotten.' Citing William Blake's 'I look through the eye, not with it' (p. 66), Cope suggests that 'In a Gnostic sense, the British scene is untrod' (Cope, web site).

As an avowedly national publication on the British scene, *The Modern Antiquarian* has to attend to some cultural complexities of historical scale. Is this ancient Britain British? Or are we dealing with things ancient and universally human which speak directly to us despite or whatever our nationality? Cope holds to a national landscape, yet at the same time mixes a neo-tribal language of communal universalism into attention to difference within Britain: 'This book aims to give something back to the culturally dispossessed of Britain, be they white, black or green; Welsh, English or In-between' (Cope, 1998, p. ix). Ancient Brits have, says Cope, not been given their due – 'When our ancestors are constantly ridiculed, we ourselves become ridiculous' (p. 134) – and the book becomes an act of cultural reclamation:

> I grew up believing that British history began with the Romans and that anything before didn't really count. The Flintstones and the Stone Age Raquel-Welch-on-a-pterodactyl movies conspired to create a vast and seemingly impenetrable prehistoric 'Then'. All of that changed the first time I saw

Avebury and my quest became clear: to discover all that had been withheld from me. (p. ix)

*The Modern Antiquarian* is devoted to a specific element of prehistory, the megalithic: standing stones, circles, burial chambers. Hill forts only get in here if they share a space with something older and stonier. A series of historical and thematic essays set out Cope's argument, the only ancient site getting a bad press being Stonehenge, presented as a later 'political monument' signifying an authoritarian power. Earlier sites make up a peaceful megalithic landscape, remnants of a sophisticated ancient culture coexisting harmoniously with earth through a feminine principle. Cope's ancient Britain is a great mother landscape, with megalithic sites positioned in relation to natural topographic recumbent female figures. Place names also signal the female landscape principle: Cope constructs etymological tables of key place name elements, finding that Britain goes back to the goddess Bridgit. Cope traces 'Bridgit Landscapes' in Dorset, Yorkshire and the Isle of Man, goddess signs resilient in the long historical face of patriarchy. Cope's essays mix spiritual transformation and material cultural change, with the development of metal and the deification of masculine individual power in a Roman Christian God finally sealing the fate of an ancient goddess culture.[8] Roman (though not Celtic) Christianity subjects the landscape to linear rather than cyclical principles of time (Cope, 1998, pp. 15, 142), and space is realigned through Roman military roads such as the Fosse Way: 'this great paranoid military barrier bisected the ancient landscape like a modern motorway' (p. 135). The Roman church sends its 'technocrats abroad' to colonize and Romanize ancient sites whose power still lingers: 'The unremitting violence and paranoid intolerance of the Roman church was largely due to its urban nature' (pp. 138–9). Cope may echo Richard Jefferies' *Story* in his flights on earthworks, but he would hardly echo Jefferies' views on Julius Caesar: 'He comes nearest to the ideal of a design-power arranging the affairs of the world for good in practical things' (Jefferies, 1979, p. 77).

Cope follows his essays with a gazetteer of 300 sites, each 'visited and verified by the author', and divided by region, with a regional map opening each section. Each site is given one page, with an account of its history, a map reference, photographs and field notes. The notes displace bookish knowledge as their author communes and connects through wind, rain, sun, elements crossing over time, blowing and raining and shining on him, seemingly, just as they did on the Ancients: 'These gibbering notes have been here included to give at least some of the flavour of these sacred sites when their powers show-out for the visitor' (Cope, 1998, p. 155). The first line of the first field note, at Boleigh Fogou in west Cornwall, reads: 'There's soil all over my book' (p. 155). At one point Cope recommends open-air song, though this is hardly *Land of the Ridge and Furrow*: 'Let us now reawaken the Great Goddess from this nameless prison by calling out to her in the landscape and chanting at her stones' (p. 147).

Much of *The Modern Antiquarian* would not, however, seem strange to the straightest geographer. The cultures of prehistory are characterized less by clear oppositions between 'orthodox' and 'unorthodox' outlooks than by intertwined ways of seeing and being in landscape, and in Cope's work antiquarian gnosis comes in part from very standard field practices. Cope's gazetteer begins with a country code on 'How To Behave in the Country' which would not be out of place in any conventional leisure or fieldwork manual, and the proximity of *The Modern Antiquarian* to an orthodox geographical culture of landscape is underlined by the prominence of the map. The Ordnance Survey Landranger series is recommended: 'Each of these maps will deluge you with sites and information, and whole holidays will be planned around one £4.25 Ordnance Survey map' (Cope, 1998, p. 152). Cope notes that: 'the Ancients placed their monuments with such care and accuracy that the traveller seemed always guaranteed to end up in some pastoral award-winning scenery' (p. 91). In both method and humour this could almost be landscape historian W.G. Hoskins, who is indeed quoted at the beginning of *The Modern Antiquarian* to the effect that 'The English landscape itself, to those who know how to read it right, is the richest historical record we possess.' Cope's off-beam scholarliness leads him, however, to quote Hoskins at secondhand while speculating that: 'Hoskins appears to have been an early 20th-century scholar with a strong mystical bent' (Cope, 1998, p. 3, citing Hoskins, 1955, p. 14, from Watts, 1993).

Cope's method is in large part to walk and look. The eye is here given transhistorical capacity, with looking ultimately the same the world and history over, part of a human phenomenological place in the world. Cope's eye, moving through landscape in his walking body, can thus function as a gnostic device for both historic and transhistoric questions. *The Modern Antiquarian* here carries echoes not only of the walking–knowing practices of land artists such as Richard Long, walking in lines and circles across England via sites such as Silbury and Glastonbury (Long, 1991), but also the phenomenological archaeology of Christopher Tilley, whose *A Phenomenology of Landscape: Places, Paths and Monuments* is almost Cope-without-the-Goddess. In Tilley's work moderns can dialogue with ancients, step into their shoes, by virtue of being looking-animals moving over the same 'bones of the land' in a 'perpetually shifting visual experience' (Tilley, 1994, pp. 71, 74), with topographies and monuments making sense through intervisibility. Just as Tilley looks for patterns of intervisibility as he walks the Dorset downland, so does Cope as he walks the Ridgeway towards Avebury. Cope identifies and plays 'the Silbury Game', spotting Silbury Hill popping up over the ridge of Waden Hill, playing visual tricks on the traveller, ancient or modern: 'Imagine the Neolithic Silbury, then pure white chalk, gliding sunrise-like along the Waden back. Travellers for the great ceremonies would be in total awe for the last forty-five minutes of the pilgrimage.' Cope maps 'The Silbury Game & Its Field Of Influence', sightlines radiating for miles along the ancient track (Cope, 1998, pp. 192–3). The modern antiquarian can be both modern and antiquarian through spatial practices deemed to be transhistoric, eyes and bodies linking over time, walking in landscape.

## Concluding remarks

To conclude we return to some themes raised at the outset, and which have run through these various subjects/objects.

In terms of attention to the historicity and spatiality of experience, the essay has addressed different forms of sensory geography. We have considered the visual geographies of modern antiquarianism, circus photography and future health illustration, and encountered visionary geographies which may or may not be tied to the workings of the physical eye. Jefferies' visionary geographies, like Cope's, also privilege a tactile connection to environment. The essay has also entailed attention to sonic geographies produced through a vocalization of place, and geographies of music also run alongside Cope's book-based antiquarianism. Geographies of sensory taste too operate through the Order of the Golden Age's fruit philosophy. As with the essay's attention to different genres of book, song, picture, etc., the intention with regard to different senses has been to highlight their interconnection and mutual constitution while respecting specificity; to discriminate between but not against.

The essay has also entailed attention to various mutually constituted scales of identity: individual, local, regional, national, imperial, cosmic. Whether it is Jefferies or Cope envisioning the cosmos on an earthwork, or Smith and Hinde documenting national nomadism, or the people of Leicester singing their city and county, we have considered subjects and objects enfolded through and generative of scales of identity. Likewise the essay has been concerned with subjects and objects enfolded through and generative of moral geographies of landscape and environment, whether through particular modes of being on pre-historic sites, or ways of moving, eating and singing in place. If work on space and subjectivity increasingly addresses relations of human and non-human, as in work on animal geographies, here is another sense in which the non-human makes up the human, in this case through the vegetable or mineral.

Lastly, the five objects/subjects considered in this essay all work through and are generative of relations of authority between performer, reader, author, and this of course not only at the time of their production but in their subsequent existence: in survival, disappearance, re-emergence, etc. These books, song-sheets and journals form, in part simply through their being bound between covers, what Harvey refers to in another context as contingent 'permanences' (Harvey, 1996, pp. 261–2), which may take on properties of momentum or inertia following their relational bringing into being. The implication here of course is that a 'relational materialism' operates temporally as well as spatially, and the act of constructing an essay such as this is itself an intervention in such processes, bringing disparate subjects/objects into conjunction. The material here has not been linked by any running historical thesis, but rather by a set of empirical and theoretical concerns, and if the five accounts are now linked it is perhaps in the manner of a constellation. Like many constellations this may end up resembling nothing more than five points claiming an inapparent shape, but to approach these five points less as representations or texts than as sites where matters of subject and object become enfolded, should have produced a different sense of their cultural geography.

## Acknowledgements

Thanks to Stephen Daniels and James R. Ryan for comments on earlier drafts of this essay, and to staff at the Leicestershire Record Office for help in tracing pageant material.

## Notes

1. In the preface to Volume 2 of *The History of Sexuality* Foucault writes of the need 'to consider the very historicity of forms of experience' (Foucault, 1986, p. 334). This approach in this essay is also informed by the Foucauldian approach to questions of the subject taken by Patrick Joyce (1994) and Nikolas Rose (1996), both of whom emphasize the spatial production of the self through the relationship to various human and non-human subjects/objects.

2. Hoskins worked at University College Leicester from 1931; on his work see Matless, 1993. Historical geographic work on relic landscapes did not itself shy from engaging with local culture. Hoskins wrote a number of guides to Leicestershire, and co-edited the first two issues of the *Leicestershire and Rutland Magazine* in 1948–9. The third issue included an essay by fellow landscape historian Maurice Beresford on 'Ridge and Furrow', suggesting this as the typical local landscape (Beresford, 1949).

3. Goodacre's collection also included verse on God, cricket, flowers and scouting. Goodacre also published village histories, plays with local themes, and in 1913 edited A.H. Dyson's history of Lutterworth, in which he highlighted his own family history; the Goodacres had come to Lutterworth Hall with banking money in 1825 (Goodacre, 1913). Goodacre also published as a numismatist (Goodacre, 1922).

4. Parker was also an ardent Wagnerian, inspired by regular attendance at Bayreuth. His parrot, Koko, could repeat all the leading motifs of the Ring Cycle (Parker, 1928, pp. 264–5).

5. If Wallis shows that for the Left pageantry could be deployed as a medium appropriate to a socialist message, a number of literary treatments of pageants deploy them as a device whereby the far from cohesive values of place are revealed through an exercise purporting to bring people together. See in particular Virginia Woolf's *Between the Acts* (1941), which centres on a village pageant, and John Cowper Powys's *A Glastonbury Romance* (1933), where a town pageant stirs ideological and religious controversy.

6. Cope first became known in the late 1970s as the singer with The Teardrop Explodes, and has produced material under his own name since the early 1980s. From the beginning Cope's work drew together interests in psychedelia and poetic senses of vision, not only in his own music but in his championing of neglected figures such as Scott Walker. Archaeology becomes prominent in his music with the 1992 LP *Jehovahkill*, which carried a plan view of the Callanish stones on the Isle of Lewis on the cover. The accompanying booklet told of his discovery of the ancient and why it was worth singing about. Subsequent LPs and side projects have developed the theme. The best source for information on Cope's work past and present is the web

site of his Head Heritage organization, whose logo is the Cerne Abbas giant with guitar.

7.  Cope does not share an enthusiasm for ley lines, finding these 'wearisome and endless' (1998, p. 29), in part one suspects because their linearity conflicts with his principle that movement should be serpentine rather than along straight tracks. On earlier debates on leys and landscape geometry, and on counter-cultural archaeology, see Matless, 1998.

8.  Cope's historical argument is often complex. Megalithic culture, he suggests, does not denote an unconscious ancient oneness with the earth, as the propensity to mark the landscape monumentally suggests some detachment from the earth being venerated. Megaliths come therefore to signal perhaps a last moment of human–natural balance.

# References

Abbott, C.C. and Bertram, A. (eds) (1955) *Poet and Painter: Being the Correspondence between Gordon Bottomley and Paul Nash*, Oxford University Press, Oxford.

Bender, B. (1998) *Stonehenge: Making Space*, Berg, Oxford.

Beresford, M. (1949) 'Ridge and Furrow', *Leicestershire and Rutland Magazine* 1 (3) 115–20.

Bingham, N. (1996) 'Object-ions: from technological determinism towards geographies of relations', *Environment and Planning D: Society and Space* 14, 635–57.

Borrow, G. (1851) *Lavengro: The Scholar, The Gipsy, The Priest*, Harrap, London.

Campbell, J. (1983) *F.E. Smith, First Earl of Birkenhead*, London, Jonathan Cape.

Chippindale, C., Devereux, P., Fowler, P., Jones, R. and Sebastian, T. (1990) *Who Owns Stonehenge?*, Batsford, London.

Cope, J. (1994) *Head-On: Memories of the Liverpool Punk-scene and the Story of the Teardrop Explodes: 1976–82*, Magog Books, London.

Cope, J. (1995) *Krautrocksampler*, Head Heritage, London.

Cope, J. (1998) *The Modern Antiquarian*, Thorsons/HarperCollins, London.

Cope, J. 'Headheritage' web site, www.headheritage.co.uk/antiquarian/index.html.

Cresswell, T. (1996) *In Place/Out of Place*, University of Minnesota Press, Minneapolis.

Fitter, R. (1949) *London's Birds*, Collins, London.

Foucault, M. (1986) 'Preface to *The History of Sexuality*, Volume II', in P. Rabinow (ed.) *The Foucault Reader*, Penguin, Harmondsworth, 333–9.

Goodacre, H. (ed.) (1913) *Lutterworth: John Wycliffe's Town*, Methuen, London.

Goodacre, H. (1922) *Bronze Coinage of the Late Roman Empire* (reprinted from *The Numismatic Circular*), Spink, London.

Goodacre, H. (1943) *Land of the Ridge and Furrow and other poems*, Leicester.

Harvey, D. (1996) *Justice, Nature and the Geography of Difference*, Blackwell, Oxford.

*The Herald of the Golden Age* (1916) vol. 19, no. 4.

Hinde, J. (1993) *Hindesight*, Irish Museum of Modern Art, Dublin.

Hoskins, W.G. (1955) *The Making of the English Landscape*, Hodder and Stoughton, London.

Jefferies, R. (1948) *The Old House at Coate*, Lutterworth Press.

Jefferies, R. (1979) *The Story of My Heart*, Quartet, London (first published 1883).

Jefferies, R. (1980) *After London*, Oxford University Press, Oxford (first published 1885).

Joyce, P. (1994) *Democratic Subjects: The Self and the Social in Nineteenth-century England*, Cambridge University Press, Cambridge.

Keith, W.J. (1965) *Richard Jefferies: A Critical Study*, Oxford University Press, Oxford.

Keith, W.J. (1975) *The Rural Tradition*, Harvester, Brighton.

Latour, B. (1993) *We Have Never Been Modern*, Harvester Wheatsheaf, London.

Law, J. and Mol, A. (1995) 'Notes on materiality and sociality', *Sociological Review* 43, 274–94.

Long, R. (1991) *Walking in Circles*, Thames and Hudson, London.

McKay, G. (1996) *Senseless Acts of Beauty: Cultures of Resistance since the Sixties*, Verso, London.

Matless, D. (1990) 'Ordering the Land', University of Nottingham, unpublished PhD thesis.

Matless, D. (1991) 'Nature, the modern and the mystic', *Transactions IBG* 16, 272–86.

Matless, D. (1993) 'One Man's England: WG Hoskins and the English culture of landscape', *Rural History* 4, 187–207.

Matless, D. (1994) 'Doing the English Village', in Cloke, P., Doel, M., Matless, D., Phillips, M. and Thrift, N., *Writing the Rural*, Paul Chapman, London, 7–88.

Matless, D. (1997) 'The geographical self, the nature of the social, and geoaesthetics', *Progress in Human Geography* 21, 393–405.

Matless, D. (1998) *Landscape and Englishness*, Reaktion Books, London.

Matless, D. (forthcoming) 'Bodies made of earth made of grass made of bodies: organicism, diet and national health in mid twentieth century England', *Journal of Historical Geography*.

Michell, J. (1982) *Megalithomania: Artists, Antiquarians and Archaeologists at the old Stone Monuments*, Thames and Hudson, London.

Pageant of Leicester (1932a) *Text of the Episodes*, Pageant Committee, Leicester.

Pageant of Leicester (1932b) *Official Handbook*, Pageant Committee, Leicester.

Parker, L.N. (1928) *Several of My Lives*, Chapman and Hall, London.

Piggott, S. (1985) *William Stukeley: An Eighteenth Century Antiquary*, Thames and Hudson, London.

Piggott, S. (1989) *Ancient Britons and the Antiquarian Imagination*, Thames and Hudson, London.

Pile, S. and Thrift, N. (eds) (1995) *Mapping the Subject*, Routledge, London.

Powys, J.C. (1933) *A Glastonbury Romance*, MacDonald, London.

Rolt, L.T.C. (1944) *Narrow Boat*, Eyre and Spottiswoode, London.

Rose, N. (1996) 'Identity, genealogy, history', in Hall, S. and Du Gay, P. (eds) *Questions of Cultural Identity*, Sage, London, 128–50.

Ryan, D. (1999a) 'Staging the Imperial City: the Pageant of London, 1911', in Driver, F. and Gilbert, D. (eds) *Imperial Cities*, Manchester University Press, Manchester.

Ryan, D. (1999b) 'The man who staged the Empire and lived in Sibford Gower: Frank Lascelles, 1875–2000', in Breward, C. and Kwint, M. (eds) *Material Memories*, Berg, Oxford.

Sack, R. (1997) *Homo Geographicus: A Framework for Action, Awareness and Moral Concern*, Johns Hopkins University Press, Baltimore.

Samuel, R. (1994) *Theatres of Memory*, Verso, London.

Samuel, R. (1998) 'The discovery of Puritanism, 1820–1914', in Samuel, R., *Island Stories*, Verso, London, 276–322.

Sarsby, J. (1998) 'Exmoor Village revisited: Mass-Observation's "Anthropology of Ourselves", the "feel good factor" in wartime colour photography and the photograph as art or social document', *Rural History* 9, 99–115.

Seaton, R. (1951) 'Leicestershire pageants', *Leicestershire Life* 1 (4), 1–8.

Segrott, J. (1999) 'Borrow on the move – self identities in George Borrow's *Wild Wales*', unpublished paper.

Sibley, D. (1981) *Outsiders in Urban Societies*, Oxford, Blackwell.

Sibley, D. (1995) *Geographies of Exclusion*, Routledge, London.

Smith, E. (1939) *Life's a Circus*, Longman, London.

Smith, E. (1943) *Caravan*, Hutchinson, London.

Smith, E. (1948) *British Circus Life*, Harrap, London.

Spencer, C. (1993) *The Heretic's Feast: A History of Vegetarianism*, Fourth Estate, London.

Thomas, E. (1909) *Richard Jefferies*, Hutchinson, London.

Thrift, N. (1996) *Spatial Formations*, Sage, London.

Tilley, C. (1994) *A Phenomenology of Landscape: Places, Paths and Monuments*, Berg, Oxford.

Wallis, M. (1994) 'Pageantry and the Popular Front: ideological production in the thirties', *New Theatre Quarterly* 38, 132–56.

Wallis, M. (1995) 'The Popular Front pageant: its emergence and decline', *New Theatre Quarterly* 41, 17–32.

Wallis, M. (1999a) 'Delving the levels of memory and dressing up in the past', in Barker, C. and Gale, M. (eds) *Inter-War Theatres*, Cambridge University Press, Cambridge.

Wallis, M. (1999b) 'Heirs to the pageant: mass spectacle and the Popular Front', in Croft, A. (ed.) *Weapons in the Struggle? Essays in the Cultural History of the British Communist Party*, Pluto, London.

Watts, K. (1993) *The Marlborough Downs*, Ex Libris Press.

Waugh, E. (1976) *The Diaries of Evelyn Waugh*, ed. Michael Davie, London, Weidenfeld and Nicholson.

Woods, M. (1999) 'Performing power: local politics and the Taunton pageant of 1928', *Journal of Historical Geography* 25, 57–74.

Woolf, V. (1941) *Between the Acts*, Hogarth Press, London.

# Placing anxieties

*David Sibley*

While heterotopic visions have had an appeal for academics since Jane Jacobs' *Death and Life of Great American Cities* (1961) and, arguably, since Kropotkin's *Fields, Factories and Workshops Tomorrow* (1899), people still try to secure their own spaces and exclude others. The realization of heterotopia – that ideal place where difference is celebrated rather than being a source of oppression – will, as Merrifield and Swyngedouw (1996, p. 6), argue 'necessarily involve turning the tide of commodity and image fetishism of assorted ruling forces and the theoretical self-indulgence of the academic left'. In such a radical reordering of society, attachments to place which receive spatial expression in class and racial divisions and which are deepened by institutions involved in housing markets and by governments which encourage parents to move their children to 'better' schools, need to be weakened. There is clearly a contradiction between the positive valuation of community, which implies belonging and not belonging, and pluralist and multiracial rhetoric because, as Iris Young (Young, 1990, p. 235) recognized, 'if existing together with others in relations of mutual understanding and reciprocity is the goal, then it is understandable that we exclude and avoid those whom we do not or cannot identify.'

I am pessimistic about the chances of moving beyond community towards less clearly bounded and accepting social spaces, partly because of the endemic insecurity which characterizes the global economy, insecurity which translates into personal anxieties about jobs and homes (obviously) but also collective anxieties about localities which are occasionally manifest in a rejection of others. There seems to be an overwhelming desire in modern, highly developed societies to feel safe. This means insulating self, family, neighbourhood or state from what are perceived as external threats.

This is not a new problem. There are parts of cities that have a long history of labour exploitation and casualization, like the East End of London, where long-established groups have repeatedly tried to erect metaphorical barricades against newcomers. This is a problem of closure which might be understood in

Weberian terms (Husbands, 1982) but Weberian closure theory, while providing a plausible account of exclusionary processes, does not seem to me to provide many clues as to why people become anxious or fearful about others. This is a problem, however, that I think can be usefully explored in psychoanalytical terms. As Michael Ignatieff (1998) has suggested, one place to start looking for an explanation of the failure of people to get on with each other is Freud's writing on 'the narcissism of minor difference'. So, in this essay, I want to take as a starting point some of Freud's arguments which touch on the question of difference but to build on these ideas by thinking additionally about constructions of place which amplify fears about human difference. In the process, I want to suggest why it is so difficult to get to heterotopia.

## Psychoanalysis and difference

I first want to consider some objections to using psychoanalytical theory in exploring ideas about cultural difference. It might seem that psychoanalysis is entirely inappropriate for the purpose because its propositions are seemingly fixed, settled and unlikely to vary. The essentialism of the practice and academic discipline founded by Freud can be appreciated from an examination of its early history although I would argue that there is nothing in psychoanalysis that makes it irredeemably insensitive to cultural difference. Freud came to psycho-analysis from medical science (his first professional appointment was as Lecturer in Neuropathology in Vienna in 1885) and he was concerned that his new sub-ject should also gain acceptance among scientists. When he and other central European Jewish practitioners, particularly Melanie Klein, went into exile in England (and in the United States), the academic respectability of the field assumed greater importance as a means of confirming the professional status of the *émigrés*. It was thus important to emphasize the nomothetic, scientific founda-tions of the emerging discipline. Furthermore, psychoanalytical associations, particularly the International Psychoanalytical Association, assumed a discip-linary role, effectively promoting western models of the self and discouraging alternative models produced by 'indigenous psychoanalysts' (Parker, 1997, p. 5). It is, thus, understandable that a psychoanalysis with its particular history and geography, based initially in early twentieth century Vienna, Budapest and Berlin, associated primarily with Jewish intellectuals, and later associated with centres of practice in London, New York and Paris, has made universal claims, failing to recognize its own cultural specificity. However, this has not discouraged a reworking of psychoanalysis, particularly by feminists and social anthropolo-gists who have recognized the considerable possibilities of the discipline as a hermeneutic for examining difference and power relations. The history of psychoanalysis suggests that cultural difference was an early, if marginal, inter-est, following Freud's *Totem and Taboo* (1913). Thus, in 1921, the Hungarian psychoanalyst, Géza Róheim, won the Freud Prize for his work on Australian (Aboriginal) totemism and further research by Roheim in Somalia, central Australia and with the Yuma Indians in Arizona, was financed by, among others,

Freud, Sandor Ferenczi (a close colleague of Freud's in Budapest) and Princess Marie Bonaparte (Wilbur and Muensterberger, 1951, p. xii). Although modern anthropologists largely reject these early interpretations of anthropological material, some of which were certainly racist, psychoanalysis continues to fascinate anthropologists (Heald and Deluz, 1994). More obviously, feminists have been deeply involved in rethinking psychoanalytical concepts (Wright, 1992) and have brought to the surface questions of gender and other salient forms of difference which were neglected by the early theorists, including Melanie Klein.

A further concern of some social scientists has been the apparent neglect of the social in psychoanalysis. Freud is known primarily for his discovery of the unconscious and his mapping of the topography of the mind. However, his conception of the self was very much a social one, as demonstrated in *Civilization and Its Discontents* (1930). The idea of repression, for example, concerns the self in a social context and he connected the unconscious with the social world through the processes of projection (of 'good' and 'bad' on to others) and introjection (the internalization of 'good' and 'bad' objects). Thus, Freud saw the self as being shaped by and held in a social web. His initial conception of object relations theory provided the grounding for subsequent 'English' psychoanalysis which has had a strong social emphasis, particularly in the work of Klein, Winnicott and Bion. I want to take some of these arguments about the self and the social emanating from psychoanalysis and spatialize them, in the process suggesting how possible relationships between the self, the social and the material world – the built form of cities, the countryside, and so on – contribute to and are affected by collective anxieties.

## Fear and anxiety

Psychoanalysts have tended to make a distinction between fear and anxiety. Thus, John Rickman (1957, p. 319), who had a penchant for scientific classification schemes, suggested that

> it is important to distinguish between fear and *Angst*, a German word which will be used henceforth in this paper for neurotic fear: the former is occasioned by a real object, the latter is characterized by an indefinite feeling of expectation *about* something but lacks an object.

As Adam Phillips (1995, p. 59) puts it: 'Fear has an object, anxiety [or what Rickman calls "neurotic fear"] has a vague location.' The objective of analysis for someone experiencing anxiety or perpetual unease is then to locate and work on the fear from which this anxiety is derived but not consciously connected. In developmental terms, the most important formative source of fear is the child's fear of the loss of love of the parent who protects him or her from the unpredictable. Phillips suggests that fear of the loss of love 'instigates a project to secure something that, by definition, cannot be secured'. As we become more aware of the uncertainties of life, we crave certainty and repetition to compensate for that early loss. This suggests that a rather diffuse anxiety about the unpredictable becomes, in varying degrees, a condition of human existence and that this condition may be exacerbated by specific fears, fears which have locations, which are bounded in space and time, manifest, for example, in moral panics.

Another way to think about anxiety is to consider the body as a site of purification, where distancing from sources of defilement becomes an imperative but, like the search for certainty, an impossible project. This is the problem of abjection which Georges Bataille (1970, p. 2) described as 'merely the inability to assume with sufficient strength the imperative act of excluding abject things'. The urge to purify the body is very strong in western societies and the child, at an early stage of development, learns to reject bodily residues with which it formerly had an easy relationship. In the cultural construction of defilement, however, bodily residues elide with despised and oppressed minorities. As both Stallybrass and White (1986) in their historical account of ethnic relations in Britain and Constance Perin (1988) in her analysis of white suburbia in the United States have argued, 'distancing from shit' is not just a bodily concern but also an aspect of socio-spatial relations. However, as Bataille's comment suggests, this distancing cannot be achieved in any final sense. Just as excretion is a necessary feature of existence, so defiled social groups are continually produced and reproduced and people may be concerned that they will be touched by the defiled, that they will move into the neighbourhood and transgress boundaries. Anxieties about the abject thus do not go away. Julia Kristeva (1982, p. 64) comments:

> Georges Bataille remains the only one, to my knowledge, who has linked the production of the abject to *the weakness of that prohibition* [and he] is also the first to have specified that the plane of abjection is that of the *subject/object relationship* [and not subject/other subject]. (Emphasis in original.)

Like the abject *thing*, bodily substances and other material sources of disgust, the abject group is objectified.

What I want to suggest is that many people construct a simplified map to reduce the chance of unpredictable encounters and to keep the abject at a distance. This map, with its clear boundaries and certainties, may compensate for the loss of parental love and protection but it also engenders anxieties that are confirmed through social experience and the experience of living in particular places.

## People and place

As a way into this problem, I first want to consider the core geographical theme of people in relationship to their environment. In my own writing (Sibley, 1995), I have put considerable emphasis on visual images as sources of both anxiety and desire. In placing the self in the social world, in Freudian and Kleinian terms, images constitute 'good' and 'bad' objects which are introjected and projected and through these complementary processes, the psyche enters the social and the social enters the psyche. These visual images are incomplete representations of people and things and spaces (stereotypes) which can be related to other culturally constructed things that either threaten the boundaries of the self, that is, they are sources of abjection, or they constitute objects of desire, but usually both. The negative (abject) is usually expressed as some variant of dirt, defilement or disorder – disorder suggesting matter out of place and uncertainty – while desire is a desire for something harmonious and untainted. People's feelings about these visual images are evident in their apprehensions about differences in others and in other places as they are articulated in their own observations and responses to daily life, like the presence of homeless people on the street, and by the written and visual media.

Freud tried to distinguish between 'minor' and 'major' differences between people in arguing that narcissism, self-love, was at the root of aversion to others whose small differences disturbed the antagonistic individual's perception of their ideal self (see Ignatieff, 1998, p. 50). Small differences, like religious affiliation, for example, could become major differences but he failed entirely to explain why some differences were minor and others major and why some minor differences became significant sources of social cleavage. What was lacking in Freud's analysis was an appreciation of the historical uses of culturally constructed difference to further material ends, something which can be understood through an historical analysis of the uses of visual images, like the racialized images of colonized people, for example. It is not only visual images which are introjected and projected in the process of the social entering the psyche and the psyche entering the social, however, and the scope of psychoanalytical theorizing needs to be extended to include other sensations which are employed in the construction of difference.

Robert Mandrou (1975, p. 50) has insisted on the primacy of touch, hearing and smell in pre-modern European cultures. He suggested that complex, non-visual sense perceptions gave way only slowly to the enlightened predominance of the eye that we take for granted when we describe a person or place: 'The eye, which organizes, classifies and orders was not the favourite organ of a period [the 16th and 17th centuries, in France] which preferred to listen.' He cites the oral, aural and tactile imagery of poets in support of this argument and suggests that clear glass windows and optical instruments, which contributed to the primacy of vision, were the preserve of the rich – the poor were still quite tactile and smelly. Sensations other than the visual have clearly declined in significance as markers of social difference since the mid-nineteenth century in industrialized societies, however, and these changes in the role of sensations have had interesting consequences for cultural geographies.

The most discussed is the production of the odourless city. As Ivan Illich (1984, p. 53) suggested,

> [many] people have lost the ability to imagine the geographic variety that once could be perceived through the nose. The cleaning up of the city and the self are associated with what Norbert Elias terms 'a civilizing shame' – a habitual fear of the humiliating put down due to one's uncleanliness.

Alain Corbin's similar and highly olfactory account of the changing role of sensations in French culture (Corbin, 1986) was explicitly psychoanalytical. He wrote, for example (p. 143), that 'the bourgeois projected onto the poor what he was trying to repress in himself.'

This projective identification, where aspects of the self, in this case primarily anxiety about the smell of bodily residues, are pushed outside the self in order to keep feelings of repulsion at a distance, demonstrates clearly the importance of abjection in relation to self-identity and the construction of an abject other. Corbin's graphic illustrations of the stenches of the nineteenth century city couple excrement, ragpickers and prostitutes. These were all sources of repulsion for the bourgeois but also sources of fascination and desire for middle-class reformers, echoing infantile fascination with bodily residues. Since the nineteenth century, however, we have become progressively more detached from the smells of residues so that they register more acutely and are arguably more likely to disturb. Thus, rather than the working class, who have been cleaned up and have the same sensitivities about smell as the bourgeoisie, it is some marginalized groups like the homeless and some nomads for whom access to water is precarious, who now constitute an olfactory other, threatening the sanitized majority.

The other aspect of cultural change which conditions object relations is change in the significance of touch. Touch, and particularly the sensation of skin touching skin, may have very specific cultural significance. Anxieties about touching others seem to be a serious problem in some modern societies, like Britain, but not in others. Who can we touch? Michael Rustin (1991) implies that touch as an aspect of interpersonal relations has become more problematic as the nuclear family has become more strongly bounded, encouraged by political rhetoric which suggests that the family provides a secure base for society in an insecure world. He argues that in physical contact between parents and children, sensations of touch, in skin contact and the feel of hair, the perception of family resemblance becomes suffused with intense feeling. Touch within the family encourages strong bonding (maybe) but people outside the family are not to be touched.

This was highlighted by a recent controversy in Britain (summer 1998) over advice to teachers not to apply sun protection cream to young children because of possible accusations of child abuse. This problem now extends to relations within families as Nicky Akehurst argued in a 1997 Channel 4 documentary on taboos associated with children's bodies. Because of prevalent anxieties about child abuse in Britain, parents, and particularly fathers, may become inhibited in contact with their young children. Children do need their own space but this space may be too firmly bounded. Autonomy and the rights of the individual may be argued to the detriment of a healthy dependency. Contact with an adult, the touch of an adult, becomes intrusive in a strongly bounded world and a taboo on contact within the family extends to others beyond the family.

As Richard Sennett argues in *Flesh and Stone* (1994, pp. 19–20), 'Through our sense of touch, we risk feeling something or someone as alien' but touch is also a source of comfort and familiarity. The primacy of vision and the possibility of vicarious satisfaction gained from observing others remotely, for example, on television or the internet, removes that risk of unwanted contact but at the same time threatens to impoverish social relations. It is easy to lament this apparent decline in the tactile element of social relations but rather more difficult to write a history of the problem. Sennett has produced an intriguing history of the Jewish ghetto in Venice, built around fears of touching, fears of the *goyim* because of an association made between Jews, syphilis and leprosy in the sixteenth century and, conversely, fears for Jewish identity because of the socializing of Jewish women outside the ghetto, touching the *goyim*. In the modern period, however, it is likely that the single family home, increasing space in the home allowing more children to have their own space, and now as Kevin Robins (1996) has argued, plugging into cyberspace, have all contributed to a separation of people, particularly parents and children, accentuating the fear of touch to the detriment of social relationships and the communal use of space. This argument might be qualified. What about the touchy, feely counselling culture we are supposed to be becoming a part of, the greater tactility of women than men, and so on? Generalizations about cultural change supposedly characteristic of a society are hazardous and, obviously, observations need to be contextualized, spatially, historically and culturally. However, I would argue that the dominance of the visual and a decline in the cultural appreciation of other senses has made a particular contribution to distanciation in the construction of cultural difference and social relations. Visual images, through stereotyping, provide that simplified map of cultures that creates the illusion of certainty, sameness and difference (which is neatly contained) but, at the same time, the map contributes to anxieties about the possible rupture of boundaries. This is one manifestation of the 'impossible project' of securing certainty described by Adam Phillips.

## People in landscapes: geographies of anxiety

Suzette Heald (1994, p. 185) has argued that 'the unconscious dynamic must be shown to have some external correlative, either in the experience of the individual or in cultural practice' (although I would say in both the individual and culture rather than either/or). Thinking first about some hypothetical 'western self' through object relations, about what positions an individual in society and space through the simultaneous projection and introjection of objects, that is, people and things, I would suggest that we are drawn inevitably into thinking about cultural difference and spaces of difference as correlates of the unconscious. In understanding anxiety about place and people, it is the elision of bodily abject things and primarily visual images of cultural difference, which is manifest in the urge of people to distance themselves from others constructed negatively as threatening (but simultaneously desired). These elisions create imagined landscapes that constitute 'objects' in the terms of object relations theory.

 This assertion will hopefully make more sense through an illustration taken from an essay by Deborah Root (n.d.) that describes how imagined people and imagined places are unsettled by contact and the real experience of place. Her essay, 'Sacred landscapes/colonial dreams: the desert as escape', is a commentary on Paul Bowles' 1949 novel, *The Sheltering Sky*. Interviewed in 1989, Bowles maintained that the novel was about the desert, not about people. 'It swats them like we swat flies. It's hard to get through the desert and come out on the other side alive.' (Root, n.d., p. 25) My reading of Deborah Root's commentary on the book, however, is that *The Sheltering Sky* is as much about the power of imaginary geographies in the construction of abject and desired others (in this sense, very much like *Edward Scissorhands*) as it is about the power of the desert over people. Three 'western selves', two American men and an American woman, travel into the Sahara, where the western subject goes to escape, to forget. The desert constitutes a 'pure outside'. This view of a pure outside radically affects the westerner's response to native people in the sense that both people and place are supposed to exist in an undefiled state and in a harmonious relationship. Two of the westerners cycle into the wilderness from a town on the desert's edge, at sunset.

> They encounter two Arabs, one praying and totally oblivious to their presence, and another who looks at them indifferently as he shaves his pubic hair with a knife. The Arabs are silent and disclose nothing to the visitors, and hence function as a *spectacle* for Kit and Port that is at once impure and quotidian. Their activities, on the one hand, prayer (Baudelaire's fanaticism) and, on the other, a particular kind of bodily hygiene – both of which have, for the European, a degraded quality and consequently take place indoors and in private – here occur outdoors and on the land. (Root, n.d., p. 30)

In the colonial imaginary, the desert Arab is a part of pure nature, but only in the imagination is the Arab a 'good object', an object of desire. This imagined Arab is a part of the desert as a pure outside. Hence, the spectacle, not an engagement with but a viewing of the other, is an utterly negative experience for the westerners because they see defilement measured against their conception of purity. The cultural reversal of public and private in relation to the pubic shaving makes the Arab abject but more particularly because this act pollutes a pure space. Distanciation from abject others, which is achieved by travel into an imagined desert as an act of purification, is frustrated by the act of travelling into the desert.

This story clearly tells us nothing about the culture of the Arabs in the desert, about their world-view, their boundaries and identity. It does reveal quite a lot, however, about western selves (a term which needs endlessly qualifying), about the ways in which representations of colonialism are introjected and projected on to others, about the importance of seeing and dreaming, and how place is implicated in this process. A similar story could be told about the imaginary geographies of romanticized and abject English gypsies, placed either in romanticized and accommodating countryside, or in a rural or urban space which is purified and excluding (Sibley, 1997). Again, close encounters jar with imagined socio-spatial relationships. Similarly, Black minorities may be in the imagination confined to a Black inner city, regardless of their actual residential distribution so that the boundary of this imagined Black space becomes the locus of anxieties about possible transgressions. In all these cases, the geographies are constructed to place and displace anxieties, the anxieties of a white, western self, a model self bounded by many marks of otherness, but the exercise is self-defeating. The geographies themselves create more unease because boundaries, once created, might be transgressed.

Anxieties are accentuated by movement. In *The Sheltering Sky*, it was travel into a colonized space and the way in which this space was imagined that caused the problem of identity for the westerners. When migration is the other way, when the western abject other moves into western territory, a desired distanciation is also frustrated. As Zygmunt Bauman (cited by Robins, 1996, p. 33) wrote:

> strangers bring the outside in and, in so doing, they seem to disturb the resonance between physical and psychical distance, the sought after co-ordination between moral and topological closeness, the staying together of friends and the remoteness of enemies.

Bauman's comment touches on the important issue of the heightening of anxiety through the creation of homogeneous and exclusionary spaces which, because of their homogeneity, are then threatened by the entry of the discrepant, those who do not fit. Binary thinking still dominates the organization of space in the public sphere. It is about insides and outsides and the clear differentiation of private and public. 'Thirdspaces' are in short supply in the postmodern city and the consequences of clear separations of public and private for groups who do not conform to mainstream values can be very serious. Benedikt Fischer and Blake Poland (1998, p. 189) note that '[some] populations existentially rely on public spaces as ends to conduct essential aspects of their private lives, rather than just using public spaces as a means to connect well-connected private spaces.' Thus, the homeless who inevitably move through public spaces are stigmatized because they have to try and organize their private lives in public and increasing purification of public space, like the smoking bans in the public spaces of Canadian cities discussed by Fischer and Poland, serve to increase stigmatization and the threat of removal. Having a fag becomes a highly deviant act when both the act and the person are deemed out of place. Strangeness and discrepancy are thus produced partly through binary thinking in regard to the organization of space.

This brings us back to Bataille and Kristeva's argument about the imperative of abjection. Consciousness of a clear and unambiguous boundary of the self and, by extension, 'the community', produces feelings of discomfort and unease about those things and social categories which have been constructed as abject. At the social level, the abject is continually being reconstituted so it is always there. Groups enter and leave the abject category and they have different degrees of membership – gypsies in Romania, for example, are more likely to have abject status than the homeless on the streets of British cities but homeless people who are alcoholics or hard drug abusers probably fare worse than young runaways. However, the abject status of 'others' can generally be connected to sources of bodily abjection, to those residues that mark the boundary of the pure and defiled body. Space then compounds the problem. Spaces may themselves be constructed as defiled, thereby accentuating the abject status of their occupants, like the residual spaces occupied by many English gypsies, for example, or the abjectness of people as a social category may be accentuated because they are discrepant in a particular space, a space where they do not belong.

## Conclusion

Adam Phillips (1995, p. xvi) stresses the importance of 'the finding of languages for what matters most to us; for what we suffer from and for; for how and why we take our pleasures'. My very partial and selective reading of the psychoanalytical literature suggests to me that cultural geography can gain from psychoanalysis because the latter suggests how feelings about people and their relationship to place can be articulated. I have emphasized anxieties and the ways in which they might be produced by and contribute to the shaping of cultural space because of their political importance. At the local level, being anxious about others is important in relationship to ideas of community and the stigmatizing and marginalization of discrepant others. Pleasures are also important but, for me, rather less than questions of social injustice.

My analysis is a pessimistic one. Despite the material changes and new cultural forms which some identify as postmodern, despite the evidence of mixing and hybridity in highly developed societies, in many places lines are still clearly drawn between those who belong, or who in Tim Cresswell's words, are 'in place' and those who are 'out of place' (Cresswell, 1996). I think that this has a lot to do with 'family values' and the way in which, in western cultures, children learn to value order and to be wary of the disordered world that lies beyond the home or the neighbourhood. The emphasis in psychoanalysis on the dominant culture in western societies is helpful in this context. Academics who are a part of the cultural mainstream, as they usually are, need to be conscious of their peculiarities and of the consequences of their own cultural practices for others. Psychoanalysis unwraps bourgeois anxieties and desires and we can do more work on this by examining connections between the unconscious and the material world, the unconscious and landscape. An appreciation of these relationships might contribute to an anarchist programme designed to break down barriers, encourage social mixing and promote heterogeneity in the built environment. This is unlikely to happen but, at least, a psychoanalytically informed cultural geography might help us to better understand our own position in the world.

# References

Bataille, G. (1970), 'L'Abjection et les formes misérables', in *Essais de Sociologie, Oeuvres Completes*, Paris, Gallimard, 2–27.

Corbin, A. (1986), *The Fragrant and the Foul: Odor and the French Social Imagination*, Harvard University Press, Cambridge, Mass.

Cresswell, T. (1996), *In Place, Out of Place: Geography, Ideology and Transgression*, Minneapolis, University of Minnesota Press.

Fischer, B. and Poland, B. (1998), 'Exclusion, risk and social control – reflections on community policing and public health', *Geoforum*, 29 (2), 187–198.

Freud, S. (1913), *Totem and Taboo*, London, Routledge and Kegan Paul, 1950.

Freud, S. (1930), *Civilization and Its Discontents*, Dover, New York, 1994.

Heald, S. and Deluz, A., eds (1994), *Anthropology and Psychoanalysis*, London, Routledge.

Husbands, C. (1982), 'East End racism, 1900–1980', *The London Journal*, 8, 3–26.

Ignatieff, M. (1998), *The Warrior's Honor*, Chatto and Windus, London.

Illich, I. (1984), *H₂0 and the Waters of Forgetfulness*, Dallas Institute of Humanities and Culture, Dallas.

Jacobs, J. (1961), *The Death and Life of Great American Cities*, New York, Random House.

Kristeva, J. (1982), *Powers of Horror*, New York, Columbia University Press.

Kropotkin, P. (1899), *Fields, Factories and Workshops Tomorrow*, Freedom Press, London, 1995.

Mandrou, R. (1975), *Introduction to Modern France, 1500–1640*, London, Edward Arnold.

Merrifield, A. and Swyngedouw, E., eds (1996), *The Urbanization of Injustice*, London, Lawrence and Wishart.

Parker, I. (1997), *Psychoanalytic Culture*, London, Sage.

Perin, C. (1988), *Belonging in America*, Madison, University of Wisconsin Press.

Phillips, A. (1995), *Terrors and Experts*, London, Faber and Faber.

Rickman, J. (1957), *Selected Contributions to Psychoanalysis*, London, Hogarth Press.

Robins, K. (1996), *Into the Image*, London, Routledge.

Root, D. (n.d.), 'Sacred landscapes/colonial dreams: the desert as escape', *Lusitania*, 1 (4), 25–32.

Rustin, M. (1991), *The Good Society and the Inner Self*, London, Verso.

Sennett, R. (1994), *Flesh and Stone*, London, Faber and Faber.

Sibley, D. (1995), *Geographies of Exclusion*, London, Routledge.

Sibley, D. (1997), 'Endangering the sacred: nomads, youth cultures and the English countryside', in P. Cloke and J. Little, eds, *Contested Countryside Cultures*, London, Routledge, 218–231.

Stallybrass, P. and White, A. (1986), *The Politics and Poetics of Transgression*, London, Methuen.

Wilbur, G. and Muensterberger, W. (1951), *Psychoanalysis and Culture*, New York, International Universities Press.

Wright, E., ed. (1992), *Feminism and Psychoanalysis*, Oxford, Blackwell.

Young, I. (1990), *Justice and the Politics of Difference*, Princeton, N.J., Princeton University Press.

# Landscape mapping and symbolic form

## Drawing as a creative medium in cultural geography

*Keith D. Lilley*

## *Introduction*

This chapter is about the subjective qualities of drawing maps and it seeks to show that mapping is a creative process that feeds the geographical imagination. To do this the chapter focuses on the mapping of landscapes, since landscape has once again become an important element of contemporary cultural geography. Landscape representation in particular has received close attention in recent years, with studies of maps, paintings, novels and diaries each providing a particular 'reading' of a landscape or landscapes (see Daniels and Cosgrove, 1988; Daniels, 1993). Recently, though, the tendency has been for landscapes to be seen less as material or substantive creations, and more as visual or textual renderings. This has caused some alarm, and in a recent paper Ken Olwig (1996) set out why it is that landscapes in their material form should continue to be scrutinized in their own right. Linked with this debate about the multiplicity of landscape has been a growing awareness amongst geographers of the politics of representation, in particular in cartography (Harley, 1989; Wood, 1993). The act of mapping as a way of 'reading' landscapes has, however, been relatively neglected by geographers. Instead, the representation of landscape through drawing has attracted the attention of those actually involved in the production of new landscapes, such as landscape architects. On this matter the work of James Corner (1992) has much to offer, and his approach to the mapping of landscapes, and his use of drawing as a creative medium, serves as a reminder that drawing maps is a way of connecting with landscape, and those who shape it.

The chapter is structured in three main parts. The first part examines 'mapping and subjectivity' by considering recent criticism of cartography and the broadening scope of 'mapping' beyond its traditional format and practice. The exposure of the 'cartographic illusion' (Harley, 1989), and the increased use of mapping metaphors, are contrasted with the apparent demise of literal mapping in contemporary cultural geography. Following this, the ideas of Corner (1992) on drawing and mapping are discussed to show that producing maps is a creative, reflective process through which the geographical 'self' is constituted and mediated. The second part of the paper uses Corner's ideas to discuss 'mapping and landscape' within the context of interpreting landscape. Using the notion of 'the morphology of landscape' it is argued that mapping provides a means of interpreting 'symbolic form' (cf. Sauer, 1925 [1963]; Panofsky, 1927 [1991]). To explore this the final part of the chapter examines the symbolic form of Grenade-sur-Garonne in south-west France, a late thirteenth century town with a designed landscape. The aim of this chapter, then, is to develop the use of drawing maps as an interpretative technique in cultural geography, as well as to consider the importance of studying the form, or morphology of landscape. Both are approaches which provide a way of understanding landscape, and engaging not only with the geographical self, but also the 'historical mind' (Harris, 1978).

# Mapping and subjectivity

Maps and mapping have continued to occupy the attention of cultural geographers over the last decade but in quite different ways compared to their predecessors. In a series of highly influential papers the 'cartographic illusion' of maps was carefully deconstructed by Harley (1988, 1989, 1992), as well as others (Woodward, 1985; Turnbull, 1989; Wood, 1993). Harley's principal concern in latter years was to expose the 'politics in maps' as well as the role of 'maps in politics' (Taylor, 1992), and in particular, by using the conceptual schemas of Roland Barthes (1972), Clifford Geertz (1975) and Michel Foucault (1977), he criticizes 'the belief that mapping is merely a technical or methodological strategy [rather] than a discourse in its own right' (Harley, 1989, p. 83). He does this by showing how subjectivity formed an important, but usually overlooked, aspect of the mapping process, within both historical geography and cartographic history.

Although Harley acknowledged 'that by disengaging ourselves as historical geographers from cartographic process we have failed to grasp the illusion of cartographic representation' (ibid.), he did not himself suggest that geographers ought to involve themselves in actually drawing maps of their own. His concern was with the influence of maps and claims of cartographic 'objectivity', rather than with using mapping as a creative medium. The question that troubled Harley (ibid., p. 82) was 'how are we representing the world when we make maps of some aspects of its "reality"?', rather than, what can the process of *drawing* a map reveal to us as geographers? For Harley, then, the map was a product that had flaws, and these he rigorously exposed by viewing the 'map as text' (ibid., pp. 84–5). But the map is more than a product: it is itself a vehicle for constructing and developing ideas, an imaginative tool, a means to an end and not necessarily an end in itself. It is this aspect of subjectivity and mapping that requires further thought.

In recent years 'mapping' in cultural geography has become more popularly conceived as a metaphorical device. Indeed, the 'literary turn' of the late 1980s not only provided Harley (1992) with a vocabulary and conceptual apparatus necessary to 'deconstruct the map', it also provided a new way of writing geography and of conceiving 'mapping'. A series of books and articles in the early 1990s made use of spatial and geographical metaphors to acknowledge the geographer's 'speaking position' (Pratt, 1992; Barnes and Duncan, 1992). Playing with metaphors makes it possible to view 'mapping' much more broadly than if the map is simply taken literally. In particular, the notion of 'mapping' has been used as a linguistic device to convey the qualities and complexities of defining geographies of sexuality (Bell and Valentine, 1995) and cultural trans-formation (Pile and Thrift, 1995). Some, however, have started to question whether the use of metaphors in geographical writing is compromising clarity, by pointing out the difficulties and ambiguities caused by conflating metaphor-ical and literal interpretations of particular spatial and geographical expressions (Keith and Pile, 1993; Farrar, 1997). The trend towards metaphorical mapping in cultural geography also raises the question of whether geographers are losing sight of the value of mapping in its literal sense. Indeed, does (literal) mapping have a future in the 'new cultural geography' (cf. Mitchell, 1995)? And if it does, what does mapping have to offer?

Producing a map, like writing a text, is a discursive practice that is both creative and imaginative. A recent call to acknowledge the subjectivities in geo-graphical writing, and to recognize the creative dimension of writing geography (Cosgrove and Domosh, 1993), provides a basis for also arguing that drawing maps may be equally creative. In this respect the work of James Corner (1992) is particularly helpful. In his paper, 'Representation and landscape: drawing and making in the landscape medium', Corner considers landscape from an 'archi-tectural point of view', as 'something to be made or designed', and that draw-ing is a way of actually producing a landscape. Essentially, his argument is that drawing provides a creative medium to develop ideas. In his case, as a landscape architect, Corner is primarily concerned with how drawing stimulates thoughts for creating landscapes. For cultural geographers, drawing provides a means of fashioning thoughts about landscapes and reflecting upon them.

Drawing landscapes, whether as sketches, paintings or maps, has been an essential part of writing geography since the nineteenth century, and these representational, 'paper' landscapes have recently been the focus for detailed and critical study by cultural geographers (Godlewska, 1995; Martins, 1998; Söderström, 1996). Yet it would seem that few geographers today consider that *the act* of drawing can still play a critical role in attempting to understand and make sense of the world. Corner's (1992, p. 243) view is that, 'drawing – a textual medium which is secondary to the actual landscape – can never be simply and alone a case of reflection and analysis'. There is more to drawing a map, then, than simply communicating information cartographically. Indeed, drawing 'is more fundamentally an eidetic and *generative* activity, one where the drawing acts as a producing agent or ideational catalyst' (ibid., pp. 243–4). Drawing a map therefore provides a way of engaging with what is 'out there' (the make-up of a landscape for example), as well as what is 'inside us' (the geographical self). It is worth considering this in a little more detail here.

First, producing a map is a physical, tactile process. It involves material, tangible things such as pens, tracing paper, sticky tape and, of course, the information that is being mapped, perhaps, as in the context of this paper, part of a landscape. Second, drawing a map is also imaginative and serves as a creative medium. This is recognized by Corner who sees drawing as an 'eidetic' activity, or process, by which he means drawing 'pertains to the visual forma-tion of ideas' (ibid., note 3). By bringing together pen and paper, and actually tracing out a map of a particular landscape, ideas are formed about the land-scape itself. As one continues drawing the map new ideas emerge and further thoughts are stimulated. The process of creating a map is thus a dialogue, an interaction 'arising from both a reactive response to the medium [of landscape] and from an imaginative source from deep within':

> Here the body and the imaginal are joined, inextricably involved with one another in a concentrated and creative, yet unselfconscious, unity. The making itself is a dialogue, a perceptive conversation between the medium and the imagination that cannot be intellectualized or thought of as external to experience. (ibid.)

Drawing a map therefore *involves* the subjective (self), making mapping a personal, reflective experience. As a result the map is a 'map' of experiences, and of course can be read as such. Its materiality, its texture and feel, is a crit-ical part of the map's ability to provoke dialogue, and as a result, in turn, that dialogue becomes imprinted on the map. The course of this 'conversation' stems from the eidetic activity of drawing. For me, drawing is a peaceful, calming and engaging process, where ideas are literally (and metaphorically) *drawn out*. Drawing a map thus has a role in constituting and mediating the geographical self. When we draw our maps, however, just as 'when we write our geograph-ies, we are creating artifacts that impose meaning on the world' (Cosgrove and Domosh, 1993, p. 37).

As well as having the capacity to generate ideas, maps are also a means through which intentionality and subjectivity can be acknowledged. Using the work of two contemporary architects, Carlo Scarpa and Mario Ridolfi, Corner (1992, pp. 270–1) notes how their 'dynamic drawings . . . are made neither for construction nor presentation, but rather for the disciplined *work* of the architect'. He goes on to explain that drawings embody decisions and actions ('utterances'), and that these can be expressed in the drawing itself. For this Corner uses 'the term deictic' from the Greek '*dikononei*, meaning to show, to make evident' (ibid.). Drawing acts not only as a catalyst for the creation of ideas, therefore, but also charts the 'work' of the geographer. In this context, drawing actually capitalizes on subjectivities, and drawing maps enables the geographer to undertake journeys of self-realization and discovery.

## Mapping and landscape

In Corner's view drawing 'holds the possibility of forming a field of revelation, prompting one to figure previously unforeseen landscapes of a richer and more meaningful dimension' (Corner, 1992, p. 245). One crucial component of this process of revealing landscape is the study of form, or morphology, as this provides a way of making sense of the complexity of landscape however it is represented, whether in art or sculpture, or on a painting or map; or however it is viewed, whether from above or from the ground. Form, it is suggested here, is a basis for ordering what we see and interpreting the meaning of landscape, while mapping provides a basis from which to study the morphology of landscape. 'Landscape', however, is itself a contested term, and needs some consideration here.

In his essay Corner emphasizes the multiplicity of 'landscape'. Landscape lives through experience and interaction, and is 'constructed' and configured by representation. That is, 'the actual lived landscape is the medium of both construal and construction; the representation is not only encoded in various related textual media, such as literature and painting, but is more significantly embodied in the constructed landscape' (ibid., p. 243). In recent years cultural geographers have intensely debated the nature of 'landscape'. Some have suggested that 'a landscape is a cultural image, a pictorial way of representing, structuring or symbolizing surroundings', in which, for instance, 'a landscape park is more palpable but no more real, nor less imaginary, than a landscape painting or poem' (Daniels and Cosgrove, 1988, p. 1). Others have tried to argue for a more 'substantive' landscape which exists outside of (but is partly constituted by) representations of it (Olwig, 1996). Whichever view one chooses, maps provide a way of making sense of landscapes. A map may have been drawn by someone else, perhaps as a plan for an intended landscape, or as a means of conveying the topography of a landscape; or the map may have been self-drawn, as an eidetic activity, and to communicate a particular set of ideas about a landscape.

During the first half of the twentieth century, mapping was the principal medium for interpreting a landscape, and in cultural geography this interpretation usually involved the study of landscape form ('morphology'). Today, morphology is sometimes thought of as a rather descriptive activity, but this is far from how it was first seen by highly influential geographers, such as Carl Sauer. He and others shared the view that morphology provided a way of understanding the evolution of landscapes, as well as a method for interpreting the cultural factors that shaped the landscape. Morphology, it was argued, revealed how 'a rejuvenation of the cultural landscape sets in, or [how] a new landscape is superimposed on [the] remnants of an older one' (Sauer, 1925 [1963], p. 343).

Studying the form of landscape makes it possible to study cultural change both temporally and spatially. In Europe, before the Second World War, the 'morphologic method' (ibid., p. 326) was used by a number of scholars in the study of urban landscapes, including Otto Schlüter, whose ideas subsequently played a critical part in the shaping of Richard Hartshorne's (1939, 1959) 'nature of geography' and, by implication, the work of other English-speaking cultural geographers in the post-war period (see Elkins, 1989, pp. 25–8).

The work of Otto Schlüter (1899, 1903), as well as other geographers of the period, such as Walter Geisler (1918, 1924), focused on the morphology of the cultural landscape (*Kulturlandschaft*) and in particular the urban landscape (*stadtlandschaft*). Here, the word *landschaft* was meant 'in the sense for which we would use "region"' (Elkins, 1989, p. 26), and Schlüter saw the urban land-scape 'as a distinct category of the cultural landscape and as such a regional unit in its own right' (Whitehand, 1981, p. 2). In his research Schlüter found that differing settlement forms in the Unstrut valley emerged during the middle ages from the contrasting cultures of German and Slav settlers. Instrumental in this work was the use of maps to analyze the form of individual settlements, and also to show regional variations in different types of settlement form (Schlüter, 1899). This approach was soon taken up by geographers and historians in Germany and Austria, not least for political reasons, during the early decades of the twentieth century (for example, Geisler, 1924 and Höenig, 1921), but as Whitehand (1981, p. 4) has noted, what tended to happen was that they 'allowed themselves to be pushed by the enormous scope of their projects into merely morphographic classification, producing profuse nomenclature with little meaning' (Figure 1). This was an 'unfortunate failure of purpose . . . in spite of Schlüter's clearly and repeatedly expressed concept of *Kulturgeographie*' (ibid.), and despite his emphasis on 'causal-genetic explanation' (Elkins, 1989, p. 27).

In his essay on the 'morphology of landscape', Sauer (1925 [1963], p. 342) was quite taken with Geisler's work, particularly his study *Die deutsche Stadt*, subtitled, 'a contribution to the morphology of the cultural landscape' (Geisler, 1924). But for the advance of morphological study it is actually Geisler's thesis on Danzig (Gdansk), supervised by Schlüter at Halle, and 'investigating specific aspects of urban form in much greater detail than had been done before', that Whitehand (1981, p. 3) sees as the more innovative piece of work. During the 1950s and 1960s, Geisler's study of Danzig attracted the attention of certain English-speaking urban geographers, such as Dickinson (1951) and Conzen (1960, 1968), and thus the study of form came to influence the work of geo-graphers in Britain (see Whitehand, 1981, p. 13). Studies of settlement form by rural historical geographers in Britain sought to classify settlement forms (Thorpe, 1949; Roberts, 1972, 1987), whilst those geographers working in the developing subdiscipline of 'urban morphology' focused more on processes of morphological change, or 'morphogenesis' (see Whitehand, 1981, pp. 7–18), using analyses of building fabric and urban layout (built form) as the basis for reconstructing aspects of the historical evolution of towns and cities (Slater, 1990; Whitehand and Larkham, 1992; Vernez-Moudon, 1997).

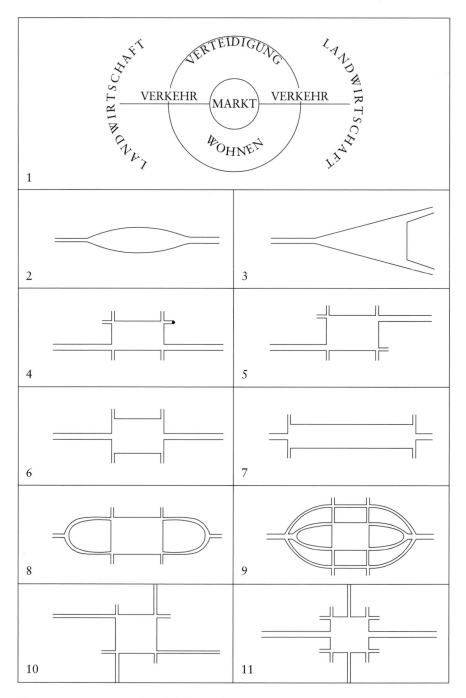

**Figure 1** Street systems classified by Anton Höenig in his study of German settlement in Bohemia.
Source: Höenig (1921)

Yet, despite providing a wealth of new information on how medieval and modern urban landscapes are formed and transformed, this genre of landscape study, so reliant on mapping as a mode of analysis and representation, has recently become viewed by some cultural geographers as 'a somewhat narrow perspective' for interpreting landscape in view of the variety of other ways in which landscapes can be 'read' (Driver, 1995, p. 769). This criticism derived from new studies of landscape by cultural geographers in the 1980s and 1990s, in which the focus of study was on representations of landscapes, through imagery and text (see for example, Cosgrove and Daniels, 1988; Cosgrove 1993; Daniels, 1993; Hemingway, 1992). Instead of seeing landscape as an object for studying 'its cycle of development', as Sauer (1925 [1963], p. 343) did, landscape in the 'new' cultural geography came to be seen as 'an idea', and the main consideration became 'what it means to imagine "a landscape"' (Driver, 1995, p. 764). This reconsideration of 'landscape' as a cultural image and mode of human signification was the subject of a highly influential volume of collected papers, brought together under the appealing title of the *Iconography of Landscape* (Cosgrove and Daniels, 1988).

However, although the 'iconographic' study of representations of landscape and the 'morphologic' study of residual past landscapes may appear to be at odds with each other (in conceptual and methodological terms), the two genres do actually share common ground, and if taken together provide a potential basis for future research on landscape in cultural geography. A connection between these two genres of landscape study is provided by the work of Erwin Panofsky, an art historian writing during the mid-twentieth century and whose ideas on iconography and iconology were taken up by Daniels and Cosgrove (1988, pp. 2–7). By linking morphology and iconography both conceptually and methodologically, I will now show how morphology can provide an important, critical approach for interpreting landscape as a cultural image.

## Symbolic form

In their introduction to the *Iconography of Landscape*, Daniels and Cosgrove (1988, pp. 2–3) draw attention to the approach adopted by Panofsky (1970, pp. 51–2) to read ' "what we see" according to the manner in which objects and events are expressed by forms under varying historical traditions'. To interpret the iconography (the symbolic value) of landscape through art and writing, they argue, relies to some extent on the recognition and interpretation of form (see also Cosgrove, 1985). However, how this is done is not made altogether clear, and the vexed issue of how to 'read' landscape proves to be elusive. There are pointers however, and these relate to the writing of Erwin Panofsky and his colleague, Ernst Cassirer. Both Panofsky and Cassirer, as Daniels and Cosgrove (1988, p. 2) point out, were concerned with *symbolic form*, 'the notion of a core symbolizing activity', where 'the different spheres of human creativity were the "forms" produced by this activity' (Wood in Panofsky, 1925 [1991], p. 14). In art, as Panofsky showed, form could be used as a basis for reading 'what we see', and Daniels and Cosgrove (1988, pp. 3–5) put this forward as the basis for interpreting landscape painting. But the work of Panofsky and Cassirer on form has a much wider relevance in the context of interpreting landscape than studying past representations of landscape. Their work can also be used to interpret the morphology of landscape.

In *Philosophie der symbolischen Formen*, Cassirer (1923 [1953], p. 53) set out to show that 'in every case "symbolic form" is a condition either of the knowledge of meaning or of the human expression of meaning'. In doing so he focused on language, myth, art and science. With these he concerned himself particularly with space, time and number, and in the second volume of his *Philosophie*, Cassirer (1925 [1955]) used 'mythical thought' in order to follow through the spatial, temporal and numerical dimensions of symbolic form. Specifically, he considered how, in the context of ancient Rome, for example, 'every new boundary established in space by mythical thinking and mythical-religious feeling became an ethical and cultural boundary' (ibid., p. 101). In doing this, Cassirer made connections between the symbolic form of ancient cosmologies and the ground plans of Roman camps and cities, and Christian churches (ibid., pp. 101–2), spelling out how 'everywhere such sacral conceptions are bound up with the general view of space and distinct spatial boundaries'. Concrete forms are thus earthly realizations of mythical forms. Both are symbolic, he argued, and 'both go back to a basic sensuous-spatial idea' (ibid., p. 103). Cassirer's notion of symbolic form was quickly and explicitly taken up by Panofsky (1927 [1991]) in his essay on *Perspective as Symbolic Form* (*Die Perspektive als 'symbolische Form'*).

For Panofsky, perspective was 'one of those "symbolic forms" in which "spiritual meaning is attached to a concrete, material sign and intrinsically given to this sign" ' (ibid., p. 41, citing Cassirer). Panofsky specifically deals with this in the third section of his essay which begins with 'a general morphology of medieval art . . . conducted in terms of framing devices, surface values, the binding power of the plane, coloristic unity, the homogeneity of space [and] the emancipation of bodies from mass' (Wood in Panofsky, 1927 [1991], p. 21). In using morphology, Panofsky aimed to dig beneath the surface of representational space (ibid., pp. 51–5). He continued to do this in his later work, too. On medieval architecture (Panofsky, 1957) 'he argued that designers of gothic cathedrals "began to conceive of the forms they shaped, not so much in terms of isolated solids as in terms of a comprehensive 'picture space' " ' (Daniels and Cosgrove, 1988, p. 3). Furthermore, 'while acknowledging its status as building', Panofsky 'found it fertile to regard gothic architecture as text, not just "a way of seeing – or designing", but as a mode of literary representation, a treatise in stone, an architectural scholasticism' (ibid.).

Recently, some use of this idea has been made in cultural geography to study urban landscapes, principally by Duncan (1990, 1993) whose work on the built form of nineteenth century Kandy blended 'the morphology and symbolism of the city' (Duncan, 1993, p. 236). Duncan demonstrates how 'the city was composed of two rectangles, the sacred shape of the cities of the gods', and how 'the temples of the four gods . . . mark out the four cardinal directions of the sacred rectangle and render it a representation of the heavens on top of the cosmic mountain at the centre of the universe' (ibid.). This reading is similar to the one Cassirer (1925 [1955], p. 103) undertakes of the structure and order of the Roman camp, in which he states that 'the plan of the camp was drawn up according to that of the city, while the city in turn was constructed according to the general plan of the world and the different spatial zones of the cosmos' (see also Wheatley, 1971). There remains, however, much scope for this sort of approach to studying landscape, to which morphology provides the key. To end this chapter, and to explore some of these ideas in a little more detail, I shall turn to consider the symbolic form of a European medieval town. This requires some consideration of landscape mapping, for as will be seen, mapping provides a basis from which to interpret symbolic form.

Grenade-sur-Garonne is a town in south-west France between Bordeaux and Toulouse, situated on a tributary of the Gironde. It was founded as a new town in 1290–1 by Eustache de Beaumarchais in collaboration with the abbot of Grande Selve at a time when many new towns (bastides), were being established in Gascony on behalf of the English and French kings, whose lands were divided in this region (Lauret, Malebranche and Seraphin, 1988, p. 288; see also Randolph, 1995). The study of the form of bastides has a long pedigree in France stretching back to the pioneering work of Pierre Lavedan (1926), and the towns have been a source of interest for economic and social historians, too (Beresford, 1967; Higounet, 1992). However, the case of Grenade is peculiar because it has also been the subject of careful scrutiny by architectural and art historians (Bucher, 1972; Friedman, 1988; Randolph, 1995), all of whom have seen it as an example of a town that was laid out using a system of proportion based on the ratio 1:√2. This configuration is clear if the layout of the town is considered morphologically. To gain access to the town's medieval form it is first necessary to draw a plan of Grenade's urban landscape.

In drawing a plan of Grenade certain connections are made with its medieval built form (Figure 2). The town was not depicted cartographically until some centuries after it was first laid out. In 1827, as part of a national programme of cadastral mapping, Grenade was surveyed to show the disposition of various parcels of land that made up the town at the time (see Lauret, Malebranche and Seraphin, 1988, pp. 166–7). Drawn to the scale of 1:2500, and beautifully executed in colour ink-wash, the plan of 1827 shows in immense detail the morphology of Grenade. Although this plan is nineteenth century in date the features that it depicts, the streets, the plots and many of the buildings, date back to the late thirteenth century. The cadastral plan of 1827 is thus a more or less complete plan of Grenade at the time of its foundation. A sense of the town's long straight streets, its rigid orthogonality, is mediated through drawing off the morphological features from the 1827 plan and re-presenting them: as one's pen follows the lines of medieval streets and plots, and the new plan appears as an outline on a piece of tracing paper, the morphology of Grenade becomes all the more visible and all the more sensible (Figure 2). The physical act of tracing out Grenade's medieval layout of streets and plots helps to stimulate thoughts about the meaning of the town's plan, its symbolic form.

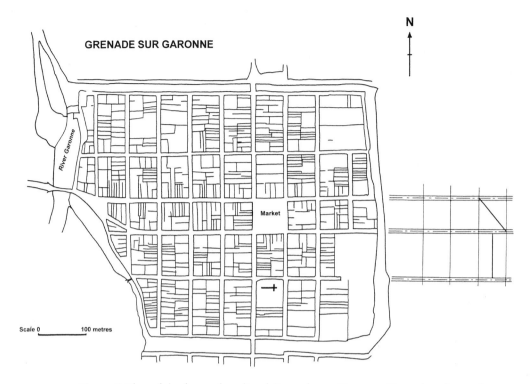

**Figure 2** Plan of the form of medieval Grenade-sur-Garonne (France), redrawn from a cadastral survey of 1827, and showing the proportional relationship between axial streets.

The drawing and redrawing of Grenade's plan is itself an 'eidetic activity', in the sense that the drawing process acts as a catalyst for generating ideas. Humanistic geographers have recognized the subjective qualities of this sort of experiential process, something which Cole Harris (1978) noted in his essay on 'the historical mind and the practice of geography' where he suggested that engaging bodily with historical material makes it possible to connect subjectively with the past. By drawing out Grenade's medieval built form it becomes possible to think through how and why such a plan was conceived and laid out, to reflect on its symbolic form and enter the minds of those who first devised the layout of the town.

To understand the form of medieval Grenade it is necessary not only to draw its plan but to place it in the context of medieval cosmology. In the same way that gothic cathedrals were designed according to geometry, the plan of Grenade was devised using a geometrical rule where 'the right rectangle or diagon is produced by a square whose diagonal determines the side of the rectangle' (Bucher, 1972, p. 43). Such use of proportion was an important part of medieval architectural design practice and it often appears in the ground plans of cathedral buildings (see Guidoni, 1970; Fernie, 1976). Geometry in the middle ages was used to symbolize divine order, in which God was the architect of the universe; *artifex principalis* (Frayling, 1995; Gimpel, 1983). The geometry that governs the medieval plan of Grenade placed the town (and its people) closer to God. Through the rhythmic proportion of its form the beauty of truth is articulated. In texts written at the time when Grenade was laid out, the beauty of form was taken as a sign of God. For example, in his *De Pulchro et Bono* Albertus Magnus wrote,

> just as corporeal beauty requires a due proportion of its members and splendid colours . . . so it is the nature of universal beauty to demand that there be mutual proportions among all things and their elements and principles, and that they should be resplendent with the clarity of form. (in Eco, 1986, p. 25)

The importance of geometry in medieval cosmology was articulated in concrete forms, as Cassirer (1925 [1955], p. 102) himself recognized in his comments on the configuration of church ground plans. By creating the ground plan of a town according to geometry it was possible to make the 'invisible' visible, something that was deeply believed and understood in the middle ages (see Lilley, 1997, 1998). In his commentary on the 'Celestial Hierarchy' (*In Hierarchiam Coelestium*), Hugh de St Victor noted that

> all things visible, when they obviously speak to us symbolically, that is when they are interpreted figuratively, are referable to invisible significations and statements . . . For since their beauty consists in the visible form of things . . . visible beauty is an image of invisible beauty. (in Eco, 1986, p. 58)

In terms of its geometric exactitude, the plan of Grenade can only be fully appreciated when viewed from above, which in the middle ages meant when the plan was viewed by God. Here, then, in the morphology of a medieval town is an expression of symbolic form which takes us back to the thirteenth century. By studying form it is possible to interpret the meaning of landscape, and through the medium of mapping it is possible to connect with the past and touch the minds of those whose understanding of the world, its spatial organization and symbolic form, was not that different, if we are to believe Cassirer (1925 [1955]), from those who lived before and those who came after.

## Conclusion

This chapter has suggested a new agenda for future work in cultural geography. It has been argued that there is much to be gained from the eidetic activity of drawing, and much to be learned from studying form. Both practices have in recent years fallen out of favour not just in cultural geography but human geography as a whole. Yet, both can provide geographers with a way of engaging with the world as well as feeding their own geographical imagination. In the context of drawing, the work of James Corner (1992) ought to be appreciated more widely, for it offers a theoretical and practical demonstration of the subjective qualities of representing landscape. The use of maps in the work of contemporary artists, as well as the links between mapping, art and imagination, are beginning to be discussed by cultural geographers (see Crouch and Matless, 1996; Nash, 1998), but rather than encouraging geographers to engage themselves directly in the representation of landscape by drawing maps of their own, much of this work relates more broadly to the place of mapping and art in the study of landscape representation (for example, Alfrey and Daniels, 1990). To an extent, what I am suggesting here is that geographers *themselves* could use drawing and mapping, not only as a way of understanding landscape but for *drawing out* one's 'geographical self', and gaining an experience of place and time. This suggestion is analogous to what Raphael Samuel (1994) meant when he pointed out that more historians ought to be engaging with mock battles, or driving steam locomotives, or building replica villages, for it is these sorts of activities that provide us with an experiential capacity necessary to make sense of our subject.

In the context of landscape, this chapter has put forward a case for the study of substantive, concrete landscapes. This is not to say that landscapes can be, or should be, ignored as representational spaces. It is necessary to consider landscape both as a material and representational creation. In doing so, the study of form, or morphology, becomes fundamental.

Although the use of form to interpret landscape has been accepted by cultural geographers for most of the twentieth century, latterly it has been increasingly sidelined. However, the blending of morphology and iconography deserves renewed consideration, bringing in the work of urban morphologists, for instance, alongside the work of those art historians and urban theorists who have been highly influential on landscape study in the 'new cultural geography'. As 'ways of seeing' landscape, morphology and iconography are conceptually closely connected, as is evident in the work of Erwin Panofsky (1927 [1991], 1957). Morphology thus provides a means of linking conceptions of space with concrete spatial forms, and one particular avenue where this approach has clear potential is in the study of urban form. To some extent this has been recognized in recent studies of the symbolic forms of temples, towns and cities in India, Sri Lanka and Mexico (Nitz, 1992; Duncan, 1993; Low, 1993), but within the context of Europe there seems to be a certain reluctance amongst English-speaking cultural geographers to use morphology as a 'way of seeing' landscape. This leaves open a rich and rewarding task for cultural geographers of the future; to take on board different genres of landscape study and interpretation, and use the activity of drawing maps of landscapes as a way of engaging the 'historical mind' as well as the geographical imagination.

## *Acknowledgements*

I am grateful to members of the landscape surgery for comments on material contained in an early draft of this chapter, especially Dr Jeremy Foster and Professor Denis Cosgrove. My thanks are due also to Professor Jeremy Whitehand and members of the

Urban Morphology Research Group for their views. The chapter has been written whilst in receipt of a British Academy Post-Doctoral Fellowship at the Department of Geography, Royal Holloway, University of London.

# *References*

Alfrey, N. and Daniels, S. (eds) (1990) *Mapping the Landscape: essays on art and cartography*, University Art Gallery, Nottingham.

Barnes, T. and Duncan, J. (eds) (1992) *Writing Worlds: discourse, text and metaphor in the representation of landscape*, Routledge, London.

Barthes, R. (1972) *Mythologies*, translated by Lavers, A., Cape, London.

Bell, D. and Valentine, G. (eds) (1995) *Mapping Desire: geographies of sexuality*, Routledge, London.

Beresford, M.W. (1967) *New Towns of the Middle Ages*, Lutterworth, London.

Bucher, F. (1972) 'Medieval architectural design methods, 800–1560', *Gesta*, 11, 37–51.

Cassirer, E. (1923) *Philosophie der symbolischen Formen: Die Sprache*, translated by Manheim, R., *The Philosophy of Symbolic Forms, volume one: language*, Yale University Press, London, 1953.

Cassirer, E. (1925) *Philosophie der symbolishen Formen: Das mythische Denken*, translated by Manheim, R., *The Philosophy of Symbolic Forms, volume two: mythical thought*, Yale University Press, London, 1955.

Conzen, M.R.G. (1960) *Alnwick, Northumberland: a study in town-plan analysis*, Institute of British Geographers special publication, 27, Alden Press, Oxford.

Conzen, M.R.G. (1962) 'The plan-analysis of an English city centre', in Norberg, K. (ed.), *Proceedings of the IGU Symposium in Urban Geography, Lund, 1960*, Lund, 383–414.

Conzen, M.R.G. (1968) 'The use of town plans in the study of urban history', in Dyos, H.J. (ed.), *The Study of Urban History*, Edward Arnold, London, 113–30.

Conzen, M.R.G. (1988) 'Morphogenesis, morphological regions and secular human agency in the historic townscape, as exemplified by Ludlow', in Denecke, D. and Shaw, G. (eds) *Urban Historical Geography*, Cambridge University Press, Cambridge, 253–72.

Corner, J. (1992) 'Representation and landscape: drawing and making in the landscape medium', *Word and Image*, 8, 243–75.

Cosgrove, D. (1985) 'Prospect, perspective and the evolution of the landscape idea', *Transactions of the Institute of British Geographers*, new series, 10, 45–62.

Cosgrove, D. (1993) *The Palladian Landscape: Geographical Change and its Cultural Representations in 16th Century Italy*, Leicester University Press, Leicester.

Cosgrove, D. and Daniels, S. (eds) (1988) *The Iconography of Landscape*, Cambridge University Press, Cambridge.

Cosgrove, D. and Domosh, M. (1993) 'Author and authority: writing the new cultural geography', in Duncan, J. and Ley, D. (eds) *Place/Culture/Representation*, Routledge, London, 25–38.

Crouch, D. and Matless, D. (1996) 'Refiguring geography: parish maps of common ground', *Transactions of the Institute of British Geographers*, 21, 236–55.

Daniels, S. (1993) *Fields of Vision: landscape imagery and national identity in England and the United States*, Polity Press, Oxford.

Daniels, S. and Cosgrove, D. (1988) 'Introduction: iconography and landscape', in Cosgrove, D. and Daniels, S. (eds), *The Iconography of Landscape*, Cambridge University Press, Cambridge, 1–10.

Dickinson, R.E. (1951) *The West European City: a geographical interpretation*, Routledge and Kegan Paul, London.

Driver, F. (1995) 'Visualizing landscape', *Journal of Urban History*, 21, 764–71.

Duncan, J. (1990) *The City as Text: the politics of landscape interpretation in the Kandyan Kingdom*, Cambridge University Press, Cambridge.

Duncan, J. (1993) 'Representing power: the politics and poetics of urban form in the Kandyan Kingdom', in Duncan, J. and Ley, D. (eds), *Place/Culture/Representation*, Routledge, London, 232–48.

Eco, U. (1986) *Art and Beauty in the Middle Ages*, Yale University Press, London.

Elkins, T.H. (1989) 'Human and regional geography in the German-speaking lands in the first forty years of the twentieth century', in Entrikin, J.N. and Brunn, S.D. (eds), *Reflections on Richard Hartshorne's Nature of Geography*, Annals of the Association of Human Geographers occasional publication, Washington DC, 17–34.

Farrar, M. (1997) 'Migrant spaces and settlers' time: forming and deforming an inner city', in Westwood, S. and Williams, J. (eds) *Imagining Cities: scripts, signs, memory*, Routledge, London, 104–24.

Fernie, E.C. (1976) 'The ground-plan of Norwich cathedral and the square-root of two', *Journal of the British Archaeological Association*, 129, 77–86.

Foucault, M. (1977) *Discipline and Punish: the birth of the prison*, Allen Lane, London.

Frayling, C. (1995) *The Strange Landscape: a journey through the Middle Ages*, BBC Books, London.

Friedman, D. (1988) *Florentine New Towns: urban design in the late Middle Ages*, MIT Press, Cambridge, MA.

Geertz, C. (1975) *The Interpretation of Cultures: selected essays*, Hutchinson, London.

Geisler, W. (1918) *Danzig: ein siedlungsgeographischer*, Danzig.

Geisler, W. (1924) *Die deutsche Stadt: ein Beitrag zur Morphologie der Kulturlandschaft*, Engelhorn, Stuttgart.

Gimpel, J. (1983) *The Cathedral Builders*, Pimlico, London.

Godlewska, A. (1995) 'Map, text and image: the mentality of enlightened conquerors: a new look at the *Description de l'Egypte*', *Transactions of the Institute of British Geographers*, new series, 20, 5–28.

Gregory, D. (1994) *Geographical Imaginations*, Blackwell, Oxford.

Guidoni, E. (1970) *Arte e Urbanistica in Toscana, 1000–1315*, Bulzoni, Rome.

Harley, J.B. (1988) 'Maps, knowledge, power', in Cosgrove, D. and Daniels, S. (eds), *The Iconography of Landscape*, Cambridge University Press, Cambridge, 277–312.

Harley, J.B. (1989) 'Historical geography and the cartographic illusion', *Journal of Historical Geography*, 15, 80–91.

Harley, J.B. (1992) 'Deconstructing the map', in Barnes, T.J. and Duncan, J. (eds), *Writing Worlds*, Routledge, London, 231–47.

Harris, C. (1978) 'The historical mind and the practice of geography', in Ley, D. and Samuels, M.S. (eds), *Humanistic Geography: prospects and problems*, Croom Helm, London, 123–47.

Hartshorne, R. (1939) 'The nature of geography: a critical survey of current thought in the light of the past', *Annals of the Association of American Geographers*, 29 (3 and 4), 171–658.

Hartshorne, R. (1959) *Perspective on the Nature of Geography*, Association of American Geographers, Murray, London.

Hemingway, A. (1992) *Landscape Imagery and Urban Culture in Early Nineteenth-Century Britain*, Cambridge University Press, Cambridge.

Higounet, C. (1992) *Villes, Sociétés et Economies Médiévales*, Fédération historique de Sud-Ouest, Bordeaux.

Höenig, A. (1921) *Deutscher Städtebau in Böhmen*, W. Ernst, Prague.

Keith, M. and Pile, S. (eds) (1993) *Place and the Politics of Identity*, Routledge, London.

Lauret, A., Malebranche, R. and Seraphin, G. (1988) *Bastides: villes nouvelles du moyen age*, Éditions Milan, Toulouse.

Lavedan, P. (1926) *Histoire de L'urbanism: volume II antiquité – moyen âge*, H. Laurens, Paris.

Lilley, K.D. (1997) 'Geometry, urban planning and town design in the high Middle Ages', *Planning History*, 20(1), 4–12.

Lilley, K.D. (1998) 'Taking measures across the medieval landscape: aspects of urban design before the Renaissance', *Urban Morphology*, 2(2), 82–92.

Low, S.M. (1993) 'Cultural meaning of the plaza: the history of the Spanish-American gridplan-plaza urban design', in Rotenberg, R. and McDonogh, G. (eds), *The Cultural Meaning of Urban Space*, Bergin and Garvey, London, 75–93.

Martins, L. (1998) 'Navigating tropical waters: British maritime views of Rio de Janeiro', *Imago Mundi*, 50, 141–55.

Mitchell, D. (1995) 'There's no such thing as culture: towards a reconceptualization of the idea of culture in geography', *Transactions of the Institute of British Geographers*, new series, 20, 102–16.

Nash, C. (1998) 'Mapping emotion', *Environment and Planning D: Society and Space*, 16(1), 1–9.

Nitz, H.-J. (1992) 'Planned temple towns and Brahmu villages as spatial expressions of the ritual polities of medieval kingdoms in south India', in Baker, A. and Biger, G. (eds), *Ideology and Landscape in Historical Perspective*, Cambridge University Press, Cambridge, 107–24.

Olwig, K. (1996) 'Recovering the substantive nature of landscape', *Annals of the Association of Americal Geographers*, 86(4), 630–53.

Panofsky, E. (1927) *Perspective as Symbolic Form*, translated by Wood, C., Zone Books, New York, 1991.

Panofsky, E. (1957) *Gothic Architecture and Scholasticism*, World Publishing Co., New York.

Panofsky, E. (1970) 'Iconography and iconology: an introduction to the study of Renaissance art', in Panofsky, E. (ed.), *Meaning in the Visual Arts*, Penguin, Harmondsworth, 51–81.

Pile, S. and Thrift, N. (eds) (1995) *Mapping the Subject: geographies of cultural transformation*, Routledge, London.

Pratt, G. (1992) 'Spatial metaphors and speaking positions', *Environment and Planning D: Society and Space*, 10, 241–4.

Randolph, A. (1995) 'Bastides in south-west France', *Art Bulletin*, 77, 290–307.

Roberts, B.K. (1972) 'Village plans in County Durham: a preliminary statement', *Medieval Archaeology*, 16, 33–56.

Roberts, B.K. (1987) *The Making of the English Village*, Longman, London.

Samuel, R. (1994) *Theatres of Memory: past and present in contemporary culture*, Verso, London.

Sauer, C. (1925) 'The morphology of landscape', in Leighly, J. (ed.), *Land and Life: a selection from the writings of Carl Ortwin Sauer*, University of California Press, Berkeley, 1963, 315–50.

Schlüter, O. (1899) 'Bemerkungen zur Siedlungsgeographie', *Geographischer Zeitschrift*, 5, 65–84.

Schlüter, O. (1903) *Die Siedlungen im nordöstlichen Thüringen: ein Beispiel für Behandlung siedlungsgeographischer Fragen*, Mitler, Berlin.

Slater, T.R. (ed.) (1990) *The Built Form of Western Cities*, Leicester University Press, Leicester.

Söderström, O. (1996) 'Paper cities: visual thinking in urban planning', *Ecumene*, 3, 249–81.

Taylor, P.J. (1992) 'Politics in maps, maps in politics: a tribute to Brian Harley', *Political Geography*, 11, 127–9.

Thorpe, H. (1949) 'The green villages of County Durham', *Transactions of the Institute of British Geographers*, 15, 155–80.

Turnbull, D. (1989) *Maps are Territories: science as an atlas*, Deakin University Press, Geelong, Victoria (Australia).

Vernez-Moudon, A. (1997) 'Urban morphology as an emerging interdisciplinary field', *Urban Morphology*, 1(1), 3–10.

Wheatley, P. (1971) *The Pivot of the Four Quarters: a preliminary inquiry into the origins and character of the ancient Chinese city*, Edinburgh University Press, Edinburgh.

Whitehand, J.W.R. (1981) 'Background to the urban morphogenetic tradition', in Whitehand, J.W.R. (ed.), *The Urban Landscape: historical development and management. Papers by M.R.G. Conzen*, Institute of British Geographers special publication, 27, Academic Press, London, 1–24.

Whitehand, J.W.R. and Larkham, P.J. (eds) (1992) *Urban Landscapes: international perspectives*, Routledge, London.

Wood, D. (1993) *The Power of Maps*, Routledge, London.

Woodward, D. (1985) 'Reality, symbolism, time, and space in medieval world maps', *Annals of the Association of Americal Geographers*, 75, 510–21.

# Index